AF716154

Geographies of Migration

Migration is an enormously broad topic of academic enquiry engaging researchers from many different social science disciplines. A wide variety of contributors from across the globe capture some of the methodological and conceptual range of migration research in the discipline of Geography today. This volume covers a large area geographically and in the expanse of subject areas involved: eighteen chapters investigate migration from, to, or within at least fifteen countries, with several sections spanning multiple places and scales. Many chapters are deeply concerned with vulnerable populations, which is not only a characteristic of much immigration scholarship but also one that connects with other areas of geography. The study of geographical assertions of sovereign power via the discourses of disorder, chaos, and crisis, shows that in these transnational times, national power is being violently reasserted, on, within, and beyond international borders. Other important topics covered include migration and climate change, 'illegality', security, government policy, labor, family, and sexual orientation.

This book was previously published as a special issue of *Annals of the Association of American Geographers*.

Richard A. Wright holds the Orvil E. Dryfoos Chair in Public Affairs and has been a Professor of Geography at Dartmouth College since 1985. With grant support from the Guggenheim Foundation, National Science Foundation, the Social Science Research Council, and the Russell Sage Foundation, he has authored more than 70 scholarly papers. His research and teaching focuses on race, residential segregation, and migration.

Geographies of Migration

Edited by
Richard A. Wright

Routledge
Taylor & Francis Group
LONDON AND NEW YORK

First published 2016 by Routledge

2 Park Square, Milton Park, Abingdon, Oxfordshire OX14 4RN
711 Third Avenue, New York, NY 10017

Routledge is an imprint of the Taylor & Francis Group, an informa business

First issued in paperback 2018

British Library Cataloguing in Publication Data
A catalogue record for this book is available from the British Library

ISBN 13: 978-1-138-66714-3 (hbk)
ISBN 13: 978-1-138-39307-3 (pbk)

Typeset in Goudy
by RefineCatch Limited, Bungay, Suffolk

Publisher's Note
The publisher accepts responsibility for any inconsistencies that may have arisen during the conversion of this book from journal articles to book chapters, namely the possible inclusion of journal terminology.

Contents

Citation Information vii
Notes on Contributors xi

1. Migration: An Introduction 1
Richard A. Wright

2. Moving "Out," Moving On: Gay Men's Migrations Through the Life Course 3
Nathaniel M. Lewis

3. Being CBC: The Ambivalent Identities and Belonging of Canadian-Born Children of Immigrants 12
Audrey Kobayashi and Valerie Preston

4. Diasporic Families: Cultures of Relatedness in Migration 21
Joanna C. Long

5. Migration, Urbanization, and Political Power in Sub-Saharan Africa 31
Clionadh Raleigh

6. Following Migrant Trajectories: The Im/Mobility of Sub-Saharan Africans en Route to the European Union 40
Joris Schapendonk and Griet Steel

7. North Korean Women's Narratives of Migration: Challenging Hegemonic Discourses of Trafficking and Geopolitics 49
Eunyoung Choi

8. Environmental Hazards as Disamenities: Selective Migration and Income Change in the United States from 2000–2010 58
J. Matthew Shumway, Samuel Otterstrom, and Sonya Glavac

9. Migration Amidst Climate Rigidity Traps: Resource Politics and Social–Ecological Possibilism in Honduras and Peru 70
David J. Wrathall, Jeffrey Bury, Mark Carey, Bryan Mark, Jeff McKenzie, Kenneth Young, Michel Baraer, Adam French, and Costanza Rampini

10. The Amenity Principle, Internal Migration, and Rural Development in Australia 83
Neil Argent, Matthew Tonts, Roy Jones, and John Holmes

11. "Under the Radar": Undocumented Immigrants, Christian Faith Communities, and the Precarious Spaces of Welcome in the U.S. South 97
Patricia Ehrkamp and Caroline Nagel

12. Enclaves of Rights: Workplace Enforcement, Union Contracts, and the Uneven Regulatory Geography of Immigration Policy 107
Virginia Parks

13. On the Work of Urbanization: Migration, Construction Labor, and the Commodity Moment 116
Michelle Buckley

14. Spaces of Immigrant Advocacy and Liberal Democratic Citizenship 126
Helga Leitner and Christopher Strunk

15. On Distance and the Spatial Dimension in the Definition of Internal Migration 135
Thomas Niedomysl and Urban Fransson

16. The Tactics of Asylum and Irregular Migrant Support Groups: Disrupting Bodily, Technological, and Neoliberal Strategies of Control 151
Nick Gill, Deirdre Conlon, Imogen Tyler, and Ceri Oeppen

17. Chaos and Crisis: Dissecting the Spatiotemporal Logics of Contemporary Migrations and State Practices 160
Alison Mountz and Nancy Hiemstra

18. Living the Way the World Does: Global Indians in the Remaking of Kolkata 169
Pablo S. Bose

19. In the "Service" of Migrants: The Temporary Resident Biometrics Project and the Economization of Migrant Labor in Canada 179
Rebecca Pero and Harrison Smith

Index 191

Citation Information

The chapters in this book were originally published in the *Annals of the Association of American Geographers*, volume 104, issue 2 (March 2014). When citing this material, please use the original page numbering for each article, as follows:

Chapter 1
Migration: An Introduction
Richard A. Wright
Annals of the Association of American Geographers, volume 104, issue 2 (March 2014) pp. 223–224

Chapter 2
Moving "Out," Moving On: Gay Men's Migrations Through the Life Course
Nathaniel M. Lewis
Annals of the Association of American Geographers, volume 104, issue 2 (March 2014) pp. 225–233

Chapter 3
Being CBC: The Ambivalent Identities and Belonging of Canadian-Born Children of Immigrants
Audrey Kobayashi and Valerie Preston
Annals of the Association of American Geographers, volume 104, issue 2 (March 2014) pp. 234–242

Chapter 4
Diasporic Families: Cultures of Relatedness in Migration
Joanna C. Long
Annals of the Association of American Geographers, volume 104, issue 2 (March 2014) pp. 243–252

Chapter 5
Migration, Urbanization, and Political Power in Sub-Saharan Africa
Clionadh Raleigh
Annals of the Association of American Geographers, volume 104, issue 2 (March 2014) pp. 253–261

Chapter 6
Following Migrant Trajectories: The Im/Mobility of Sub-Saharan Africans en Route to the European Union
Joris Schapendonk and Griet Steel
Annals of the Association of American Geographers, volume 104, issue 2 (March 2014) pp. 262–270

Chapter 7
North Korean Women's Narratives of Migration: Challenging Hegemonic Discourses of Trafficking and Geopolitics
Eunyoung Choi
Annals of the Association of American Geographers, volume 104, issue 2 (March 2014) pp. 271–279

Chapter 8
Environmental Hazards as Disamenities: Selective Migration and Income Change in the United States from 2000–2010
J. Matthew Shumway, Samuel Otterstrom, and Sonya Glavac
Annals of the Association of American Geographers, volume 104, issue 2 (March 2014) pp. 280–291

Chapter 9
Migration Amidst Climate Rigidity Traps: Resource Politics and Social–Ecological Possibilism in Honduras and Peru
David J. Wrathall, Jeffrey Bury, Mark Carey, Bryan Mark, Jeff McKenzie, Kenneth Young, Michel Baraer, Adam French, and Costanza Rampini
Annals of the Association of American Geographers, volume 104, issue 2 (March 2014) pp. 292–304

Chapter 10
The Amenity Principle, Internal Migration, and Rural Development in Australia
Neil Argent, Matthew Tonts, Roy Jones, and John Holmes
Annals of the Association of American Geographers, volume 104, issue 2 (March 2014) pp. 305–318

Chapter 11
"Under the Radar": Undocumented Immigrants, Christian Faith Communities, and the Precarious Spaces of Welcome in the U.S. South
Patricia Ehrkamp and Caroline Nagel
Annals of the Association of American Geographers, volume 104, issue 2 (March 2014) pp. 319–328

Chapter 12
Enclaves of Rights: Workplace Enforcement, Union Contracts, and the Uneven Regulatory Geography of Immigration Policy
Virginia Parks
Annals of the Association of American Geographers, volume 104, issue 2 (March 2014) pp. 329–337

Chapter 13
On the Work of Urbanization: Migration, Construction Labor, and the Commodity Moment
Michelle Buckley
Annals of the Association of American Geographers, volume 104, issue 2 (March 2014) pp. 338–347

Chapter 14
Spaces of Immigrant Advocacy and Liberal Democratic Citizenship
Helga Leitner and Christopher Strunk
Annals of the Association of American Geographers, volume 104, issue 2 (March 2014) pp. 348–356

Chapter 15
On Distance and the Spatial Dimension in the Definition of Internal Migration
Thomas Niedomysl and Urban Fransson
Annals of the Association of American Geographers, volume 104, issue 2 (March 2014) pp. 357–372

Chapter 16
The Tactics of Asylum and Irregular Migrant Support Groups: Disrupting Bodily, Technological, and Neoliberal Strategies of Control
Nick Gill, Deirdre Conlon, Imogen Tyler, and Ceri Oeppen
Annals of the Association of American Geographers, volume 104, issue 2 (March 2014) pp. 373–381

Chapter 17
Chaos and Crisis: Dissecting the Spatiotemporal Logics of Contemporary Migrations and State Practices
Alison Mountz and Nancy Hiemstra
Annals of the Association of American Geographers, volume 104, issue 2 (March 2014) pp. 382–390

Chapter 18
Living the Way the World Does: Global Indians in the Remaking of Kolkata
Pablo S. Bose
Annals of the Association of American Geographers, volume 104, issue 2 (March 2014) pp. 391–400

Chapter 19

In the "Service" of Migrants: The Temporary Resident Biometrics Project and the Economization of Migrant Labor in Canada
Rebecca Pero and Harrison Smith
Annals of the Association of American Geographers, volume 104, issue 2 (March 2014) pp. 401–411

Notes on Contributors

Neil Argent, Professor in the Division of Geography and Planning at the University of New England, Armidale, Australia.

Michel Baraer, Ecole de technologie superieure, Universite du Quebec, Canada.

Pablo S. Bose, Associate Professor in the Department of Geography at the University of Vermont, Burlington, USA.

Michelle Buckley, Assistant Professor in the Department of Human Geography at the University of Toronto Scarborough, Canada.

Jeffrey Bury, Associate Professor in the Department of Environmental Studies at the University of California Santa Cruz, and Faculty Director at the Center for Integrated Spatial Research, USA.

Mark Carey, Associate Dean and Associate Professor of History at Clark Honors College, University of Oregon, Eugene, USA.

Eunyoung Choi, Independent Scholar, Seoul, South Korea.

Deirdre Conlon, Lecturer in Critical Human Geography in the School of Geography at Leeds University, UK.

Patricia Ehrkamp, Associate Professor in the Department of Geography at the University of Kentucky, Lexington, USA.

Urban Fransson, School of Business, Economics and Law, Department of Business Administration, Gothenburg University, Sweden.

Adam French, Energy and Resources Group, University of California, Berkeley, USA.

Nick Gill, Associate Professor in the Department of Geography at Exeter University, UK.

Sonya Glavac, Lecturer in the School of Behavioural, Cognitive and Social Sciences at the University of New England, Australia.

Nancy Hiemstra, Assistant Professor in the Department of Cultural Analysis and Theory, State University of New York at Stony Brook, USA.

John Holmes, Emeritus Professor in the School of Geography, Planning and Environmental Management at the University of Queensland, Australia.

Roy Jones, Emeritus Professor in the School of Built Environment at Curtin University, Perth, Australia.

Audrey Kobayashi, Professor in the Department of Geography and Planning at Queen's University, Canada.

Helga Leitner, Professor in the Department of Geography at the University of California, Los Angeles, USA.

Nathaniel M. Lewis, Lecturer in Human Geography, Geography and Environment at the University of Southhampton, UK.

Joanna C. Long, Independent Researcher, Manchester, UK.

Jeff McKenzie, Associate Professor in the Department of Earth and Planetary Sciences at McGill University, Montreal, Canada.

Bryan Mark, Associate Professor in the Department of Geography at The Ohio State University, Columbus, USA.

Alison Mountz, Professor in the Balsillie School of International Affairs at Wilfrid Laurier University, Waterloo, Canada.

Caroline Nagel, Associate Professor in the Department of Geography at the University of South Carolina, Columbia, USA.

Thomas Niedomysl, Associate Professor in the Department of Human and Economic Geography and the Centre for Innovation, Research and Competence in the Learning Economy, Lund University, Sweden.

Ceri Oeppen, Lecturer in Human Geography in the Department of Geography at Sussex University, UK.

Samuel Otterstrom, Professor in the Department of Geography at Brigham Young University, Provo, USA.

Virginia Parks, Associate Professor in the School of Social Service Administration at the University of Chicago, USA.

Rebecca Pero, PhD Candidate in the Department of Geography at Queen's University, Canada.

Valerie Preston, Professor in the Department of Geography at York University, Canada.

Clionadh Raleigh, Professor in the Department of Geography, University of Sussex, UK.

Costanza Rampini, PhD candidate in the Department of Environmental Studies at the University of California Santa Cruz, USA.

Joris Schapendonk, Assistant Professor in the Department of Geography, Planning and Environment, Radboud University Nijmegen, Netherlands.

J. Matthew Shumway, Professor in the Department of Geography at Brigham Young University, Provo, USA.

Harrison Smith, PhD student, Faculty of Information, University of Toronto, Canada.

Griet Steel, Associate Researcher, Institute of Development Policy and Management, University of Antwerp, Belgium.

Christopher Strunk, Assistant Professor in the Department of Geography at Augustana College, Rock Island, USA.

Matthew Tonts, Head of the School of Earth and Environment at the University of Western Australia, Crawley, Australia.

Imogen Tyler, Professor in the Sociology Department at Lancaster University, UK.

David J. Wrathall, Senior Researcher at the Institute for Environment and Human Security, United Nations University, USA.

Kenneth Young, Professor in the Department of Geography and the Environment at the University of Texas, Austin, USA.

Migration: An Introduction

Richard A. Wright

Department of Geography, Dartmouth College

This is the sixth annual special issue of the *Annals of the Association of American Geographers*. These special issues allow the four section editors to highlight themes of international significance that showcase the breadth and depth of the discipline of geography in a format accessible to a range of readers. Bruce Braun will edit the next special issue scheduled for Spring 2015 ("Futures: Imagining Socio-Ecological Transformation"). In 2016, Mei-Po Kwan will assemble the eighth, the theme being "Geographies of Mobilities."

More than 140 scholars or teams of scholars submitted abstracts for consideration in the Migration special issue. Three members of the *Annals* editorial board—Tom Cooke, Anna Secor, and Monica Varsanyi—joined me in selecting thirty-four full submissions from which the final eighteen published papers emerged. Given the tighter word limits, *Annals* special issues trade off between empirical detail and conceptual nuance and the diversity of topics. We tried to ensure that the collection would capture some of the methodological and conceptual range of migration research in the discipline today. We also sought to publish a set of articles that covered a lot of ground geographically: The eighteen contributions investigate migration from, to, or within at least fifteen countries, with several articles spanning multiple places and scales. I had hoped to include at least one paper on the recent massive migrations within China, but that expectation did not materialize.

This brief introduction situates the contributions in various ways. Several of the articles are deeply concerned with vulnerable populations. This is not only a characteristic of much immigration scholarship but also one that connects with other areas of geography. I am reminded that the first special issue on peace and armed conflict attempted to place exactly such marginalized subjects on center stage (Kobayashi 2009). Whether those populations are, for example, women exiting North Korea (Choi), migrants facing climate change impacts on local environments (Wrathall et al.), or hotel employees in the United States laboring without work authorization (Parks), many of the article feature migrant agency and resiliency. Two articles specifically complement and augment this subtheme and analyze the work performed by various immigrant and asylum support groups who contest governmental policies and practices (Gill et al.) or the proliferation of exclusionary immigration policies in the United States (Leitner and Strunk). Another explores some of the limits to immigrant advocacy wherein the practice of hospitality in certain faith communities fails to meet, in the authors' own words, "biblical ideals" (Ehrkamp and Nagel).

In 2013, Edward Snowden released tens of thousands of documents to the press, revealing the extent of the surveillance program of the U.S. National Security Agency. Geographers increasingly intervene in debates about surveillance, privacy, rights, and securitization, and immigration scholars are particularly well placed in these respects. Two articles in this issue tackle issues of sovereignty and security. The idea of transnational migration sometimes suggests a diminution of the importance of national borders. Alison Mountz and Nancy Hiemstra's study of geographical assertions of sovereign power via the discourses of disorder, chaos, and crisis, however, shows that in these transnational times, national power is being violently reasserted, on, within, and beyond international borders. Becky Pero and Harrison Smith's study also ranges across borders and argues that Canada's deployment of biometric screening is part of a global program designed to promote and manage temporary, short-term labor.

An issue devoted to migration that did not feature some articles dealing with identity would be incomplete. Valerie Preston and Audrey Kobayashi's article explores how and where second-generation Chinese Canadians express their identities. Nathaniel Lewis's research extends the concept of gay migration to include more than just the disclosure or development of a gay identity, showing how gays use migration to negotiate a variety of life circumstances and transitions.

As Joanna Long says, "family is central to migration." Writing against the grain of some of the traditions in family migration research in geography (e.g., tied migration, family decision making, gender), Long chooses to explore what she calls the "idiosyncratic cultures of relatedness" through which families are produced in diasporic contexts. All readers (transnational or not) will relate to the central idea in this article that contrasts what a family is versus how it is imagined to be.

This Migration special issue includes three very different articles that foreground the role of migrants in urbanization and development. Michelle Buckley's study of migrant construction work in Dubai paves new pathways toward understanding urbanization processes, and Pablo Bose deconstructs the discourse of what he calls the global Indian through an analysis of luxury condominium developments in the Indian city of Kolkata. The third article assesses how African political institutions direct migration and shape community fortunes and shows how migration is implicated in the acquisition and possession of political power in Africa (Raleigh).

The special issue also features three articles that focus on the environment. In addition to Wrathall et al.'s study of climate change and migration in Central and South America, Argent et al. find that in much of rural Australia amenity migration is changing particular rural areas. Access and coastal location are keys, but their study also shows that amenity migration intersects with more traditional rural land uses, leading them to shy away from notions of rural gentrification and move toward a more nuanced model of "multifunctional rural transition." Shumway et al. show that the disamenities associated with environmental hazards produce selective intercounty flows in the United States, such that those places least able to adjust to net losses of the relatively wealthy suffer from the outmigration of precisely that population.

Geographers' interests in migration and mobility reach back to Zelinsky (1971), whose work anticipated both the emergence of mobility studies as well as mass migration to the global north (cf. King 2012). Schapendonk and Steel's study of the evolution of Sudanese and Nigerian migration trajectories builds on the mobilities paradigm and south–north migration. They highlight the spatial frictions (the immobilities) inherent in almost all transnational connectivities.

Niedomsyl and Fransson explore an even older concern of geographers: distance and migration. Ravenstein (1885, 191, 198) observed that the majority of migrants move over relatively short distances. Niedomsyl and Fransson compare actual distances moved with those derived from administrative boundaries. They confirm that the vast majority of moves are short but also find that the size of administrative units in migration analysis can confuse short and long migration.

Sincere thanks go to Tom, Anna, and Monica for all their hard work. Not only did they help in the initial selection, but they also served as referees on more than one submission. About ninety other colleagues served as reviewers; they provided terrific feedback and commentary and significantly elevated the quality of every published submission. Robin Maier and Miranda Lecea ensured that the selection and production processes ran smoothly. I would like to thank everyone who submitted an abstract and who made the task of distilling the pool to final published papers very difficult. Finally, thanks go to the authors, all of whom adhered to some pretty strict deadlines and the speeded up Just-in-Time production regime of an *Annals* special issue.

References

King, R. 2012. Geography and migration studies: Retrospect and prospect. *Population, Space and Place* 18 (2): 134–53.

Kobayashi, A. 2009. Geographies of peace and armed conflict. *Annals of the Association of American Geographers* 99 (5): 819–26.

Ravenstein, E. G. 1885. The laws of migration: I. *Journal of the Statistical Society* 48:167–227.

Zelinsky, W. 1971. The hypothesis of the mobility transition. *Geographical Review* 61:219–49.

Moving "Out," Moving On: Gay Men's Migrations Through the Life Course

Nathaniel M. Lewis

School of Geography, University of Nottingham

Research in the field of sexuality and space has begun to explore the relationships between gay and queer sexual subjectivities and migration. Much of this research examines the regulation and policing of queer international migrants or identity formation processes among younger queer people migrating within countries. This study, although located partially within the second category, broadens and deepens existing accounts of gay men's migrations within countries by focusing on life circumstances and events beyond an initial coming-out process and considering the migration experiences of gay men at multiple points in the life course. This study uses life course theory to contextualize the migration narratives of 48 self-identified gay men in Ottawa, Ontario, Canada, and Washington, DC. The findings lend credence to recent claims that migration is central in the lives of gay men and other queer people but extends the concept of gay migration to include more than just the disclosure or initial development of a gay identity. They reframe migration as a tool used to negotiate a variety of life circumstances and transitions (e.g., establishing careers, creating meaningful community identities) rendered challenging by variegated landscapes of stigma and inclusion in North America.

性慾与空间的场域之研究, 已着手探讨同性恋及酷儿的性主体和迁徙之间的关联性。该研究多半检视对于酷儿国际迁徙的管制与维安, 抑或是年轻的酷儿在国内迁徙的身份认同形构过程。儘管本研究部分属于第二类的范畴, 但却透过聚焦超越最初出柜过程的生命境遇与事件, 并考量男同性恋者在其生命历程的多重时间点中的移居经验, 扩展并深化对于男同性恋者在国内进行迁徙的既有解释。本研究运用生命历程理论, 脉络化四十八位来自加拿大安大略省渥太华以及美国华盛顿特区、自我认同为男同性恋者的迁徙叙事。研究发现印证以下主张: 迁徙对男同性恋者及其他的酷儿人们的生活而言至关重要, 但也延伸了男同性恋迁徙的概念, 用以包含不只是同性恋身份认同的揭露或是最初的发展。它们将迁徙再概念化为用来协商北美的社会污名与包容使之具有挑战性的各种生命境况与变迁地景 (例如建立事业、创造有意义的社群认同) 之工具。

La investigación en el campo de sexualidad y espacio ha empezado a explorar las relaciones entre las subjetividades homosexuales y la migración. Un gran volumen de esa investigación se ha dedicado a examinar las regulaciones y el control policial de este tipo de migrantes internacionales, o los procesos de formación de identidad entre los homosexuales más jóvenes que migran entre países. Aunque ubicado parcialmente dentro de la segunda categoría, el presente estudio amplía y profundiza los registros existentes de las migraciones internas de homosexuales varones, concentrándose en las circunstancias de vida y eventos que ocurren tras el proceso inicial de dejar de esconderse y considerando las experiencias migratorias de hombres gay en múltiples puntos del curso de la vida. El estudio utiliza la teoría del curso de la vida para contextualizar las narrativas migratorias de 48 hombres auto-identificados como *gays* en Ottawa, Ontario, Canadá, y en Washington, DC. Los hallazgos aportan credibilidad a recientes afirmaciones de que la migración es central en la vida de hombres gay y gente de similar condición, pero extiende el concepto de migración gay para incluir otras cosas además de la revelación o el desarrollo inicial de una identidad gay. Estos hallazgos replantean la migración como una herramienta que se usa para negociar una variedad de circunstancias de la vida y transiciones (e.g., estableciendo carreras, creando identidades comunitarias significativas) revestidas como retos por variados paisajes de estigmatización e inclusión en Norteamérica.

Life course theory considers how individual subjectivities such as age, gender, and ethnoracial identity converge with particular place-embedded networks of relationships and institutions to shape the timing and nature of life events such as labor force entry, childbirth, marriage, and relocation (Elder 1985; Mulder and Hooimeijer 1999). The life trajectories of individuals who identify as gay, lesbian, bisexual, transgendered, or queer, however, are examined less frequently.[1] The burgeoning literature on geographies of sexualities, perhaps in an effort to avoid the type of structuration and categorization critiqued by queer theory (Sullivan 2003), has also been hesitant to choreograph events in queer lives. This article consequently seeks to establish some of the patterns of one life event, migration, which has been described as central in the lives of queer people (Cant 1997; Knopp 2004).

The literature on queer migration focuses mainly on heteronormative national immigration policies (Mountz 2010), diasporic transformations of queer cultures across international borders (Manalansan 1995), HIV/AIDS diagnosis and care (Ellis and Muschkin 1996), and, to a lesser extent, early-in-life, within-country, usually rural-to-urban moves associated with coming out (Weston 1995; Cant 1997; M. Brown 2000). Less is known, however, about how migration might be employed by queer people throughout life and across a variety of historic and geographic contexts. This article addresses this gap by examining the migration narratives of forty-eight gay-identified men living in Ottawa, Ontario, Canada, and Washington, DC, and situating the narratives within the participants' broader life histories. The narratives suggest that gay men migrate to both navigate specific life events (e.g., labor market entry, partnering) and to facilitate lifelong processes of identity development—not just processes of coming out. The findings reveal, on the one hand, that gay men's migrations are as diverse and everyday in nature as those observed elsewhere in life course research. On the other hand, they also demonstrate that migration is particularly central in gay men's lives because of the precarious social and institutional contours they must navigate.

Applying a Life Course Lens to Gay Men's Migrations

Social scientists have increasingly employed the life course as an alternative means of examining life events and changes frequently explained through pregiven, biologically driven, stage-sequential models of human development or life cycle (e.g., Levinson 1978). The life course acknowledges that differing social, institutional, and geographic contexts shape the progressions of individual lives. These contexts, coupled with the ways in which an individual's age, gender, or ethno-racial subjectivities affect their experience, create significant variations in the timing, patterning, and experience of the life course—both across groups (e.g., men vs. women) and within those groups (Bailey, Blake, and Cooke 2004; Hopkins and Pain 2007; Cooke 2008; Bailey 2009). Similarly, migration is now understood less as a universal response to economic or demographic pushes and pulls and more as an individualized process contingent, for example, on gender, class, age, and ethno-racial subjectivities and identities (Silvey 2004). Thus, researchers have begun to study migration as spatial tool through which individuals negotiate life events and individual identities by shifting or switching—at a given point in time—the social or institutional contexts in which they are located (Kobayashi and Preston 2007; Bailey 2009; Finney 2011).

Research on the timing and sequencing of life events such as childbirth and labor market entry (Elder 1985), as well the impact of migration on those events (Bailey, Blake, and Cooke 2004), has tended to focus on age and sex groups or on heterosexual couples and families. Increasingly, however, life course research has considered the gendered and ethnically differentiated social dynamics that inform the timing and impact of life events (Cooke 2008; Finney 2011). Sexual identity, which is not easily categorized and frequently absent from large-scale data sets, is only rarely considered in life course research (but see Hostetler and Cohler 1997; Kertzner 2001; Barker, Herdt, and de Vries 2006) and even less frequently with regard to migration specifically (but see Cooke and Rapino 2007).

The positioning of "homosexual" as a deviant identity in the Anglo-American world (Halperin 2000) created distinct differences between the life courses of gay men and their heterosexual counterparts. Gay men have not historically had rights to marry or raise children, and the legal and social impact of anti-gay stigma has been and continues to be uneven across space and time (Hostetler and Cohler 1997; Kertzner 2001; Smith 2008). It follows, then, that gay men might relocate to shift, facilitate, or speed up events (e.g., coming out, partnering) that might be impossible or more difficult in a different location (Barker, Herdt, and de Vries 2006). The remainder of this article investigates how the variations in gay men's experiences of individual development, social support, and institutional infrastructure—both across places and as distinct from heterosexual men—inform their migration decisions across the life course.

Methods

This study includes the life narratives of forty-eight self-identified gay men living in Ottawa, Ontario, Canada, and Washington, DC (Cohler and Hostetler 2003). Using two participant-driven snowball samples (Frank and Snijders 1994) that began with personal contacts or service agency referrals, the author conducted twenty-four semistructured interviews in each city between October 2009 and May 2010. Participants were twenty-four to fifty-nine years old; had moved between six months and twenty-eight years prior to being interviewed; worked mostly in professional sectors such as education, health, and government; and identified as white, Anglo-Saxon, Hispanic, African-Canadian, Aboriginal, and other ethno-racial signifiers. Interviews lasted between forty-five and seventy-five minutes and were transcribed in full and coded by the author. In line with a constructivist grounded theory approach allowing for flexibility in the creation and ordering of codes and themes (Charmaz 2006), each transcript was analyzed once to establish broad themes, such as "coming out" and "employment," and then recoded for subthemes. The life course emerged as a theoretical thread that connected all of the themes and explained differences in migration experiences and decision-making processes for men at different ages and life stages.

This study has two notable limitations. First, the focus on men who have moved excludes the experiences of those who have stayed relatively stationary during the life course. This is not to predetermine the importance of migration but to further understand the relationships among gay men's identities, life events, and migrations across the life course. Second, the study sites are not necessarily the most historically or culturally significant destinations for gay men in North America. At the same time, studying migrations to these midsized "government towns" offers a unique opportunity to examine how gay men's migrations are influenced by both urban hierarchies and the institutions of specific cities (see Lewis 2012a, 2012b).

Individual Development: Maturation and Mobility

Research in psychology and psychiatry has long positioned coming out, or disclosing one's sexual identity to others, as a central but singular transition marked by identity confusion and same-sex experimentation, identity assumption, and finally self-identification as gay (Troiden 1989). Much of the work on gay men's migrations within countries (Weston 1995; M. Brown 2000; Gorman-Murray 2007) has focused on coming out, perhaps contributing to an emergent discourse that "it gets better" once gay identities are disclosed and homophobic places are left behind (Gay, Lesbian, and Straight Education Network 2012). From a social perspective, however, coming out—and the relocation that might be associated with it—is not a straightforward antidote to the long-term impact of social stigma on gay men's lives. Due to both legal prejudice in many jurisdictions and informal prejudice in family homes, workplaces, and social settings, gay men might continue to feel both "out of place" (Cresswell 1996) and "off-schedule" (Kertzner 2001) throughout the life course. Certain environments might force them to conceal their sexuality or to delay or forego relationships; in others they continue to experience significant stigma and discrimination.

Participants' narratives demonstrate that gay men's concerns about their individual development vary across the life course. For men earlier in life who were initiating a process of coming out, moving was a more urgent, singularly focused act motivated by frustrations over "treading water," anxieties over "getting out," and desires to galvanize or realign life courses that felt stunted (see also M. Brown 2000). Sebastian, forty-two, who moved to Ottawa from a town in Nova Scotia about twenty years earlier, said that relocating was a way to ensure that he was "doing things that [were] age-appropriate at an age-appropriate time."[2] Later in life, however, men were less concerned about forming a newly "gay" sexual identity or making contact with "the gay community." Sam, forty-four, who moved to Ottawa from the known gay center of Montreal in his early thirties, said, "I think [my gay identity] is becoming less prominent . . . than my Montreal days, just because I wasn't in that coming out phase . . . I think it's sort of a natural life cycle, I guess, in sort of the coming out process that you, um, start to wean yourself away from the community." As the next sections show, however, relocation does not cease after coming out; rather, it is used to navigate a variety of life events rendered challenging by the geographically and institutionally variable nature of both anti-gay stigma and gay and queer inclusivity in North America.

Social Support and the Gay Community

Social support plays a crucial role in gay men's lives, not only in terms of facilitating identity formation but also as a source of lifelong security for men who have broken ties with families or might not have the opportunity to create similar family structures of their own (Hostetler and Cohler 1997; Kertzner 2001; Barker, Herdt, and de Vries 2006). As the participants' narratives of their own identity development suggest, younger men's concerns often center on making first contact with an imagined gay community and finding initial opportunities to date and socialize. As Gorman-Murray (2009) suggested, men engaging in "coming-out migration" are also seeking comfort in the ability to define themselves vis-à-vis an accessible group of imagined gay compatriots. Often, the urban gay "scenes" in cities such as San Francisco, New York, and Toronto, acting as beacons for young gay men moving from elsewhere (Cant 1997), have been positioned as the central spaces in gay lives (Knopp and Brown 2003; Valentine and Skelton 2003). Recent studies have suggested, however, that large cities and the established gay scenes within them become less desirable, or even restrictive and disenfranchising, to gay men beyond an initial coming-out process (Waitt and Gorman-Murray 2011a, 2011b; Annes and Redlin 2012).

The participants who had moved as part of a coming-out process referenced the established concept of gay migration as a form of starting over that involves exchanging one set of social networks for another and inserting themselves into an imagined urban gay community (Fortier 2001, 410). For these men, the goal of moving was to dislocate themselves from relationships that they feared would become irreparably altered or damaged by coming out as gay and to gravitate toward communities that they felt would provide a security net (Gorman-Murray 2009). Daniel, twenty-eight, who moved to Ottawa from rural eastern Ontario in the early 2000s, said, "There's definitely a transition when you cut off your established connections to a community and then move to a new one where you have to rebuild it." Consequently, participants who had moved earlier in life perceived the critical mass of the gay community, alternately defined in terms of "density," "options for meeting people," or a "large enough base," as a crucial factor in their decision to move.

Both relative location and absolute location were central to these earlier-in-life migration decisions. While coming out, many participants felt it necessary to move a significant distance from home to disentangle themselves from existing social and familial networks with sometimes far-reaching tendrils (see also Annes and Redlin 2012; Lewis 2012b). At the same time, many purposefully chose a specific urban center that would provide a critical mass of other gay men (e.g., for dating and friends) and a community within which they could define themselves. Among younger single men, the perceived size of the broader gay community was typically described as an important factor. Among the thirty-three men who moved to Ottawa or Washington, DC, under age thirty, twenty-five of them (75 percent) moved from a locale smaller than their destination city, sometimes substantially (see Figure 1). Rick, twenty-seven, suggested that whereas Ottawa, a midsized city, might be initially attractive for younger men moving from rural areas (i.e., as a coming-out stopover), their subsequent relocations—at least early in life—would continue to move up the urban hierarchy: "Ottawa's seen as a transition city for a lot of gay men," he said.

In contrast, among gay men in midlife, identity might be defined more in relation to accumulated "families of choice" rooted in place and created at will, rather than the perceived size or social options of a pregiven urban gay community (Kertzner 2001; Weeks, Heaphy, and Donovan 2001; Waitt and Gorman-Murray 2011a; Annes and Redlin 2012). For two participants, this identity transition was analogous to moving away from an urban gay village. For Pete, forty-one, who left Montreal in his early thirties, the social life he had in Montreal's gay village had become restrictive: "Being in a big city like Montreal, you always like some people [but] you're forced to hang out with [others] . . . [in Ottawa] you're . . . not necessarily more exclusive, but [you have] more friends that are suitable to your lifestyle." For Raymond, thirty-one, who had just recently chosen to relocate from Toronto to Ottawa rather than Montreal said, "I sort of was afraid that I'd get sucked back into that life that I had in Toronto where it was very sort of village-based . . . and I didn't necessarily want that again."

Many of the participants in midlife also intended to stay in Ottawa or Washington, DC, because stable cultures of employment facilitated the types of long-term relationships that were crucial for social support and potentially for later-in-life networks of care provision (Barker, Herdt, and de Vries 2006). Their narratives show that migration triggers such as retirement, particularly for gay men whose accrued sources of social support extended well beyond the traditional family unit, were "necessary but not sufficient" conditions for movement (Bailey, Blake, and Cooke 2004, 1620; Mulder and Hooimeijer 1999). Ashton, forty-seven, described the logic accordingly:

> We all have that discussion, so when you retire, are you going to stay in Ottawa or are you gonna move on? And

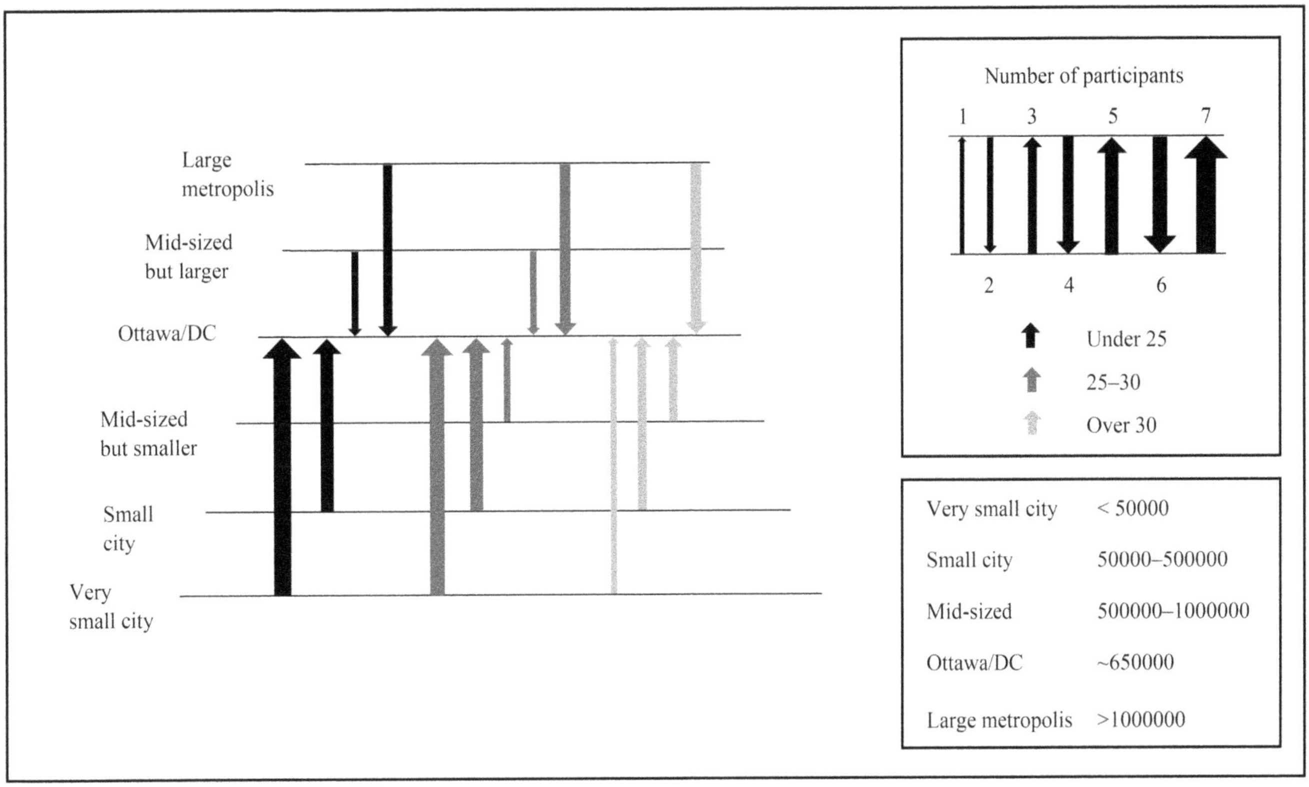

Figure 1. Place moved from, forty-eight self-identified gay men, Ottawa, Ontario, Canada, and Washington, DC.

> a lot of people were like . . . I thought I'd be going somewhere else but I look around and I think, I have all my network of friends here . . . if I move away I've gotta build it all from scratch.

Declining preferences for large urban communities—and for moving in general—among men in midlife were also observed in the data for participants' next anticipated migrations. Among the twenty-nine participants who were under forty at the time of the interview, twenty-one (72 percent) anticipated moving to a larger city and another six (21 percent) planned to stay (two were unsure).[3] Among the sixteen men over forty, only four (25 percent) anticipated moving to a larger city, six (35 percent) planned to stay, and the other six (35 percent) planned to move to somewhere smaller (see Figure 2). The mixed responses show that although migration continues throughout many gay men's lives, affinity for the liminal spaces of the scene or the repeated moves associated with "queer identity quests" (Knopp 2004) might erode as the life course progresses.

A gradual weaning from the gay community was not necessarily a universal, seamless transition for all men in or approaching midlife. Some were reluctant to leave the safety of the gay villages they had come out in. For others, reaching midlife without having attained a coupled, stationary, "homonormative" lifestyle (Duggan 2002) rendered a midsized city like Ottawa a socially restrictive place where singlehood felt like failure (Hostetler and Cohler 1997). Sebastian, forty-two, observed: "Ottawa has a good infrastructure to support a long-term relationship lifestyle . . . but . . . after five years of being single in this city . . . the infrastructure to support going out isn't there." Sebastian's comment suggests that a bricks-and-mortar gay scene—something Ottawa was observed to lack—continues to be important for men in midlife who have not "caught up" (or choose not to catch up) to a homonormative life course.

Participants' narratives also showed that the perceived importance of locating oneself within a known urban gay community depends on historical context. For the few men who had moved fifteen or twenty years prior to being interviewed, their social supports were often gained from participating directly in gay rights and HIV/AIDS movements after arriving in Ottawa or Washington, DC. One participant who moved from Halifax, Nova Scotia, in the 1980s said, "What I sought [in moving to Ottawa], of course, was, you know, acceptance and a community . . . but at that time when I

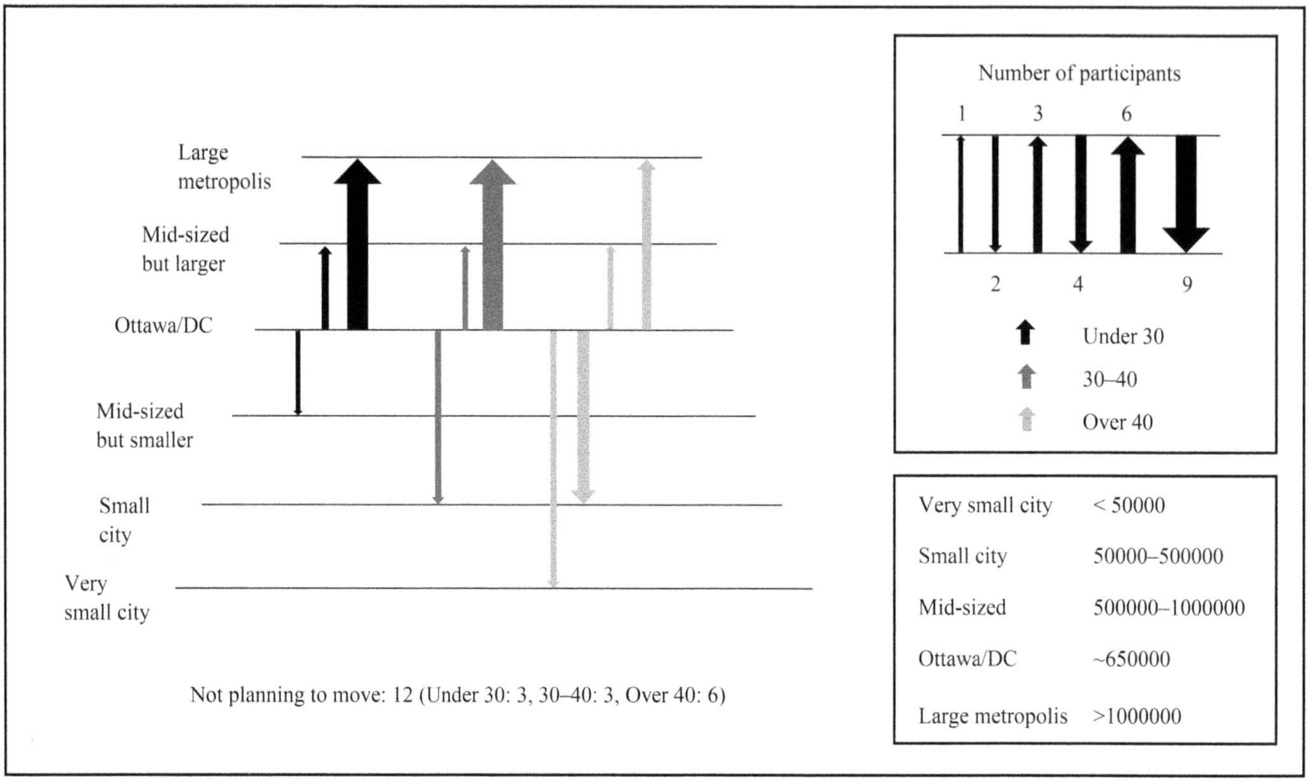

Figure 2. Destination of next anticipated migration at time of interview, forty-eight self-identified gay men, Ottawa, Ontario, Canada, and Washington, DC. *Note:* Does not include two "not sure" responses.

moved here . . . it was starting to grow. So I grew with it, and the things that I probably expected or sought were the things that I was creating or helping to create" (Adrian, fifty-six). His portrayal of HIV/AIDS activism as an early "point of entry" into urban gay communities provides a corollary to existing studies that focus on HIV/AIDS diagnosis as an impetus for return migration, sometimes from a large city to a smaller one (see Ellis and Muschkin 1996). Yet many men's narratives also suggest that points of entry into the gay community are now more diffuse than they were in the past. The Internet, for example, has not only made it easier for gay men to research potential destinations but has also perhaps decentered the role of the urban gay community in providing social support (Bryson et al. 2006). As James, forty, said, "As a general rule, we don't need to go to clubs as much to meet people; there's the Internet, there's those kinds of things." Such narratives suggest that visible, geographically concentrated villages and scenes play a temporarily limited role in gay men's migration decisions and, possibly, a slowly declining role in the provision of social supports to gay men and other queer people (but see Lewis 2013).

Institutional Infrastructure and Community Identities

As previous life course research has suggested, the social and spatial trajectories of individuals' lives are shaped not only by shifts in individual priorities and desires but also by the institutions they encounter across the life course (Hopkins and Pain 2007; Bailey 2009). For gay men, access to the institutions that might facilitate different life events (e.g., coming out or establishment of careers and community roles) is more historically and geographically uneven than those for the general population (Hostetler and Cohler 1997). Early in life, participants frequently distanced themselves physically and psychically from places where they felt marginalized or excluded, such as high schools, summer factory jobs, and communities centered on the heteronormative family and highly gendered work–home divisions. As previous narratives suggested, the institutions sought elsewhere were those that would provide safety and facilitate making first contact with the gay community. "One of the first things I did when I came to Ottawa was go to Pink Triangle Youth Meetings and just kind of met some people," said Derek, thirty-one.

Later in life, however, as "the social and existential meaning of homosexual identity becomes less clear" (Kertzner 2001, 88), gay men's decisions about where to live might be based on career security and the desire to impact their communities in a meaningful way. Although both Ottawa and Washington, DC, have distinct histories of regulation and oppression among gay civil servants (e.g., through systematic expulsions), the civil service sectors in both cities now promote inclusion through clear antidiscrimination protocols, the addition of gay and lesbian professional groups to most key agencies, and (in some cases) the provision of same-sex partner benefits (Lewis 2012a). Many men therefore targeted the capital city as an environment that might help overcome the insecurities they experienced elsewhere. Mark, forty-nine, who moved from Newfoundland to Ottawa in the early 1990s, feared being outed and potentially fired while working as a Catholic school teacher. Randall, forty-five, who moved from Halifax, Nova Scotia, also in the early 1990s, said that his racialization had both affected his entry into the labor market and rendered coming out impossible because it would have been yet another identifier of difference: "The double whammy of being both Aboriginal and Black meant that no matter how clever I was, no matter how good my portfolio was, um, how well I presented myself." Adam, thirty-four, who had only recently sought out a transfer from an Ohio branch of a U.S. government agency to the DC headquarters, said that the institutional (although not legal) embeddedness of gay inclusion in federal government offices made it "very comfortable to have [gay men] working without any issues in just about every federal agency . . . [he] didn't get a sense of that in Ohio." These comments not only show the importance of institutional cultures for gay men in the labor force but also suggest that migration decisions among gay men are not about consumption (Duggan 2002). Rather, they concern economic production and security.

For other participants, moving was connected to opportunities to create identities through participating in advocacy, service, and community building (although not always gay-related). Some had left behind places where they felt unable to contribute meaningfully to their communities. Pete, forty-one, for example, had quit coaching youth baseball in his town in Quebec for fear of becoming the target for "easy assumptions" about his conduct with the players. Rick, twenty-seven, discussed the difficulty of doing gay advocacy work in London, Ontario, a city that he felt was suffused with "3M [corporation], London Life [insurance company] moneyed sort of blueblood conservatives . . . and a serious history with homophobia." After moving to Ottawa or Washington, DC, participants had frequently joined gay and lesbian professional organizations and performed pro bono work for national and local gay media. Many had also attended local-cum-national events such as equal rights protests on Parliament Hill in Ottawa and fundraisers sponsored by DC's Human Rights Campaign (HRC)—the chief U.S. national gay rights organization. In this way, particularly among men in midlife, the institutions of the capital city provided opportunities to create a secure livelihood and make meaningful contributions to advocacy and community—what Shane, twenty-seven, called "an important part of [the gay] experience"—even as the allure of traditional gay homelands like Montreal and San Francisco had declined over historical time and across their individual life courses.

Conclusions

This exploration of gay men's migration decisions through the life course, although not exhaustive, reveals complex motivations and considerations for migration contingent on age, life stage, and other subjectivities. In particular, the narratives reveal important patterns in the motivators for migration, the relationships between identity and migration, and the changing landscape of potential migration destinations for gay men. For men who had moved at younger ages, relocating was bound up with plans to explore their sexual identity or disclose it to others and to exchange (at least temporarily) a previous home for a new, usually larger one. For them, "the big city" had provided evidence of potential social and romantic contacts (e.g., high numbers of local profiles on gay dating sites) and institutions that might facilitate coming out or meeting others (e.g., support organizations). For men in midlife, moving—or the decision to not move—was more related to desires for secure economic livelihoods and meaningful community identities. Over time, however, these divisions have perhaps become less stark. With the advent of the Internet, younger men are perhaps less likely to make a "leap of faith" coming-out journey to the nearest gay homeland than they are to carefully research the sexual as well as community and professional potentials of a variety of places or to come out entirely in an online context before moving. By the same token, some men might feel that they should wean themselves from the "scene" at a certain age, yet yearn for it after they have moved to a smaller city with a more limited social infrastructure or when their relationship status changes from partnered to single.

Migration and identity formation are still strongly interlinked for gay men beyond the coming-out process, particularly because the channels of social and institutional support and, alternately, exclusion, remain particularly variegated for this population. The aspects of identity developed through relocation also tend to change over the life course. Early-in-life migrations often entail establishing a sexual identity itself by finding romantic partners or friends to go out with. Later-in-life moves might involve seeking out places where inclusive workplaces and organizations are dense enough to foster meaningful professional and community identities, particularly because gay men might not have the opportunity to raise children or participate in the child-related activities that often ground heterosexual men's identities later in life (Kertzner 2001). Consequently, a secure and meaningful gay identity is not something attained in a singular journey of moving out and coming out but through lifelong (and multijourney) process of "becoming" (Worth 2009).

Finally, the narratives suggest a landscape of potential destinations for gay men altered radically from the homeland–hinterland binary suggested by Weston (1995). Amidst an improved but still uneven landscape of gay rights and protections (Smith 2008) and more geographically diffuse, politically institutionalized modes of advocacy (Smith 2005), the declining profile of homelands such as San Francisco might create preferences for more "ordinary" cities (G. Brown 2008). These cities might lack historically significant and geographically concentrated "scenes" and "villages" but offer institutional configurations that promote secure and meaningful livelihoods.

These three observations suggest that gay migration, like all migration, is highly interdependent with and contingent on life course events (Cooke 2008, 262) and not always foregrounded in gay-specific events (e.g., coming out, HIV/AIDS, gay neighborhood formation) that have long dominated research in this area. New research on gay and queer migration could perhaps focus on the ongoing relationships between sexuality and more mundane processes (e.g., job-seeking and economic security) that—because of the long-term stigmatization of gay identity—can be more challenging than, or at least geographically distinct from, those of the general population. Continued use of a life course approach in both sexuality and space studies and migration can thus promote more fulsome understandings of gay and other queer identities and how individual places—and movement between those places—might facilitate the health, well-being, and inclusion of queer people in a range of life stages and locational contexts.

Notes

1. In this article, *queer* is used to denote individuals who identify as gay, lesbian, bisexual, or transgendered. Although *queer* has also been framed as a political position that rejects categorizing sexual identity, it is used here to refer to identification with any sexual subjectivities different from or outside of heterosexual norms. *Gay*, in contrast, refers to gay-identified men and the self-ascribed identity of the forty-eight men considered in this study.
2. Pseudonyms are used to protect the identities of the participants.
3. Participants in the thirty-to-forty and over-forty categories demonstrated similar patterns of preference. Percentages do not add up to 100 due to "not sure" responses.

References

Annes, A., and M. Redlin. 2012. Coming out and coming back: Rural gay migration and the city. *Journal of Rural Studies* 28:58–68.

Bailey, A. 2009. Population geography: Lifecourse matters. *Progress in Human Geography* 33 (3): 407–18.

Bailey, A., M. K. Blake, and T. J. Cooke. 2004. Migration, care, and the linked lives of dual-earner households. *Environment and Planning* A 36:1617–32.

Barker, J. C., G. Herdt, and B. de Vries. 2006. Social support in the lives of lesbians and gay men midlife and later. *Sexuality Research & Social Policy* 3 (2): 1–23.

Brown, G. 2008. Urban (homo) sexualities: Ordinary cities and ordinary sexualities. *Geography Compass* 2 (4): 1215–31.

Brown, M. 2000. *Closet space: Geographies of metaphor from the body to the globe*. London and New York: Routledge.

Bryson, M., L. MacIntosh, S. Jordan, and H. Lin. 2006. Virtually queer? Homing devices, mobility, and un/belongings. *Canadian Journal of Communication* 31:791–814.

Cant, B. 1997. *Invented identities? Lesbians and gays talk about migration*. London: Cassel.

Charmaz, K. 2006. *Constructing grounded theory: A practical guide through qualitative analysis*. London: Sage.

Cohler, B. J., and A. J. Hostetler. 2003. Linking life course and life story: Social change and the narrative study of lives. In *Handbook of the life course*, ed. J. Mortimer and R. Shanahan, 555–78. New York: Kluwer Academic/Plenum.

Cooke, T. J. 2008. Migration in a family way. *Population, Space and Place* 14:255–65.

Cooke, T. J., and M. A. Rapino. 2007. The migration of partnered gays and lesbians between 1995 and 2000. *The Professional Geographer* 59:285–97.

Cresswell, T. 1996. *In place/out of place*. London: University of Minnesota Press.

Duggan, L. 2002. The new homonormativity: The sexual politics of neoliberalism. In *Materializing democracy: Towards a revitalized cultural politics*, ed. R. Castronovo and D. Nelson, 175–94. Durham, NC: Duke University Press.

Elder, G., ed. 1985. *Life course dynamics*. Ithaca, NY: Cornell University Press.

Ellis, M., and C. Muschkin. 1996. Migration of persons with AIDS—A search for support from elderly parents? *Social Science and Medicine* 43:1109–18.

Finney, N. 2011. Understanding ethnic differences in the migration of young adults within Britain from a lifecourse perspective. *Transactions of the Institute of British Geographers* 36 (3): 455–70.

Fortier, A.-M. 2001. Coming home: Queer migration and multiple evocations of home. *European Journal of Cultural Studies* 4:405–24.

Frank, O., and T. Snijders. 1994. Estimating hidden populations using snowball sampling. *Journal of Official Statistics* 10:53–67.

Gay, Lesbian, and Straight Education Network. 2012. The Trevor Project. http://www.thetrevorproject.org/content/glsen (last accessed 28 November 2012).

Gorman-Murray, A. 2007. Rethinking queer migration through the body. *Social & Cultural Geography* 8 (1): 105–21.

———. 2009. Intimate mobilities: Emotional embodiment and queer migration. *Social & Cultural Geography* 10 (4): 441–60.

Halperin, D. M. 2000. How to do the history of male homosexuality. *GLQ: A Journal of Gay and Lesbian Studies* 6 (1): 87–124.

Hopkins, P., and R. Pain. 2007. Geographies of age: Thinking relationally. *Area* 39:287–94.

Hostetler, A., and B. Cohler. 1997. Partnership, singlehood, and the lesbian and gay life course: A research agenda. *Journal of Gay, Lesbian, and Bisexual Identity* 2 (3–4): 199–230.

Kertzner, R. 2001. The adult life course and homosexual identity. *Annual Review of Sex Research* 12:75–92.

Knopp, L. 2004. Ontologies of place, placelessness, and movement: Queer quests for identity and their impacts on contemporary geographic thought. *Gender, Place and Culture* 11:122–34.

Knopp, L., and M. Brown. 2003. Queer diffusions. *Environment and Planning D: Society and Space* 21:409–24.

Kobayashi, A., and V. Preston. 2007. Transnationalism through the lifecourse: Hong Kong immigrants in Canada. *Asia Pacific Viewpoint* 48:151–67.

Levinson, D. J., with C. Darrow, E. Klein, and M. Levinson. 1978. *Seasons of a man's life*. New York: Random House.

Lewis, N. M. 2012a. Gay in a "government town": The settlement and regulation of gay-identified men in Ottawa, Canada. *Gender, Place & Culture* 19 (3): 291–312.

———. 2012b. Remapping disclosure: Gay men's segmented journeys of moving out and coming out. *Social & Cultural Geography* 13 (3): 211–31.

———. 2013. Ottawa's Le/The Village: Creating a gaybourhood amidst the "death of the village." *Geoforum* 49:233–42.

Manalansan, M. 1995. In the shadows of Stonewall: Examining gay transnational politics and the diasporic dilemma. *GLQ: A Journal of Gay and Lesbian Studies* 2: 425–38.

Mountz, A. 2010. *Seeking asylum: Human smuggling and bureaucracy at the border*. Minneapolis: University of Minnesota Press.

Mulder, C. H., and P. Hooimeijer. 1999. Residential relocations in the life course. In *Population issues: An interdisciplinary focus*, ed. L. J. G. Wissen and P. A. Dykstra, 159–86. New York: Kluwer.

Silvey, R. 2004. Power, difference, and mobility: Feminist advances in migration studies. *Progress in Human Geography* 28 (4): 490–506.

Smith, M. 2005. Resisting and reinforcing neoliberalism: Lesbian and gay organizing at the federal and local levels in Canada. *Policy & Politics* 33 (1): 75–94.

———. 2008. *Political institutions and lesbian and gay rights in the United States and Canada*. London and New York: Routledge.

Sullivan, N. 2003. *A critical introduction to queer theory*. Armadale, Australia: Melbourne.

Troiden, R. R. 1989. The formation of homosexual identities. *Journal of Homosexuality* 17 (1–2): 43–73.

Valentine, G., and T. Skelton. 2003. Finding oneself, losing oneself: The lesbian and gay "scene" as a paradoxical space. *International Journal of Urban and Regional Research* 27 (4): 849–66.

Waitt, G., and A. Gorman-Murray. 2011a. "It's about time you came out": Sexualities, mobility and home. *Antipode* 43:1380–1403.

———. 2011b. Journeys and returns: Home, life narratives and remapping sexuality in a regional city. *International Journal of Urban and Regional Research* 35: 1239–55.

Weeks, J., B. Heaphy, and C. Donovan. 2001. *Same sex intimacies: Families of choice and other life experiences*. London and New York: Routledge.

Weston, K. 1995. Get thee to a big city: Sexual imaginary and the great gay migration. *GLQ: A Journal of Lesbian and Gay Studies* 2:253–77.

Worth, N. 2009. Understanding youth transition as "becoming": Identity, time, and futurity. *Geoforum* 40 (6): 1050–60.

Being CBC: The Ambivalent Identities and Belonging of Canadian-Born Children of Immigrants

Audrey Kobayashi* and Valerie Preston†

*Department of Geography, Queen's University
†Department of Geography, York University

Immigrants from Hong Kong are one of the largest recent cohorts of newcomers in Canada. We examine the processes through which the children of Hong Kong immigrants construct a sense of belonging in Canada, their only home. In focus group discussions, the participants express an unequivocal sense of Canadian citizenship and belonging. Yet their identities project an idealized Canadian body that wears "Canadian" clothing, plays "Canadian" sports, and, above all, speaks "Canadian" English. The projection is always relational and spatialized, variable according to place, and especially according to whether the other is viewed as family, as another "Canadian-born Chinese" (CBC), as a recent Chinese immigrant, or as a dominant Canadian. Often viewed as not Chinese enough by recent immigrants, in turn, they denigrate recent immigrants as not Canadian enough and resent them for reinforcing a concept of otherness held by dominant Canadians that often includes CBCs. Many participants feel more "Chinese" in family settings where parents uphold Chinese traditions. Their sense of belonging is a set of paradoxes, of between-ness, that they negotiate through a sense of place-ness: place as the homes in which they belong but also as a set of shifting public and private contexts in which they express their identities and relationships in variable ways.

来自香港的移民，是加拿大晚近为数最多的新移民群体之一。我们检视香港移民的子女在加拿大——对他们而言是唯一的家——的归属感建构过程。研究参与者在焦点团体讨论中，表达了对加拿大公民身份及归属的明确感。但他们的身份认同投射出一个理想的加拿大身体，该身体穿着 "加拿大的" 服饰、从事 "加拿大的" 运动，尤其是说 "加拿大的" 英文。此般投射永远是关係性且空间化的，随着地方有所不同，并特别取决于他者被视为家庭的一分子、另一个 "加拿大出生的华人" (CBC) 、华人新移民，还是主流的加拿大人。在加拿大出生的华人经常被华人新移民视为不够具有华人特色，反之，他们则视华人新移民不够加拿大人，并怨怼这些新移民深化了主流加拿大人的他者概念，因为这个 "他者" 经常包含了加拿大出生的华人。许多参与者在父母维护华人传统的家庭环境中，感觉自己更像 "华人"。他们的归属感是透过地方感协商的一组矛盾与介乎中间性 (between-ness)：包含他们视为家的地方，以及一系列以不同方式表达其身份认同及关係的改变中的公共及私人脉络。

Los inmigrantes procedentes de Hong Kong son una de las nuevas mayores cohortes de recién llegados al Canadá. Examinamos los procesos mediante los cuales los hijos de inmigrantes de Hong Kong construyen un sentido de pertenencia en Canadá, que es su única patria. En discusiones grupales, los participantes expresan un inequívoco sentido de ciudadanía y pertenencia canadiense. Sin embargo, sus identidades proyectan una individualidad canadiense idealizada, que usa ropa "canadiense", juega deportes "canadienses" y, sobre todo, que habla inglés "canadiense". La proyección siempre es relacional y espacializada, variable según el lugar y, especialmente, según que el interlocutor sea visto como familiar, o como otro "chino nacido en Canadá" (CBC, por su acrónimo inglés), o como un inmigrante chino reciente, o como un canadiense nato dominante. Mirados a menudo como no lo suficientemente chinos por inmigrantes recientes, ellos a su vez denigran de los inmigrantes re cientes como no lo suficientemente canadienses y los resienten por reforzar un concepto de otredad que tienen los dominantes canadienses, que a veces incluye a los CBCs. Muchos participantes se sienten más como "chinos" en escenarios familiares donde los padres mantienen las tradiciones chinas. Su sentido de pertenencia es un conjunto de paradojas, o de indefinición, que se negocia mediante un sentido de lugar-idad: lugar como los hogares a los cuales ellos pertenecen, pero también como un conjunto de cambiantes contextos públicos y privados donde ellos expresan sus identidades y relaciones de variables maneras.

Between 1987 and 1996, over 242,000 Hong Kongers, 14 percent of all immigrant arrivals, settled in Canada (Skeldon 1994; Mar 2005). With the 1997 handover of Hong Kong to China, the migration flow dropped abruptly, but not before these migrants transformed both Hong Kong and their Canadian destination cities of Vancouver and Toronto (Mitchell 2004; Preston, Siemiatycki, and Kobayashi 2006; Lo 2008; Ley 2010; Kobayashi, Preston, and Murnaghan 2011).

Hong Kong immigrants to Canada represent one of the most studied groups of transnationals, with strong ties to family, friends, and businesses in Hong Kong. Many established a trans-Pacific lifestyle, with some members of the family remaining in Hong Kong for work or business and others living in Canada, especially for the purpose of ensuring an international education for children, who later carry their cultural capital to Hong Kong (Waters 2003, 2006). International migration is a family strategy, with key decisions made along the life course according to demographic transitions from school to work to retirement (Kobayashi and Preston 2006). A return to Hong Kong was always planned, particularly for those who migrated as children. At the turn of the century, estimates of returnees living in Hong Kong ran over 100,000 (Ley and Kobayashi 2005). For these transnationals, migration involves constituting a sense of belonging in a new place while maintaining ties to the place of origin.[1] Research shows that this group has a strong attachment to Canada notwithstanding their high rate of return to Hong Kong (Skeldon 1994; Preston, Siemiatycki, and Kobayashi 2007; Kobayashi, Preston, and Murnaghan 2011).

Much less work has directly studied the Canadian-born children of Hong Kong emigrant parents. Here, we focus on the sense of belonging articulated by "Canadian-born Chinese" (CBCs), the children of Hong Kong emigrants who were born in Canada or who migrated to Canada before the age of sixteen.[2] As young adults who have grown up in Canada, we expect that their sense of belonging is different from that of their parents, shaped by different places and experiences (Aitken 2010; Aitken and Plow 2010). Research on second-generation Canadians from a variety of backgrounds has shown that they often have a sense of two cultures—one "Canadian," and the other that of their parents—but that they are "capable of interpreting and even re-framing the dominant narrative in a variety of ways to assert their own sense of identity and Canadianness" (Kobayashi 2008, 5). Racialized second-generation Canadians often feel the effects of everyday racism keenly, especially in educational and work contexts, and adjust their actions accordingly (Reitz and Somerville 2004; Reitz and Banerjee 2007; Brooks 2008). Our analysis explores these two processes: on the one hand, development of a sense of Canadian identity and reaction to acts of exclusion against themselves and, on the other, turning against recently arrived immigrant youth. We address the discursive production of belonging through focus groups designed to explore understandings of "Canadian" culture and identity and to find out how the CBCs experience and react to racism and discrimination.[3] Ours is not, therefore, a study of integration but of strategies of inclusion within the participants' own country.

Methodology

The eighty-six participants included in this study ranged in age from eighteen to twenty-six (Table 1), with a median age of twenty years. Approximately 45 percent were currently university students. Their

Table 1. Demographic characteristics of focus group participants

	1.5 generation	Canadian-born
Year of birth		
1975–1979	14	3
1980–1984	43	26
Place of birth		
Hong Kong	52	0
Canada	0	29
Other (China)	5	0
Year of arrival		
Pre-1985	3	—
1985–1989	15	—
1990–1994	32	—
1995–1999	4	—
No response	3	—
Not applicable	0	29
Citizenship		
Canadian only	31	25
Canadian dual	14	1
Landed immigrant	6	0
No response	6	3
Education		
Less than high school	1	0
High school graduate	11	9
College/trades	0	1
Some university	31	17
Bachelor's degree	9	0
Postgraduate	1	0
Not stated	4	2
N	57	29

parental homes were across Canada, but the largest number came from the suburbs of Toronto and Vancouver, especially the municipalities of Richmond Hill and Markham, Ontario, and Richmond, British Columbia. They were recruited with the assistance of Chinese-Canadian service groups and student clubs, and all conversations were in English. Discussions yielded more than 200 pages of rich conversation, which we recorded and transcribed before coding with N-Vivo software.[4]

We chose focus groups as the medium for collecting information because the interaction allows individuals to express, share, amplify, and sometimes even change their understanding. The focus group is not simply a group interview but a site of discursive production (Pratt 2000; Hopkins 2007). For this reason, we have selected a relatively small number of extended excerpts from focus group transcripts to illustrate in depth the interactions that occurred throughout these conversations.

We have adopted the term Canadian-born Chinese, or CBC, to refer to all our participants, because this terminology represents a socially constructed self-identity, not a literal statement of place of birth. The term is fraught, contested, variable, and in constant transformation, but early in the research we were told that many in the 1.5 generation also identify themselves as CBC, rather than as new immigrants from Hong Kong, for whom they have another acronym: FOB, or "fresh off the boat." Unlike FOB, which we discuss later, CBC does not have negative connotations either inside the community or outside. We stress, however, that all of the participants born in Canada and most of the 1.5 generation are Canadian citizens, although some of them hold dual citizenship (Table 1). Participants often emphasized that their Canadian identity derives from their status as citizens. Discussing their sense of belonging, however, they often expressed ambivalence, as a state of being in between their overseas-born parents and the dominant Canadian majority. Our analysis addresses that ambivalence.

Being CBC and Belonging Where?

As Yuval-Davis (2006, 2011) has emphasized, relationships to people and the material world, particularly family relationships, establish and reaffirm a sense of belonging. Indeed, the family, not the nation-state, might well be the primary unit for belonging (Waters 2006; Ho 2009; hooks 2009). For many contemporary migrants who actively maintain social connections and relationships across national borders, belonging is complicated, influenced by formal relations to multiple nation-states and economic, social, and political ties in several places (Preston, Siemiatycki, and Kobayashi 2007; Ho 2009).

Almost all of our CBC participants had thought about moving to Hong Kong for work after graduation, and a few who had parents in Hong Kong (known as "satellite kids"; see Waters 2003) were fairly certain that they would move there, but the majority believed that they would start their working lives in Canada. The focus group conversations nonetheless reveal a tension between feeling "Canadian," feeling "Chinese," and the places in which those feelings emerge. CBCs typically express belonging first to their Canadian childhood homes, the only homes they name when asked where they are "from." The from-ness of place has an inevitable and uncontested quality. As many of our participants were university students, from-ness is also an identity anchor to their families, a place of security against which the many social and cultural changes of university life were measured.

Despite their secure ties to Canadian citizenship, they also described their sense of belonging as contested and contradictory. Straddling what they often described as two cultures, they described never being fully accepted as Canadian or Chinese and therefore never fully at home. The following comment occurred during a focus group in Vancouver:

> CBC1-2: I think it kind of goes back to the whole roots idea, like she said, y'know, you're born there so that's where you belong in a way. You still feel like, oh my home's back there because I was born there and that's where—I was brought up there.

A discussion among Queen's University students from across Canada began with a round of stating from-ness—Vancouver, Edmonton, Toronto, and so on—and then moved on to what it means to be tied to place:

> KY2-5: . . . The thing I noticed since I've come to Queen's, like back in Vancouver, most of my friends were Caucasian, but here it's more difficult to make friends because they didn't really know me, right? 'Cause they just saw an Asian face. It's kind of weird, like now, . . .
> KY2-4: I think it depends on where you're from. I'm from Edmonton right? There's less Asians there. So people just see you as part of the regulars. I feel this way, I don't know if it's true or not. Like in Toronto or like Richmond, Vancouver, like you can be in somewhere and you won't see a white person for the whole day. And then Canada,

> like in Toronto or Vancouver, that's when people start labeling you.
> KY2-11: I think white cultured people, they do label us. Yeah.
> KY2-2: For sure. . . . It's just by how they live in, the style or culture that they live in. Say, they go to Richmond Hill, they're the only white guys. And like you go back to downtown and they're all like white guys. The Richmond Hill territory is kind of like Hong Kong.

A few minutes later, after more discussion of what it means to be from a place where identity is contested:

> KY2-9: Well, we're not blend-in multicultural. . . . Blend-in. Like everyone seems like their own little. . . . Say in Toronto right? Woodbridge is all Italians. And Toronto, Richmond Hill and places like that, it's all Cantonese, Hong Kongnese. And Downtown I'd say is mostly Vietnamese. We just have our own areas, our own places. It's not really blended into one culture.
> KY2-9: . . . Because it takes a lot of time for you to really be a part. Like to [KY2-4] how she said how it was different in Edmonton, right? Because there are so few immigrants there? So you don't see any sources of, you can't enter into your old lifestyle right? So I'm not saying you're forced to blend in, but then your core environment is different, so you have to adapt. So you slowly go into that, whereas in Toronto, the sources are there for you, so you don't see as much need to go into the mainstream.
> KY2-7: . . . When you say mainstream, do you mean Canadian-Canadian? Like a piper person? [laughter and joking; the reference is to the "Scottish" heritage of Queen's University]. . . . I think everyone has the potential to get into this mainstream, but that would involve letting go of a lot of things which people, whether it's for themselves or for others, aren't willing to give up. When I say this, I mean like traditions, and maybe some benefits that being Chinese holds. That being Canadian might not hold. And that's the biggest barrier I see.

Sense of belonging is spatially differentiated just like the relationships from which it is constituted, not only by city and neighborhood but also by context, where participants act out place-based identities. Many feel more "Chinese" in family settings where parents uphold Chinese traditions. Going to Chinese restaurants and supermarkets on a Sunday, usually with one or both parents, was mentioned regularly. Participants emphasized that activities with their parents in the homes of family and friends and in Chinese commercial and residential enclaves reinforce and confirm their sense of belonging to a Chinese population, culture, and heritage, but even more so to a particular family.

Outside of what they view as Chinese cultural places, however, CBCs are very concerned about how they represent themselves in public places where many young adults are aware of the potential for discriminatory actions from other Canadians, including others from Hong Kong, who might label them as either CBC or recent immigrant. This topic came up in every focus group. Participants described manipulating their appearance by wearing different clothes in different places. One spoke of the clothes that her aunt in Hong Kong sent, which she refused to wear because she would be marked as foreign. Others described cars, even cell phones, as markers of difference. Language is a potent marker separating Canadians and non-Canadians, posing a challenge for CBCs who conduct different aspects of their lives in different languages. Their public identity is constructed in and through fluent English spoken with a Canadian accent, although many might speak Cantonese with their parents. Participants talked about speaking "Chinglish," a dialect that combines Chinese and English, with their friends to express connections with their Chinese and Canadian roots. Language as a place-marker is fraught, however, for they are aware of the differences between themselves and recent immigrants:

> CBC1-7: I don't know, there's one class where people think I don't know English well, or I won't understand it well. And there's another group who think that I won't know Cantonese well.

> CBC1-3: I can remember back in high school I could like eavesdrop. Like I could hear Chinese people saying stuff about me, and I would just turn around and start yelling at them because I knew what they were saying. But they always saw me as someone who was only able to speak, understand English, so it just really surprised them of like how much we actually know.

Despite their claims of being at home in Canada, CBCs are very aware that parents and some older siblings feel differently:

> CBC1-2: I feel the same way because the schools are better and the air is better and Hong Kong is now like my second home, but my parents aren't like that or even my brother because they came at an older age.

> CBC1-4: . . . because they [older siblings born in Hong Kong] don't adjust as well as we do. I was here back in elementary so it's a lot easier. Especially now that there are so many Chinese people here it's harder to see it [is] an English-speaking place because everyone, like majority, is speaking Cantonese, right. It seems like Cantonese is more of a first language than English even though in school the teachers speak in English, they don't talk throughout the whole class. If they're in the classroom and you're working

by yourself so you'll be talking in Cantonese with your own friends.

For our participants, however, the advantages of fluency in English, one of Canada's two official languages, are paramount, a shield against discrimination and exclusionary processes of othering:

KY2-5: But I grew up a Canadian so I think I would probably experience less racial prejudice than other people that immigrated here and stuff. But I still think to some degree that when I go to the work force one day, I will probably face it, too. But to a lesser extent because I could probably relate to them more, and I could probably try to slide into that role where they realize that in a lot of ways I am just like them, right? 'Cause they could probably relate to me, too, 'cause I play a lot of sports. I could socially hang out with them. And I think that's the way that I could break through that barrier. But I also see a lot of people not being able to do that too, right?
Facilitator: For what reasons?
KY2-5: I think that one of the keys is language. I think that's a big barrier. And maybe that's why some of our parents are finding it difficult here because of the language. And language, I think that's a big factor in trying to fit into the society, whatever society that you're living in.

Speaking of Being Racialized

Racialization is also a set of discursive relationships that often take distinctive linguistic forms (Essed and Trienekens 2008). The earlier comments show that CBCs often view racism as something that can be avoided by being the right kind of Canadian, speaking correctly, and thereby avoiding stereotypes. They also spoke of the hurtful experiences of everyday racism, which Essed (2002) described as the experiences that one recognizes on the basis of patterns of interactions with others. Conversations invariably turned to unpleasant public encounters, often involving classmates, store clerks, and strangers.

KY2-10: . . . Yeah, I remember once I was in grade eight or something, I was sitting in class, just sitting. And one guy came up to me and said, "Why don't you . . . go back where you came from?" Okay . . .
KY2-4: Yeah, . . . somebody told me that he was shopping at Cat Centre and some white guy just went up to him and goes, "Why don't you go back home?" Like out of the blue. He was just shopping. I think as a whole, we can't say all Canadians are racist, but it exists, but sometimes in a sense, I can say that my mom, she thinks everyone is racist. She always feels that she's discriminated, but I think it's because she lost her pride. . . . Some parents are just like that.
. . . Okay, just listen to this. This is so pathetic. There's this girl, like a nice, Canadian Caucasian girl. And I was hanging out with my Caucasian friends. And I would talk to her, and she would not talk to me. She would just talk to all my Caucasian friends. And there's like one or two Caucasian, and there's an Indian individual, and whatnot, and only talk to the Caucasian people. I ask them questions, they answer to them. And we're talking about discrimination. I've had enough discrimination. Since grade school, . . . these kids would come up to me and teach me this song like, "Me Chinese, Me Know Dumb, Me Know . . ." I was like yo man, kid, I'm smacking your face. Sorry. Seriously, I had this kid in grade six, he used to pop the air out of my Nike Air's just because I was Chinese. He was like, "Go home you fool." And I'm like, "Where? Thornhill?" [laughter]
I don't know, I have to say that I've seen a lot of it and I felt a lot of discrimination, but I know for sure that not all Canadian people are like this. And just quite recently, on our school trip, we had a club trip to Montreal. And we're walking in Chinatown, there's this guy and he was like, "All you China men, you're all the same. You guys smell." And I'm like, "Yeah, Uncle Tom." I'm like, "Good luck with your pickup. That's a nice lumberjack jacket." [laughter]
KY2-5: Yeah, but there are like a lot of instances. I guess my mom had to deal with it too, 'cause like she'd come home and she'd be like, if this was Hong Kong you know, I would let somebody totally have it, right? But she just couldn't communicate to give them hell, you know what I mean? The guy, I don't know if it was a guy or a girl, but like if it was me, I would have totally just went nuts on them, just like verbally right? I would have told them off, like you know, I would have stood up for myself. But I guess my mom just couldn't 'cause she didn't have the language. She couldn't defend herself verbally, you know what I mean? And I think a lot of people have to deal with that, that can't speak English that well.
Facilitator: How do they deal with it?
KY2-5: They don't. They take it and they go home most of the time.
KY2-1: I think name-calling and telling people to their face that, I don't know, to go back on the boat or something, these are very direct forms of racism or discrimination, and you can stand up for yourself if someone does call you to your face. But let's say, I would go shopping and then, you know how at the cashier they would say, "Oh hi, how are you?" And they would start having a conversation with the person in front of you who's Caucasian, and they're really, really friendly. And then you're next in line, all you do is pay and then go, right? I mean I would consider that discrimination, too, I mean, but how are you going to pinpoint it? Like yell at the person, like,

> why aren't you making a friendly conversation with me? I mean you can't do that. If someone does call you a name, then sure, you can stand up for yourself, but there are many subtle forms of discrimination, which I guess you just have to take it 'cause you can't yell at someone for no reason. Yeah, that's how I feel. Being called in the face, you can say something about that, but other forms, it's quite hard 'cause you know that you're being discriminated against and I guess you can see that with other people of color or other immigrants. . . . I think one of the best ways to combat it is to know Canadian culture so that when someone from the Canadian culture says something to you, you know you can just hit them back with it. And then they don't have anything else to say.

Agency and control are the main focus of these discussions. Participants spent very little time talking about the racists and their practices, although their descriptions highlight the painful nature of these experiences and their banality. Outright, overt racism is easier to address than the everyday experience of being treated differently and unfairly. The patterns of disfluency, especially the frequent insertion of the filler term "like," occur most often when making direct reference to racism that cannot be easily controlled. The discussion focuses on how participants attempt to control situations, through dress, language, and knowing what they view as Canadian culture. Their cultural comfort and fluency with English are major advantages that CBCs feel equip them to deal with racism. For example, one participant argues that a quick response that emphasizes his residence in Canada, home is Thornhill, a Toronto suburb, deflects comments aiming to label him as a foreigner. The last participant also emphasizes that familiarity with Canadian culture allows him to respond effectively. CBCs consider themselves more equipped to deal with racism than their parents, who have less fluency in English and less knowledge of Canada. The discussions underscore the belief that having grown up in Canada, they have the power to effect change with their own actions and words and to dispel the words of others.

Being a CBC and Not Being a Recent Immigrant

The CBCs also deflect their experiences of racism by transferring their own prejudices to recently arrived immigrants, whom they label as FOBs. In nearly every group, the topic of relationships with recent immigrants came up, and it was always uncomfortable. The first time was at the end of a discussion about experiences of racism, when someone declared, "It's *their* fault that we experience racism." This comment invoked an immediate and vociferous response against "FOBs." Some people were visibly uncomfortable with this linguistic turn, literally squirming in their seats, avoiding eye contact with the speakers. But as the conversation shifted to follow up on why they did not want to be taken for recent immigrants, participation and consensus grew. The consensus was that they did not want to be taken for "Chinese" rather than "Canadian." What they viewed as the insistence of recent immigrants on creating a "Chinese" landscape consisting of certain kinds of cars, clothing, and, of course, language, made the CBCs vulnerable to racism and othering.

Some of the conversations were thoughtful, if fraught. The following conversation turns on calling recent immigrants "FOB," positioning the CBC self in relation to the foreign other:

> KY5-10: . . . I did have a kind of reverse racism against more recent immigrants because I didn't understand when I was younger, why they couldn't assimilate into the Canadian culture. Like if you come to Canada, why aren't you learning to speak English properly, or make friends as opposed to being a little clique?
> KY5-5: . . . I'm like, what's wrong with these people? Why can't they just assimilate into like everyone else? And stop being so segregated, 'cause that's what bothered me the most when it came down to it. Like do you guys always have to display that you guys are this group? And it really bothered me.
> Facilitator: Why do you think it bothered you?
> KY5-5: I guess partly because just the exposure to the Chinese culture. The only Chinese culture I've ever gotten was within my home, and I guess being bombarded with like so many different kinds of Chinese culture? Like you see the CBCs, you see the complete white-washed, you see the hard core Hong Kong people, and it's just like whoa, these people are all Chinese? . . . It's like, whoa, what's going on? . . .
> KY5-1: I have to say, I never knew the term [FOB], like, like that's what people call [them]?
> KY5-5: Yeah, I didn't know either until I came [here].
> KY5-1: I didn't know until someone started, who didn't even know that it was bad, to tell me what it was. I didn't understand why people would like call people that, because originally all your parents are from the same place. And these students are coming here, and I find in a way, kind of racism. . . . So when I see people coming here, I kind of think, I understand why they're all together, and a lot of them are going to go back to Hong Kong, too? I kind of accept it.
> Facilitator: Do you think that's a racist term?
> KY5-1: I did initially, but it's kind of like now, it's used like more like . . .

KY5-5: Description.
KY5-1: Yeah, it's not like I see that people aren't trying to be racist when they say it. Not try, but they don't mean it to be like that? They say, uh . . .
Facilitator: It just becomes part of regular conversation?
KY5-1: Yeah, at first when I first heard it, I thought it was so weird. And I didn't understand how people could say it, because we're all like first-generation children of immigrants, and all our parents probably had the same . . .
KY5-4: Yeah, last year, . . . I made a bunch of Asian friends, and I was so shocked at how often they used that word to describe like people that just came from Hong Kong. They just like, I remember we were at bubble tea one time, and there was a group of like people just from Hong Kong, and I was sitting with my friends, and they were just saying that word out loud, and I was so embarrassed. And I was just like, I can't believe they were just saying that in front of them, . . .
KY5-5: Yeah, when I first moved to Toronto, my cousin had to explain all the different terminology, and I was like, why would they ever make up something like that? And it just kind of, made me think, like is that right or not right . . .

This four-way conversation contains many subtleties. It was one of the more sensitive and thoughtful interactions, revealing how CBCs learn and participate in racialization, how ubiquitous the denigration of recent immigrants is among their peers, the significance of coded language to differentiate the CBCs from the others, and the ambiguous guilt that accompanies language they know to be othering, the transition from shock to normal in a conversation about the other that is actually about the self.

Conclusions

CBCs are not immigrants. They are certainly not, as common parlance and even some scholarly literature (e.g., Portes and Rumbaut 2001) describes them, "second-generation immigrants." They express an unequivocal sense of Canadian citizenship and belonging. They believe in a multicultural Canada in which they have a legitimate place. Yet their identities project an idealized Canadian body, one that wears "Canadian" clothing, plays "Canadian" sports, and, above all, speaks "Canadian" English. The projection is always relational and spatialized, variable according to place, and especially according to whether the other is viewed as family, as another CBC, as a recent Chinese immigrant, or as a dominant Canadian. Notwithstanding some recognition of the complexity of multiculturalism in Canada, their multiculturalism has only two dimensions: Chinese and Canadian. They are often viewed as not Chinese enough by recent immigrants. In turn, they denigrate recent immigrants as not Canadian enough, and they resent recent immigrants for reinforcing an otherness held by dominant Canadians that often includes them. They go to considerable lengths to differentiate themselves from recent immigrants. Yet they feel conflicted over what they see as a clash of cultures between the traditional expectations of their parents and their desire to be recognized within the Canadian mainstream. Their sense of identity and belonging is therefore a set of paradoxes of between-ness that they negotiate through a sense of place-ness: place as the homes in which they belong but also as a set of shifting public and private contexts in which they express their identities and relationships in variable ways.

Thomas (2005) described multiculturalism for American high school students as "banal," an uncontested, unexplained, and uncontextualized idealization that "racializes subjects within a narrow definition of cultural expression" (5). Our subjects echo this concept. CBCs are deeply racialized, in a relational, discursive project that involves not only the directly racist speech acts of the dominant group but also a range of conformist acts through which they position and create themselves. It is not only the attributes of speech, clothing, and the like but also the sense of agency, of controlling the situation through a secure sense of citizenship and knowledge of themselves as Canadian, that allows them to be confident of their identity and their place. Our findings are drawn from conversations with Canadian-born Chinese of Hong Kong ancestry, but additional research that investigates the strategies of inclusion deployed by children of immigrants from other origins, particularly those of Chinese background from Taiwan and the People's Republic of China, is warranted to understand how these children place themselves in Canadian society.

Acknowledgments

This research was supported by a strategic grant from the Social Sciences and Humanities Research Council of Canada, No. 829-99-1012, to Drs. A. Kobayashi (Principal Investigator), D. Ley, G. Man, V. Preston, and M. Siemiatycki. We are grateful to Janine Rose, Susanne Cliff-Jungling, and Ann Marie Murnaghan for invaluable research assistance, The Rockefeller Foundation Bellagio Centre where Valerie Preston worked on this article, and two reviewers and the editor whose comments improved the article.

Notes

1. This research is part of a larger project investigating the transnational experiences of Hong Kong Canadians in Vancouver, Kingston, and Toronto, involving 257 participants in a total of thirty focus groups, conducted in Cantonese and English. Participants were recruited from a range of social and service organizations, most notably SUCCESS in Vancouver and CCIS in Toronto. All participants received reimbursement for travel costs to meetings that took place in various public places, including community centers and libraries. The social and economic characteristics of the Hong Kongers from Toronto and Vancouver were similar (Kobayashi, Preston, and Murnaghan 2011), and the same was true for the CBCs included in this analysis.
2. By combining these two groups, we follow Portes and Rumbaut (2001), who also group together the American-born children of immigrants and immigrants who arrive as children and receive some primary or secondary education in the United States; however, we disagree with labeling this group as second-generation immigrants. The second generation is by definition Canadian because they are born or naturalized in Canada.
3. Most focus groups included men and women; however, we conducted a few groups segregated by gender. Comparison of the transcripts revealed no consistent differences in the discussions.
4. Our codes are in two parts: The first is the number of the focus group; the second is the identifier given to each individual.

References

Aitken, S. C. 2010. Not bad for a little migrant working kid. *Children's Geographies* 8:363–71.

Aitken, S. C., and V. Plow. 2010. Overturning assumptions about young people, border spaces and revolutions. *Children's Geographies* 8:327–33.

Brooks, M. 2008. Imagining Canada, negotiating belonging: Understanding the experiences of racism of second generation Canadians of colour. *Canadian Diversity/diversitè canadienne* 6:75–78.

Essed, P. 2002. Everyday racism: A new approach to the study of racism. In *Race critical theories: Text and context*, ed. P. Essed and D. T. Goldberg, 176–94. Malden, MA: Blackwell.

Essed, P., and S. Trienekens. 2008. "Who wants to feel white?" Race, Dutch culture and contested identities. *Ethnic and Racial Studies* 31:52–72.

Ho, E. L. E. 2009. Constituting citizenship through the emotions: Singaporean transmigrants in London. *Annals of the Association of American Geographers* 99:788–804.

hooks, b. 2009. *Belonging: A culture of place*. London and New York: Routledge.

Hopkins, P. 2007. Thinking critically and creatively about focus groups. *Area* 39:528–35.

Kobayashi, A. 2008. A research and policy agenda for second generation Canadians: An introduction. *Canadian Diversity/diversitè canadienne* 6:3–6.

Kobayashi, A., and V. Preston. 2006. Transnationalism through the life course: Hong Kong immigrants in Canada. *Asia Pacific Viewpoints* 48:151–67.

Kobayashi, A., V. Preston, and A. M. Murnaghan. 2011. Affect and transnationalism through the voices of Hong Kong immigrants to Canada. *Social and Cultural Geography* 12:871–88.

Ley, D. 2010. *Millionaire migrants: Trans-Pacific life lines*. Malden, MA: Wiley-Blackwell.

Ley, D., and A. Kobayashi. 2005. "Back to Hong Kong": Return migration or transnational sojourn? *Global Networks* 5:111–28.

Lo, L. 2008. DiverCity Toronto: Canada's premier gateway city. In *Migrants to the metropolis: The rise of immigrant gateway cities*, ed. M. Price and L. Benton-Short, 97–127. Syracuse, NY: Syracuse University Press.

Mar, P. 2005. Unsettling potentialities: Topographies of hope in transnational migration. *Journal of Intercultural Studies* 26:361–78.

Mitchell, K. 2004. *Crossing the neoliberal line: Pacific Rim migration and the metropolis*. Philadelphia: Temple University Press.

Portes, A., and R. Rumbaut. 2001. *Legacies: The story of the immigrant second generation*. New York: Russell Sage Foundation.

Pratt, G. 2000. Research performances. *Environment and Planning D: Society and Space* 18:639–51.

———. 2010. Listening for spaces of ordinariness: Filipino-Canadian youths' transnational lives. *Children's Geographies* 8:343–52.

Preston, V., M. Siemiatycki, and A. Kobayashi. 2006. Transnational urbanism: Toronto at a crossroads. In *Negotiating borders and belonging: Identities and practices in Canada*, ed. L. Wong and V. Satzewich, 91–110. Vancouver, Canada: University of British Columbia Press.

———. 2007. The dual citizenship of Hong Kong Canadians: Convenience or commitment? In *Dual citizenship in global perspective: From unitary to multiple citizenship*, ed. T. Faist and P. Kivisto, 203–26. New York: Palgrave Macmillan.

Reitz, J. G., and R. Banerjee. 2007. Racial inequality, social cohesion, and policy issues in Canada. In *Belonging? Diversity, recognition and shared citizenship in Canada*, ed. K Banting, T. J. Courchene, and F. L. Seidle, 489–545. Montreal, Canada: Institute for Research on Public Policy.

Reitz, J. G., and K. Somerville. 2004. Institutional change and emerging cohorts of the "new" immigrant second generation. *Journal of International Migration and Integration* 5:385–415.

Skeldon, R. 1994. *Reluctant exiles?: Migration from Hong Kong and the new overseas Chinese*. Armonk, NY: M. E. Sharpe.

Thomas, M. 2005. "I think it's just natural": The spatiality of racial segregation at a US high school. *Environment and Planning A* 37:1233–48.

Waters, J. L. 2003. "Satellite kids" in Vancouver: Transnational migration, education and the experiences of lone children. In *Asian migrants and education*, ed. M. W. Charney, B. S. A. Yeoh, and C. K. Tong, 165–84. Dordrecht, The Netherlands: Kluwer Academic.

———. 2006. Geographies of cultural capital: Education, international migration and family strategies between Hong Kong and Canada. *Transactions of the Institute of British Geographers* 31 (2): 179–92.

Yuval-Davis, N. 2006. Belonging and the politics of belonging. *Patterns of Prejudice* 40:197–214.

———. 2011. *The politics of belonging, intersectional contestations*. London: Sage.

Diasporic Families: Cultures of Relatedness in Migration

Joanna C. Long

Independent Researcher, Manchester, UK

Family is central to migration. It can be the reason and mechanism for movement, as well as a key context in which the effects of migration are felt. Recent work has demonstrated that families are not necessarily destroyed by migration but rather that migrants develop new ways to maintain a sense of family over distance. Focusing on the doing rather than the definition of family, this article explores the idiosyncratic cultures of relatedness through which people make family in diasporic contexts and the associated tension between what family is and should be. The article examines how family is deployed as both an enduring and elastic migration resource and adds to arguments that distance does not foreclose familial intimacy. It also questions the presumed correlation between proximity and intimacy, in which geographical and emotional closeness are conflated, and the making of family over larger distances is ignored. Drawing on in-depth interviews with Palestinians living in Britain, the article examines practices of making family through visiting relatives in different parts of the world, in conjunction with broader opinions and experiences of family. It explores how migration can reveal the tensions of geography and kinship—tensions between the perceived incontrovertibility of family relationships and the need to continually construct those relationships across space. I conclude that migration precipitates a reworking of family relationships and a reimagination of the very notion of family itself but that these new formations remain tied to conventional ideas of "proper" family as a spatially and temporally situated unit, with potentially painful implications for feelings of family relatedness.

家庭是移民的核心。家庭可以是迁徙的理由和机制, 亦是迁徙的影响被感知的关键脉络。晚近的研究已証实, 家庭并不一定会因迁徙而被摧毁, 反之, 迁徙则建立了远距离维繫家庭感的崭新管道。本文聚焦家庭的实作, 而非家庭的定义, 探讨人们在离散的脉络中打造家庭的亲缘关係之特殊文化, 以及 "何为家庭" 与 "家庭应该如何" 之间的紧张关係。本文检视家庭如何同时部署做为持久与弹性的移民资源, 并补充有关距离并不会消解家庭亲密性的主张。本文亦质问邻近性与亲密性之间的预设关联, 其中地理与情感的亲近被合而为一, 而较为远距的家庭打造则被忽略。本文对居住于英国的巴勒斯坦家庭进行深度访谈, 藉此检视透过探访世界各地亲戚的家庭打造实践, 及其与更广阔的家庭概念和经验的结合。本文探讨移民如何得以揭露地理和亲属关係之间的张力——在所感知的清楚家庭关係, 与不断跨越空间建构这些关係的需求之间的张力。我将在结论中主张, 移民促使了家庭关係的再造, 以及家庭观念本身的再想像, 但这些新构成, 仍然与做为座落在同一时空单位的传统 "适当" 家庭概念密切相关, 并可能对家庭亲缘关係的感觉产生痛苦的意涵。

La familia es un elemento central de la migración. Asuntos familiares pueden ser la razón del movimiento y de los mecanismos involucrados, lo mismo que el contexto clave dentro del cual se sienten los efectos de la migración. El trabajo reciente demuestra que las familias no necesariamente se destruyen por la migración sino por el contrario que los migrantes desarrollan nuevas maneras de conservar y enaltecer el sentido de familia por encima de la distancia. Enfocando más la definición de familia que las acciones relacionadas, este artículo explora las culturas idiosincráticas de pertenencia familiar con las cuales la gente define el ámbito familiar en contextos diaspóricos y la tensión asociada entre lo que la familia es y lo que debería ser. El artículo examina cómo la familia es desplegada como recurso de migración perdurable y elástico, contribuyendo a la argumentación de que la distancia no obstaculiza la fortaleza familiar. Cuestiona también la supuesta correlación entre proximidad e intimidad, donde la cercanía geográfica y la emocional se fusionan, y la construcción de familia sobre distancias más grandes es ignorada. Con base en entrevistas de profundidad con palestinos residentes en Gran Bretaña, el artículo examina las prácticas para construir familia por medio de familiares visitantes en diferentes partes del mundo, en conjunción con opiniones más amplias y experiencias familiares. El artículo explora cómo la migración puede sacar a flote las tensiones entre geografía y parentesco—tensiones entre el carácter incontrovertible percibido de las relaciones familiares y la necesidad de edificar continuamente esas relaciones a través del espacio. Concluyo que la migración precipita el replanteamiento de las relaciones familiares y una re-imaginación de la propia noción de familia, pero que estas nuevas formaciones se mantienen atadas a ideas convencionales de la "propia" familia, como una unidad situada espacial y temporalmente con implicaciones potencialmente penosas en lo que al sentimiento de parentesco se refiere.

Family is central to migration. Whether it is the responsibilities to protect and provide for families that act as motivation to move, the practical and supportive networks of relatives that enable migration, or the trail of family members following a leading person, family can be the reason and mechanism for movement, as well as a key context in which the effects of migration are felt. Interest in families and migration has developed considerably in recent years and has contributed much to our understandings of how lives on the move are practiced. Special issues of *Global Networks* have explored transnational marriages (Charsley and Shaw 2006) and communication practices (Panagakos and Horst 2006), and contributors to the *Journal of Intercultural Studies* have drawn attention to the dynamics of transnational family belonging and intimacy (Skirbiš, Baldassar, and Poynting 2007; Svašek 2008). An important argument of this literature is that distance does not necessarily rupture or destroy families and family relationships; rather, people find new ways of connecting with one another, often using a range of communication technologies, and so maintain senses of mundane intimacy that are so central to family life in many contexts (Wilding 2006). Far from writing off families, migration opens them up to new ways of operating and of imagining themselves, fostering new cultures of relatedness over sometimes great distances. Moreover, by taking families out of their normative contexts, migration can expose ever-present but previously unnoticed aspects of family relationships. This is not to say that conventional or normative ideas about family and place, proximity, and intimacy have been left behind by those living transnational or diasporic lives. As Brah (1996) argues, diaspora space is "where the permitted and the prohibited perpetually interrogate; and where the accepted and the transgressive imperceptibly mingle even while these syncretic forms may be disclaimed in the name of purity and tradition" (208).

These cultural and emotional geographies are the latest iteration in a longer history of research into family migration. Led by economists and sociologists in the 1970s and 1980s, researchers explored the dynamics of family migration first through the lens of human capital, subsequently modifying this to account for gender in migration decision making (Cooke 2008b). Geographers waded into this debate in the 1990s, bringing with them an enhanced appreciation of intersecting social forces, geographical context, and change over time (see, e.g., Clark and Huang 2006; Mulder 2007; Cooke 2008b). Although this spatial turn in family migration still has further to go, family is coming to be recognized as a factor in almost all forms of migration and in other spatial processes, such as gentrification, studentification, and counterurbanization (Smith 2011). Family matters within cultural geography, too, although it has taken some time for it to be recognized as something more than the context in which other concerns about gender, sexuality, home, childhood, and so on, play out (Valentine 2008; see also Harker and Martin 2012). Indeed, as recent work has shown, taking family seriously means that there is much more to say about ancestry, place, and belonging (Nash 2005, 2008); about state immigration and international employment policies (Lee and Pratt 2012; Martin 2012); and about global labor markets (Pratt 2012). In each of these contexts family is symbolically, materially, and affectively enrolled in circuits of power operating at and between multiple scales, from the household, the body, and blood, to race, nation, and humanity. Others have homed in on the more intimate micropolitics of relatedness both within single households and across the world, opening up families as sites for the reproduction of certain social practices (Valentine, Jayne, and Gould 2012), as well as the politics of care and impact of separation on relationships between migrant parents and their left-behind children (Graham et al. 2012). Moreover, what these two latter studies highlight is the disjuncture between ideas of family and family-in-practice; between those ideals and aspirations we live *by* and the people we live *with* (Valentine, Jayne, and Gould 2012).

This article is concerned with what it means to live with this disjuncture, how it is shaped by diasporic life, and what are its effects are on family relatedness. In short, I am concerned with how people *do* family in the context of migration; that is, the various ways in which people operate as, experience, conceive of, and mobilize family in a diasporic context and all of the tensions and contradictions that go with that. To do so I draw on qualitative research conducted with Palestinians in Britain between January 2008 and February 2009. This consisted of repeated, in-depth individual and group interviews with seventeen people, usually in their own houses. Each conversation had a theme: family, home, and community. This was to ensure that personal, emotional, and potentially challenging questions about home were placed in the middle of the research process, flanked by more open questions oriented toward storytelling and social life. All interviews were semistructured to create flexibility for participants to steer the interview toward topics of importance to them (Dunn 2000). Furthermore, repeat interviews enabled topics to be discussed in greater depth, getting "beneath the surface" (Thomson 2006, 580), as the recurrence of issues over the course of several meetings provided insight into their significance and allowed for the meanings of such issues to be more fully explored. In the first interview on family, participants were invited to tell the story of how they came to Britain and about life as part of a dispersed family. The second interview began with questions about domestic spaces and significant houses in participants' lives, before moving on to ask about the meaning of home both in terms of places and relationships. In preparation for this interview, participants were invited to select a significant object or room in the house to talk about. Many forgot to do this, and if nothing sprang to their minds I would ask about an object mentioned previously to prompt discussion. The final interview was often shorter—sixty to ninety minutes, compared with two to three hours for the previous interviews—and focused on social relationships, community, and the importance of conversations about family names and belonging (Long 2011).

In this article I explore cultures of relatedness over distance and the impact that these have on how people experience and imagine family. I do so by focusing on practices of making family through visiting relatives in different parts of the world, in conjunction with broader opinions and experiences of family. I do not address other practical strategies of intimacy, such as international communication technologies (ICTs), as these have been richly explored by others (see, e.g., Panagakos and Horst 2006; Valentine 2006; Wilding 2006; Baldassar 2008). My key concerns are with the ways in which people attempt to construct certain kinds of family over distance, the ideals of family on which these constructions rely, and the (potentially ambivalent) effects of these efforts on feelings of relatedness. I am interested in how these cultures of relatedness are organized in space and time and the tensions between endurance and elasticity around the meaning and practice of family. Before proceeding, I would like to take a brief tour of debates about families in migration, cultures and geographies of relatedness, and the spaces and spacings of Palestinian families.

Cultures of Relatedness

Recent work in kinship studies has sought to move beyond nature versus nurture arguments to think about families through their own cultures of relatedness; that is, their own idiosyncratic ways of being connected to one another that need not conform to conventions nor to any "pre-given analytic opposition between the biological and the social" (Carsten 2000b, 4). Anthropologists have led the way in researching such culturally specific forms of relatedness, sensitive to the intersections between biology and society in the production of relatedness (Carsten 2000a; Franklin and McKinnon 2001; Galvin 2001; Stone 2001). The geographies of the family relationships being explored by anthropologists are clear not simply because they place kinship in social, cultural, national, and ethnic contexts but because they show how family is linked to the production of place and identity and how closeness and distance among family can be a function of physical and emotional proximity (Edwards and Strathern 2000). Proponents of these new kinship studies argue that such politics of family, place, and identity in context might be productively approached by focusing on the doing rather than the definition of kinship. This recognizes the enduring power and potency of "ideas of incontrovertible bonds based on blood" but situates them alongside the "routine practices of choosing kin" within which blood is also a "flexible criteria for relatedness" (Nash 2005, 452). From this perspective, a relative who is claimed by blood might also be "disowned through lack of social interest" and likewise a lapsed relative might be reclaimed "through resurrected biological links" (Edwards and Strathern 2000, 160). Moreover, thinking kinship as a doing opens up non-normative ways of being related that go beyond conventional heterosexual arrangements of partners, parents, and children produced and legitimated by Euro-American law (Borneman 2001; Butler 2002).

Processes of settling within diasporic contexts have the capacity to bring these doings of kinship into relief as all kinds of networks are mobilized for advice and support (Creese, Dyck, and McLaren 1999). For Palestinians in transnational and diasporic contexts, kin networks provide valuable resources for survival. Indeed, "family cohesion" has been described by some as "the cornerstone in the Palestinians' ability to regroup and reestablish their network" in diasporic contexts (Ghabra 1987, 50), although conventional ideas of family and practices of kinship are also being reshaped along the way. In one sense, cohesion reinforces norms of family, as conventions of providing for one's family can be influential in the very decision to migrate in search of employment and economic opportunities (Hilal 2006). In Kuwait, maintaining certain traditions, such as marriage negotiations, ensured that connections among dispersed relatives were also maintained (Ghabra 1987). In another sense, however, the exigencies of migration and the importance of relatedness in diasporic contexts can open up new possibilities for the doing of kinship. For Palestinians in Kuwait, Ghabra suggests that important family support networks, which in Palestine would have been composed primarily of paternal relatives, now expanded to include maternal kin. Family is being deployed here as both an enduring and elastic migration resource: It is both a stable framework that imbues economic migration with cultural, social, and emotional meaning and a set of connections that evolve and expand according to changing social values and personal needs.

But what effect does this reconfiguration have on what it feels like to be family in diasporic contexts? Recent work on transnational families has demonstrated that, although proximity might indeed facilitate emotional closeness among people through seeing one another regularly, sharing daily joys and struggles, and so on, distance does not foreclose familial intimacy, and global migrants of all kinds put considerable effort into the "emotional labour" or "transnational kin-work" involved in maintaining intimate relationships transnationally (Svašek 2008, 224; see also Parreñas 2005; Skrbiš 2008). Much of this research focuses on the communication practices of transnational families, particularly the reliance on a range of established and emerging ICTs (Yeoh, Huang, and Lam 2005; Panagakos and Horst 2006; Valentine 2006; Wilding 2006). Baldassar (2008) suggests that at the heart of this emotional labor are concerns about familial absence and feelings of longing for family togetherness that are ameliorated by cultivating various forms of copresence. In doing so, Baldassar situates ICTs as just one form of copresence (virtual) among other ways of feeling and imagining distant loved ones, including objects such as letters, gifts, and heirlooms that are associated with distant people (copresence by proxy); thinking of or praying for others (imagined copresence); and visiting (physical copresence). Digging deeper into the politics of technologically-mediated transnational family relationships, however, Madianou and Miller (2012) point out the inherent mediated-ness of all relationships. Through their research with Filipina mothers working in Britain and their left-behind children in the Philippines, Madianou and Miller illuminate the triadic tension between normative ideas of family relationships, how that relationship actually functions, and the promise of communication media to resolve the discrepancy. This tension within transnational family relationships signals a broader need to question notions of family itself in the context of migration. If ICTs challenge normative ideals about the relationship between the communicating individuals, on what are such normative ideals based, what is at stake in maintaining them, and at what cost?

In the next section I investigate how certain ideas of family are promoted, organized, and contested and with what effects. I am particularly interested in the dynamics of endurance and elasticity of relatedness among Palestinians living in Britain, but I am also concerned with how these play out in practice and across different spaces and their implications for feelings of family relatedness. Following Harker (2010), my focus is on the imaginative and practical spacings of family: how family is conceived of and actively produced through particular places and moments, the uneasy relationship between these conceptualizations and newer configurations of family in a diasporic context, and how differences of opinion regarding the constitution of family reverberate through family relationships and experiences of relatedness.

Doing Diasporic Palestinian Families

In this section I explore how participants spoke about the meaning of family and the effects of migration, as well as their efforts to construct particular senses of family relatedness over distance. My concern is with the tensions between ideals and practice, memory and experience, singularity and multiplicity, mundanity and magic, as well as the spatial and temporal organization of both actual and imagined family. I begin with concrete practices of making family before moving on to wider ideas of what family means and feels like in migration.

For many participants in this study, the number of people in their everyday family circle had shrunk considerably during migration. This was often because during the *Nakba* and the Gulf War people had used whatever connections they could—siblings, cousins, in-laws, colleagues—to find safety; they have since ended up scattered across the Gulf, the Middle East, Europe, and North America. Over the decades, however, many of the fragmented family units that ended up in Britain remained small due to way that the normative nuclear family is coded into everyday life through, for example, the packaging of food in portions of four or the limited availability of affordable cars and houses for more than four people. Moreover, the lack of additional family members around to help in the caring, supervising, and disciplining of children even led one couple to consciously decide against having more than three children, even though they desired a much larger family. Visiting family in different parts of the world was therefore considered by many participants (especially those who were parents) to be important, partly because it immersed their children in the large Arab family life in which they themselves had grown up but that their children miss out on in Britain and also because it helped to nurture connections between diasporic children and their relatives, teaching them "ways" of family not experienced in their everyday context and to foster a sense of, if not home, at least cultural belonging (Ali and Holden 2006). One man, Jameel, elaborated on the importance of this in particular detail. At the time we spoke, he himself had not seen his siblings in the United States for eight years because he refused to travel there while George W. Bush was president, yet he continued to send his daughter, Layla, each summer to spend time with her cousins. Because Layla was an only child growing up in a diasporic context, Jameel said that she has not had his experience of a family life with lots of siblings, but he felt that these visits to the United States have enabled closer relationships with her multitude of aunts, uncles, and cousins, which he considers central to her cultural heritage and future:

> She missed having a brother or a sister, like that. Well, there they are [in the United States]. You know, my brothers have many of them and she has a kick out of visiting with them and they become her insurance. Life insurance. So that's Eastern, it's not Western. Definitely it's Eastern. . . . Very important. Very, very important. There's no insurance in life other than family, that's my view. So that's it, she has many homes. If something happens, she has many homes. It's simple like that.

Jameel's motivations for sending his daughter to the United States every year are based on a normative (and idealized) notion of Arab family as extended and mutually supportive. He also implicitly implies that Layla's small, British-based family cannot provide the same life insurance. Large family there thus compensates for small family here by cultivating a domestic closeness between Layla and her relatives over great distances: "she has many homes." Moreover, his notion of life "insurance" twins family and home in a manner that mobilizes ideas of home as both a scene of safety and a set of incontrovertible relationships and naturalizes the incorruptible bonds of kinship by casting family as unconditional protectors. At the same time, however, the purpose of Layla's visits is to produce these ostensibly organic bonds, which might otherwise be weakened by physical distance. Thus, although Jameel argues that security (i.e., home) is a simple (i.e., natural) matter of family, he implicitly understands that kinship is also "a social process in which the relations that matter are selectively performed" (Nash 2005, 452) and that, without these transatlantic visits, his daughter's life insurance would be compromised.

Normative ideals of what family is or was and what it could or should be had considerable currency among many participants. Tawfiq (a first-generation refugee from Palestine and now a father) claimed that, as part of a nuclear household in Britain, "there is no family life." Similarly, Alifa (the daughter of first-generation refugees from Palestine) felt that "family was taken from us" by forced migration and that the relationships they did have with their grandparents, aunts, uncles, and cousins were an approximation of what being a family could be. Others were more ambivalent about their parents' attempts to deliberately cultivate certain kinds of family relatedness through visiting. This was the case for Maryam, who declared that visits to see relatives in Jordan could be overwhelming:

> I don't like families too much over there. It's very much, like, pressurized. Where the family is everything, where there's no sort of outside or personal space because the family is so close-knit that it gets just too intense. . . . I think with a lot of, like, Arab families, especially, like, in Arab countries, it's very intense, you know, people in and out all the time and there's no personal space at all and you tend to live quite close to each other, you know, even in the same building sometimes, you know, flats above each other and things. You know, if you're always seeing family in your face and there's no other sort of social contact apart from people who are related to you in some way and it just gets really much [*laughs*].

Although Maryam acknowledged the love with which all of this attention is bestowed ("they do it 'cause they care about each other"), she is used to having her own space:

> We were raised in a sort of different environment where you do have your own space and you respect other people's boundaries and so it's very different going over there and suddenly feeling like, you know, it's fine for anyone to just

come in and out . . . it's just all, like, different sort of family [*laughs*].

As a mechanism for constructing certain kinds of family relatedness and maintaining ideals of Arab family intimacy over distance, visiting has mixed results. The stated aim for many was to cultivate emotional closeness beyond the nuclear family, something that was often seen as an Arab or, as Jameel put it, an "Eastern" thing. This exposure to other ways of doing family also constitutes immersion in Palestinian-Arab cultural practices, offering diasporic children opportunities to connect with their cultural, geographical, and genealogical roots and offering their parents a chance to temporarily recapture some aspects of family life that were lost in forced migration. These promises are unevenly realized among diasporic children. Whereas Layla was reported to get a "kick" out of visiting her cousins in the United States (feelings that were echoed by several other second-generation participants), Maryam felt suffocated by the attention and intrusion of proximate family life, which contrasted so sharply with her own expectations about domestic space and privacy growing up in Britain (for more on the ambivalence of diasporic visiting, see Wagner 2008). Thus, in diasporic contexts, experiencing multiple ways of family life can multiply ideas of family itself, adding it to the list of apparently singular things (home, homeland, identity, nation, belonging, etc.) that assume greater diversity through migration.

What is more, participants' conversations about their personal meanings and experiences of family illuminated specific ways in which ideas about family are multiplied that had important implications for family relationships and feelings of belonging. During a group interview with Faruq and his two children, Noura (aged twenty-four) and Zaki (aged sixteen), Faruq was explaining that moving to Britain had changed his entire experience of family, from one that included "grandparents, cousins, aunties, and their kids" to one that was limited to the occupants of their suburban house. Noura, his daughter, then took the opportunity to expand on her own interpretation of family as, first, her parents and siblings and, second, her aunts, uncles, cousins, and grandparents. Faruq's displeasure was obvious:

Faruq: [*quietly*] [If] you will not follow my way, you will disintegrate.
[*pause*]
Noura: [*inaudible*] I'm not saying all-!
Faruq: I am telling you. OK?

The atmosphere bristled for several minutes afterward as Noura, attempting to defuse the situation, quickly moved to rationalize her perspective as the result of generational differences and her upbringing in Kuwait and Britain, away from the extended family network that her father holds so dear:

> I think probably with our generation . . . we do have that feeling [of belonging to a larger family] . . . [but] whereas maybe with our parents and my dad's generations it'd be something that's there, twenty-four-seven, no matter what, I think with us it's something that's sort of on an occasion, be it sad or happy, that, you know, my dad's definition of family sort of more kicks in. . . . I think my dad's definition of family definitely has a lot of advantages . . . I don't disagree with him, but I think just my way's different.

On one level this disagreement dramatizes the way in which singular things can multiply in diasporic contexts. More importantly, however, Faruq is casting normative Arab and British configurations of family in opposition. He promotes an image of Arab family relatedness as wide and deep and his British-based family as small and superficial. Echoing arguments that normative family formations are the origins and future of personhood, society, and culture (Butler 2002), a large Arab family is for Faruq not just a matter of individual connection but of personal survival, and his daughter's identification with a more flexible, diasporic family relatedness is the path to existential doom. Equally interesting is the way in which different versions of family are situated by each of them in space and time. Noura's two families are, first, her parents and siblings living together in Britain and, second, her extended family in Jordan, whom she visits every few years. In other words, different kinds of family (nuclear and extended) take place in different locations (Britain and Jordan) and with different temporalities (everyday and occasional). Faruq, however, believes only in the extended network of relatives, refusing to acknowledge the validity of his smaller family unit in Britain. As such, family for him is a memory of when he lived as a child not only with his parents and siblings but with grandparents, uncles, aunts, and cousins under one roof in a "proper family house." The spatiality and temporality of family here is still split, although not as it was for Noura: his family ideal is firmly rooted in one place (a house in Palestine) and time (pre-1948 childhood), even while his daily life revolves around a smaller, immediate family in Britain.

On the face of it, this spatial and temporal organizing of family appears to follow generational lines, in which the father's ideal is based on memories of his childhood in Palestine, which differs from his daughter's experience of growing up Palestinian in Britain. Comments by other participants, however, suggest that such a straightforward interpretation is misleading. Alifa and Mai, for example, who are of a similar age to Noura and whose parents fled from Palestine to Kuwait in 1948 and moved to Britain in the early 1980s, echoed Faruq's longing for past togetherness, despite not having experienced it themselves. Specifically, they imaginatively mourned the impossibility of having all of the members of their parents' immediate families together in one place:

> Alifa: I do find it sad though that, you know, [they used to share], you know, one bedroom and now they're never going to all . . . people come over but you're never going to get seven of them . . . in one country let alone, you know, one room.
> Mai: I was just saying actually the other day, my dream has always been that I'd be living [with] all, you know, six of my uncles and aunties, my dad's, and all my cousins and my grandparents in one house or in one country, so that we could all be united.

These fantasies of cohabitation romanticize both exile and proximity. Alifa and Mai's parents lived with their parents and siblings in one bedroom because they had fled a war zone in Palestine and were struggling to set up a new life in Kuwait with little money and few possessions. Nevertheless, Alifa and Mai consider the contrast between such intense togetherness and the wide dispersal they currently experience to profoundly symbolize their loss of family. Moreover, Mai's dream of sharing a house with her extended family dramatizes a certain diasporic idealization of family togetherness. Alifa acknowledged and explained this idealization when comparing her attitude toward family to that of her husband, whose family are from and continue to live in their ancestral village (now part of Israel) where he grew up next door to his uncle and a few minutes' walk from numerous other aunts and uncles. Alifa sees her husband's family as having a much more relaxed attitude toward one another as a result of this proximity:

> I don't know if it's just they've got different personalities . . . but I do feel they were all . . . laid back about family . . . it can be more of a nuisance to them, you know. But whereas we've never had—family has [never] been a nuisance because we've never been—you know, I mean, you get to meet up on such special occasions it's only positive.

Alifa envies her husband's laid-back attitude toward family and his ability to sometimes find his family "a nuisance" because it means that they are sufficiently involved in one another's everyday lives that each encounter is not a sacred, rarefied experience. "We never really had that," said Alifa, for whom being among relatives "was really quite magical." At the same time, Alifa acknowledged that she and her sisters had a somewhat distorted perspective:

> I think the reality is not, like, you know, it's not all love and absolutely amazing [and] if you do live near them there are problems and things, but to us it always means quite a lot.

The way in which migration multiplies notions of family is organized in space and time, between family "over there" and "back then," and family here and now (as well as in the future). Drawing on ideals and experiences of what it is, was, and could or should be, some aspects or versions of family are consigned to a spatial and temporal *elsewhere*. These elsewheres, which are based on memories and rememories (Tolia-Kelly 2004), are also fantasies of previous and future cohabitation and proximity. That is, the past togetherness being mourned does not have to have been directly experienced to be remembered, missed, and desired. Moreover, the extent to which this "elsewhere family" is mourned varies from person to person, not necessarily following generational lines. Indeed, I got the sense that Alifa and her sisters lamented the loss of family more than their parents, who said that they felt liberated by their British passports and the associated freedom to visit relatives internationally, which had not been possible when they lived in Kuwait. Similarly, Noura appeared to revel in her two families, among whom she felt emotionally close but in different ways, whereas Maryam found it all a bit too much. The alternative experiences of family that emerge through migration are thus shaping alternative ideas about family for those in diasporic contexts. But these innovations can have profound implications for the self and for family relatedness. Faruq predicted existential doom for those straying from the ideal of proximate, extended family, whereas for Alifa and her sisters, the denial of such a family and the mundane annoyances associated with it distorts their family relatedness by imbuing it with an inauthentic magic.

Conclusion: Making and Shaping Relatedness over Distance

Migration clearly multiplies ideas of family and experiences of family life. For some in this study, their new, smaller unit is considered an approximation or impoverished version of the "true" family life taking place in a spatial and temporal elsewhere. Visiting is one means of reestablishing both the proximity and the intimacy associated with "proper family." Although visits can take many forms (regular, crisis, duty, tourist, etc.; Baldassar 2008), the core value is the same: the exposure to and cultivation of cultures of relatedness rooted in physical and emotional closeness. In the process, fantasies of shared domesticity, family togetherness, and "life insurance" are both indulged and challenged. Family here is being carefully organized in space and time, most obviously in the everyday domestic spatialities of family life and the temporary, idealized, domestic togetherness of visits (see Long 2013) but also in the enormous distances traveled to achieve proximity, travel that is predicated on UK citizenship and the free passage a British passport enables under international travel agreements. At the heart of the practices discussed in this article is a sense of what family should be but is not or cannot be due to forced migration and an enduring determination to sustain these ideals anyway (Valentine, Jayne, and Gould 2012). This is not a case of normative versus non-normative versions of family but, rather the replacement of one normative arrangement (large, supportive Arab family) with another (small, individualistic British family). For those in my research, submitting to a different family form is not a choice. British life is implicitly organized around the nuclear family, from the size of affordable houses and cars to the packaging of food in sets of four. The costs and work involved in having a large family in Britain, without the support network of an extended family, mean that family formations are hardly a choice at all, especially when weighed against the ability to visit family scattered across the globe on a semi-regular basis or the desire to educate one's children privately.

Migration highlights these doings of kinship as well as reconfiguring them. That so much of this kin work is centered on achieving intimacy demonstrates the considerable effort required to create the kinds of knowing, loving, and caring relationships that are beyond words (Berlant 2000). Similarly, the effort made by all participants to remain connected with kin elsewhere suggests an awareness that family relationships are made as well as given. As I have shown, however, this might be at the cost of more casual kinds of relatedness that permit irritation and argument, not one's best behavior, as well as glossing over ambivalence toward family visits. Wagner's (2008, 2011) work with European Moroccans on holiday in Morocco has begun to explore the subtle politics of such visits, but more of this kind of ethnographic research is required to better understand the coincidences and contradictions between ideals and experiences of visiting, especially among different members of the same family. Furthermore, the effort invested in constructing particular kinds of familial relationships in migration can reinforce conventional modes of family that are seen to trump other experiences of relatedness and domestic life. The deeper connections forged with friends and neighbors in the context of migration are a case in point. Although some research has touched on how such connections replace, compensate for, or stand in for family (Alexander, Edwards, and Temple 2007), more work is needed on the ways in which such relationships are measured against and valued over so-called biological family. Pursuing such research promises to deepen and widen our understanding of diasporic relatedness, particularly the endurance and elasticity of family: migration might precipitate the multiplication and reconfiguration of family, both in meaning and in practice, but at the same time it can remain tied to and curtailed by conventional ideas about what counts as intimacy and who counts as family.

Acknowledgments

I am grateful to Richard Wright and one anonymous reviewer for their valuable comments and to Chris Harker, Lauren Wagner, and Richard Phillips for their feedback on an earlier draft. The original research was funded by the ESRC (Ref: PTA-031-2005-00061) and benefited enormously from the supervision of Alison Blunt and Catherine Nash. All errors are my own.

References

Alexander, C., R. Edwards, and B. Temple. 2007. Contesting cultural communities: Language, ethnicity and citizenship in Britain. *Journal of Ethnic and Migration Studies* 33 (5): 783–800.

Ali, N., and A. Holden. 2006. Post-colonial Pakistani mobilities: The embodiment of the "myth of return" in tourism. *Mobilities* 1 (2): 217–42.

Baldassar, L. 2008. Missing kin and longing to be together: Emotions and the construction of co-presence in transnational relationships. *Journal of Intercultural Studies* 29 (3): 247–66.

Berlant, L. 2000. Intimacy: A special issue. In *Intimacy*, ed. L. Berlant, 1–8. Chicago: University of Chicago Press.

Borneman, J. 2001. Caring and being cared for: Displacing marriage, kinship, gender and sexuality. *International Social Science Journal* 49 (154): 573–84.

Brah, A. 1996. *Cartographies of diaspora: Contesting identities*. London and New York: Routledge.

Butler, J. 2002. Is kinship always already heterosexual? *Differences* 13 (1): 14–44.

Carsten, J., ed. 2000a. *Cultures of relatedness: New approaches to the study of kinship*. Cambridge, UK: Cambridge University Press.

———. 2000b. Introduction: Cultures of relatedness. In *Cultures of relatedness: New approaches to the study of kinship*, ed. J. Carsten, 1–36. Cambridge, UK: Cambridge University Press.

Charsley, K., and A. Shaw, eds. 2006. South Asian transnational marriages in comparative perspective [Special issue]. *Global Networks* 6 (4).

Clark, W. A. V., and Y. Huang. 2006. Balancing move and work: Women's labour market exits and entries after migration. *Population, Space and Place* 12:31–44.

Cooke, T. J. 2008a. Gender role beliefs and family way. *Population, Space and Place* 14:163–75.

———. 2008b. Migration in a family way. *Population, Space and Place* 14:255–65.

Creese, G., I. Dyck, and A. McLaren. 1999. Reconstituting the family: Negotiating immigration and settlement. Research on Immigration and Integration in the Metropolis Working Paper Series No. 99-10, University of British Columbia, Vancouver, Canada.

Dunn, K. 2000. Interviewing. In *Qualitative research methods in human geography*, ed. I. Hay, 50–82. Oxford, UK: Oxford University Press.

Edwards, J., and M. Strathern. 2000. Including our own. In *Cultures of relatedness: New approaches to the study of kinship*, ed. J. Carsten, 149–66. Cambridge, UK: Cambridge University Press.

Franklin, S., and S. McKinnon. 2001. *Relative values reconfiguring kinship studies*. Durham, NC: Duke University Press.

Galvin, K.-L. 2001. Schneider revisited: Sharing and ratification in the construction of kinship. In *New directions in anthropological kinship*, ed. L. Stone, 109–24. Lanham, MD: Rowman and Littlefield.

Ghabra, S. N. 1987. *Palestinians in Kuwait: The family and the politics of survival*. Boulder, CO: Westview.

Graham, E., L. P. Jordan, B. S. A. Yeoh, T. Lam, M. Asis, and Su-kamdi. 2012. Transnational families and the family nexus: Perspectives of Indonesian and Filipino children left behind by migrant parent(s). *Environment and Planning A* 44 (4): 793–815.

Harker, C. 2010. On (not) forgetting families: Family spaces and spacings in Birzeit, Palestine. *Environment and Planning A* 42 (11): 2624–39.

Harker, C., and L. L. Martin. 2012. Guest editorial. *Environment and Planning A* 44 (4): 768–75.

Hilal, J. 2006. Emigration, conservativism, and class formation in West Bank and Gaza Strip communities. In *Living Palestine: Family survival, resistance and mobility under occupation*, ed. L. Taraki, 185–230. Syracuse, NY: Syracuse University Press.

Lee, E., and G. Pratt. 2012. The spectacular and the mundane: Racialised state violence, Filipino migrant workers, and their families. *Environment and Planning A* 44 (4): 889–904.

Long, J. C. 2011. Daar al Falastini: Home, family and identity among Palestinians in Britain. PhD thesis, Queen Mary, University of London, London.

———. 2013. Diasporic dwelling: The poetics of domestic space. *Gender, Place and Culture* 20 (3): 329–45.

Madianou, M., and D. Miller. 2012. *Migration and new media: Transnational families and polymedia*. London and New York: Routledge.

Martin, L. L. 2012. Governing through the family: Struggles over US noncitizen family detention policy. *Environment and Planning A* 44 (4): 866–88.

Mulder, C. H. 2007. The family context and residential choice: A challenge for new research. *Population, Space and Place* 13:265–78.

Nash, C. 2005. Geographies of relatedness. *Transactions of the Institute of British Geographers* 30 (4): 449–62.

———. 2008. *Of Irish descent: Origin stories, genealogy, and the politics of belonging*. Syracuse, NY: Syracuse University Press.

Panagakos, A. N., and H. A. Horst, eds. 2006. Return to Cyberia: Technology and the social worlds of transnational migrants [Special issue]. *Global Networks* 6 (2).

Parreñas, R. 2005. Long distance intimacy: Class, gender and intergenerational relations between mothers and children in Filipino transnational families. *Global Networks* 5 (4): 317–36.

Pratt, G. 2012. *Families apart: Migrant mothers and the conflicts of labor and love*. Minneapolis: University of Minnesota Press.

Skrbiš, Z. 2008. Transnational families: Theorising migration, emotions and belonging. *Journal of Intercultural Studies* 29 (3): 231–46.

Skrbiš, Z., L. Baldassar, and S. Poynting. 2007. Introduction—Negotiating belonging: Migration and generations. *Journal of Intercultural Studies* 28 (3): 261–69.

Smith, D. P. 2011. Geographies of long-distance family migration: Moving to a "spatial turn." *Progress in Human Geography* 35:652–68.

Stone, L. 2001. *New directions in anthropological kinship*. Lanham, MD: Rowman and Littlefield.

Svašek, M. 2008. Who cares? Families and feelings in movement. *Journal of Intercultural Studies* 29 (3): 213–30.

Thomson, R. 2006. The qualitative longitudinal case history: Practical, methodological and ethical reflections. *Social Policy and Society* 6:571–82.

Tolia-Kelly, D. 2004. Locating processes of identification: Studying the precipitates of re-memory through artefacts in the British Asian home. *Transactions of the Institute of British Geographers* 29 (3): 314–29.

Valentine, G. 2006. Globalizing intimacy: The role of information and communication technologies in maintaining and creating relationships. In *The global and the intimate*, ed. G. Pratt and V. Rosner, 365–93. New York: Feminist Press.

———. 2008. The ties that bind: Towards geographies of intimacy. *Geography Compass* 2 (6): 2097–2110.

Valentine, G., M. Jayne, and M. Gould. 2012. Do as I say, not as I do: The affective space of family life and the generational transmission of drinking cultures. *Environment and Planning A* 44 (4): 776–92.

Wagner, L. 2008. Diasporic visitor, diasporic tourist. *Civilisations* 57:191–205.

———. 2011. Negotiating diasporic mobilities and becomings: Interactions and practices of Europeans of Moroccan descent on holiday in Morocco. PhD thesis, University College London, London.

Wilding, R. 2006. "Virtual" intimacies? Families communicating across transnational contexts. *Global Networks* 6 (2): 125–42.

Yeoh, B. S., A. S. Huang, and T. Lam. 2005. Transnationalizing the "Asian" family: Imaginaries, intimacies and strategic intents. *Global Networks* 5 (4): 307–15.

Migration, Urbanization, and Political Power in Sub-Saharan Africa

Clionadh Raleigh

Department of Geography, University of Sussex

African political geographies have been overlooked in the discussion of internal migration and urbanization. Yet institutional changes since the early 1990s altered the practices of acquiring and keeping power across Africa, and it is now possible to address how rural and urban areas have changed as a result. In this article, the focus is on how political practices interact with circular internal migration flows. "Democratic" practices depend on the manipulation of rural spatial clustering and migration, but modern political institutions effectively disenfranchise urban and migrant voters, resulting in the continued reproduction and exacerbation of spatial inequalities. Further, these are responsible for the poor conditions of urban areas, the proliferation of rural bias across states, and limited support for urban political identities and programmatic change.

有关内部迁徙与城市化的讨论，总是忽略非洲的政治地理。但自 1990 年代早期以降的制度变迁，已改变了在非洲取得及维繫权力的实践，使得当下可探讨农村与城市地区如何因而改变。本文将聚焦政治实践如何与循环性的内部迁徙流动相互作用。"民主"的实践，取决于农村的空间聚集和迁徙的操控，而当代的政治制度，有效地剥夺了城市与迁徙选举人的公民权，导致空间不均的持续再生产及恶化。此外，上述问题是城市地区境况恶劣、各国的农村偏好激增，以及对城市政治认同及实际变革的支持相当有限之导因。

En la discusión sobre migración interna y urbanización en África no se han tomado en cuenta las geografías políticas de ese continente. Es necesario anotar que los cambios institucionales ocurridos desde principios de los años 1990 alteraron en todo África las prácticas sobre cómo hacerse al poder y conservarlo; ahora es posible abocar la cuestión de las transformaciones experimentadas por las áreas rurales y urbanas como resultado de aquellos cambios. El presente artículo se centra en el modo como las prácticas políticas interactúan con los flujos circulares de la migración interna. Las prácticas "democráticas" dependen de la manipulación de la concentración espacial de la población y la migración rural, pero las instituciones políticas modernas obstruyen efectivamente los derechos electorales de los votantes urbanos y migrantes, lo cual resulta en la continua reproducción y exacerbación de las desigualdades espaciales. Aún más, éstas son responsables de las pobres condiciones existentes en las áreas urbanas, la proliferación de prejuicios rurales a través de los estados y el apoyo limitado a las identidades políticas urbanas y al cambio programático.

The plethora of theories designed to explain the form, function, and flows of migration and urbanization often assume a "fairly stable political environment in which economic and social factors influence the nature and growth of cities" (Potts 2010, 18). Few suggest that migration patterns alter political institutions or the individual political circumstances of migrants. In general, the causal direction has been to suggest that political events—notably crises—lead to migration (Foresight 2011); that migration of some groups over others is facilitated due to political connections (Collinson et al. 2003); or that, across Africa, political elites brand particular communities as "migrant" or "foreign" to restrict their political rights (e.g., Ivory Coast and Democratic Republic of the Congo).

This article argues that institutional change has had negative repercussions for the urban poor and migrants (Raleigh 2013). Democratization and decentralization alter the political calculus against urban migrants and residents, who now routinely suffer from the practice of multiparty politics. From the range of intended and unintended consequences for regimes and voters, the most critical here is the de facto disenfranchisement through limited political identities in Africa outside of rural ethno-regional affiliations and alliances; the reversal of autocratic urban bias in favor of demographically strong rural voters; and negative competition between opposition and incumbent powers. This interpretation of political power is grounded in frameworks recognizing how citizens and their spatial movements are used to bolster a political project from which they are ultimately victimized (Laclau and Mouffe 2001). In short, new patterns of mobility and settlement across Africa are at odds with the practices of power.

An ongoing debate presumes that as urbanization continues, the focus and attention of political institutions and regimes will return to urban areas to generate economic growth and placate a growing population. This neglects how the manipulation of the political process through democracy sustains a rural focus, through which spatialized and ethnic claims to power have more relevance. Manipulations centered around voting, registration, and representation make migrants pawns of political elites, kept in positions of dependence and patron subservience due to the underprovision of economic opportunities in urban areas and the lack of alternative political identities through which to support systematic change. In limiting effective political agency to those in rural areas, many African political transitions lead citizens to vote against their immediate self-interest for access to services, representation, and development. In addition, the dynamics of elite capture—including administrative rigging through decentralization, underfunding opposition areas, and creating disincentives for alternative political identities—leads to the relegation of urban migrants and residents as secondary citizens and has perpetuated decades of public mismanagement on urban issues.

The article proceeds as follows: I first present a brief general introduction to internal African mobility. Following that is a discussion of rural bias created by democracy and decentralization and the ways in which migrants and the urban poor are engaged in contests to secure power. I conclude by discussing the consequences for underrepresented migrant and urban residents.

Why Migrate? And to Where?

African migration is often a socioeconomic obligation, undertaken to distribute resources in high-risk environments and cope with high levels of production uncertainty in both rural and urban spaces (Roncoli and Ingram 2001; Tacoli 2002; Homewood, Coast, and Thompson 2004; Perch-Nielsen, Bättig, and Imboden 2008). Most socioeconomic theories of movement posit that migration sustains rural homesteads and livelihoods during difficult periods and provides urban residents with a safety net from the growing pressures in urban areas (Todaro 1969; Smit 1998; Potts 2005). The typical migrant engages in reversible movements (Dustmann 2003; Raleigh and Jordan 2010) characterized by temporary, circular,[1] and internal migration from rural areas. It is generally agreed that circular migration dominates (Potts 2010) and varies in form by direction, steps, time span, and whether family members accompany a migrant (Skeldon 2012).

The absolute and relative location of movement (urban or rural) is largely structured by group or individual social capital, preestablished links, and infrastructure linking rural and urban areas (de Haan, Brock, and Coulibaly 2002). Satterthwaite (2010) noted that the scale and direction of people's migrations change according to the geography of economic opportunities: cities, towns, and rural areas with expanding economies that attract migrants, who are responsive to income variability (Lilleør and van den Broeck 2012). Destination characteristics beyond economic gain, however, are increasingly shaping movement (Vertovec 2007; Brettell 2008; Brooks and Waters 2010). Urbanization figures in several developing countries suggest a trend of downward growth in larger cities where populations are generally stagnant or decreasing, whereas secondary and tertiary cities are growing faster than capitals. African migrants seem most attracted to proximate small to large cities (Potts 2010; see Beauchemin [2011] on West Africa).

Overall, the African continent is rapidly urbanizing[2]: 39.6 percent of Africans are urban and annual urban population growth averages over 5 percent (World Bank 2008; United Nations 2011). Yet, these figures aggregate countries with high rates of urbanization (e.g., Djibouti, Democratic Republic of the Congo) and those with the lowest (e.g., Burundi, Rwanda, Uganda, and Ethiopia). Further, urban versus rural growth figures suggest that those states that are highly urbanized have the lowest urban growth, whereas countries like Rwanda (17 percent urban growth in 2010), Burundi (6.15 percent), and Malawi (5.6 percent) are the sites of rapidly growing urban centers. Rural–urban migration significantly contributes to urbanization rates across Africa, but it is not entirely responsible for the rates of urban growth; individual cities and states might have similar rates of urban growth caused by numerous variations in contributing factors (e.g., natural increase, slowing or increasing in- or outmigration, recategorization of centers). Although the data on circular migration are particularly poor (see Gould [1995] for discussion), both inflows and outflows are considerable.[3]

In response to growing pressures, African governments have attempted to alter migration patterns, flows, and incentives. To limit urbanization, many incentivize the rural poor to remain in or return to their villages through nationwide access to primary and secondary education (Mabogunje 2007). Such policies might alter migration patterns but ultimately fail to stem rural-to-urban circular movements (Beauchemin 2011). This sort of migration occurs in all African states and involves large portions of the populations.

Governance and Migration Effects

Spatial reconfigurations and demographic changes within African states cannot be divorced from expressions of power and the means by which it is acquired and exercised. From a political geography standpoint, the questions are not how Africa's political changes have affected migration but, rather, how institutional changes have altered the urban–rural political divide, how migrants and migration have supported political change, and what the political consequences of high internal migration are. This article argues that the main results of institutional change are that rural areas are politically privileged, despite needed attention to urban issues; migrants and the urban poor vote against their own self-interest, which leads to poorer urban conditions; and decentralization has stunted an effective opposition and alternative urban political identity. In short, many African governments engage in different modes of informal disenfranchisement and demographic manipulation (Schedler 2002).

Rural and Urban Biases

Previous research into the distinctions between African governance in urban and rural areas largely focused on theories of urban bias and rural patronage and repression (see Bates 1981, 1993). Urban bias was the result of economic policies that systemically favored those living in towns and cities at the expense of the rural majority (Bates 1981, 1993). In particular, distorted commodity pricing systems for urban residents were designed to quell potential collective action against the autocratic state (Van de Walle 1989), as this was considered far easier for urban groups to organize than disenfranchised rural groups. Urban bias implies that individuals engage in and support violent collective action—including riots, protests, or coups—to maximize economic self-interest. The end to urban bias was ascribed to structural adjustment policies (SAPs); despite relatively poor adherence to those (Van de Walle 2001), the urban population bore the costs of retrenchment of the public sector, new collections of tax, and an end to favorable prices for commodities (Nugent 1999; Harding 2011). There are few references to migration practices in this literature, although Potts (2005) has argued that SAPs resulted in high poverty rates within the urban civil service and a significant return migration to rural areas.

Democracy

In addition to SAPs, concurrent, externally mandated changes in political institutional form ushered in multiparty democracy to many African states. Since the early 1990s, multiparty systems clearly position demographically strong rural areas as the main constituency (Nugent 1999; Boone 2003; Harding 2012). Simple political demography demands more efforts by regimes in rural areas, as they are often more populated than urban sectors. Indeed, smaller and more demanding urban constituencies have lost in the democratic transition (Bates 1993). In addition to a rural focus, the structure of most sub-Saharan African political systems reinforces the dominance of ethno-political-territorial identities. This is a result of both politicized ethnicity and spatial considerations: Ethnicity is a significant predictor of party support in most African societies (Norris and Mattes 2003). In response, political parties across Africa do not distinguish themselves through programmatic policy stances (Scarritt and Mozaffer 1999; Van de Walle and Butler 1999; Posner 2005) and instead represent ethno-political divisions (Salih and Mohamed 2003; Harding 2012; in specific reference to Senegal, see Galvan 2001).

Ethnic pandering is much more successful in countries with clustered ethnic communities (Ichino and Nathan 2013). African ethno-regional groups are strongly identified with specific territories (Scarritt and McMillan 1995), and group-based calculations can result in entire areas of electoral support (Boone 2003). This allows politicians to cater a message on "service delivery" without needing to emphasize an ethnic dimension.

Furthermore, incumbents enjoy multiple rural advantages: While in power, they engage in rival bargaining and consolidation of key constituencies into party alliances (Van de Walle and Butler 1999) and use state resources to strategically plan the distribution of public and private (e.g., ministry position) goods to elevate their standing across communities and court rivals. For example, Yoweri Museveni's 2011 reelection in Uganda is largely explained by massive government spending intended to rouse support during the election (some estimates suggest 85 percent of the annual state budget; Conroy-Krutz and Logan 2012). Finally, incumbents orchestrate obstacles for opposition candidates to limit their ability to educate and appeal to voters. Spatial clustering in rural areas, coupled with the creation of electoral vote districts benefiting incumbents, has allowed multiparty elections to become a fight over rural dominance. As evidence of the link among political favor, space, and ethnic demography, rural candidates support incumbent parties in higher numbers, whereas urban areas return higher than average opposition support (Conroy-Krutz 2009).

Decentralization

The political geography of African states is formalized within the institutional structure, allowing for both rural and regime dominance. For more than thirty years, the decentralization process was designed to limit centralized state power. In practice, it has been a process of regime reorganization, capture, and recentralization due to the creation of governing units in exchange for patronage (e.g., Uganda), a reluctance to relinquish significant power to the subnational level, and a refusal to fiscally support new units (e.g., Nigeria; Wunsch 2001). Consequently, urban movements, residents, identities, and programs are effectively restricted.

The creation of decentralized administrative units is not a politically neutral process (Boone 1998), as both district proliferation and the use of decentralized administrative positions allow central regimes to extend patronage networks in periods of constrained state budgets (Tripp 2010; Conroy-Krutz and Logan 2012). This is often due to the largely hierarchical and centralized patronage distribution systems operating in many African states, where the objective is to create nested power hierarchies in which "local power structures have been facilitated by the desire for rural elites to create and sustain power bases in the countryside" (Crook 2003, 77).[4] These local governance agents become intermediaries to exert control of the countryside and ensure regime support within a locality (Kasara 2007).

Central regimes create and acknowledge decentralized units but provide administrators with little to no fiscal support. In a majority of African states, the transfer of funds from the center is calculated to range from less than 5 percent of national budgets (Ndegwa 2002) to less than 15 percent (Therkildsen 1994; e.g., Kenya's rate was 3.5 percent; World Bank 2000). The reluctance of central governments to allocate sufficient funds is particularly important in urban areas with large influxes of migrants and high growth: The lack of support is possibly related to the higher than average rates of opposition support in urban areas (Resnick 2011). Given that opposition parties controlling city councils might benefit from urban programs, by starving the council of adequate funds, the central government is also managing opposition support and risk to regime stability. In summary, rural decentralization by regimes is designed to capture political bases; urban areas are decentralized but inadequately funded to effectively limit opposition appeal.

These factors go far in explaining the lack of urban agendas across parties in rapidly urbanizing areas, the poor conditions that urban residents persistently live with, the lack of incumbent support in cities, and agency of opposition parties. Further, as both migrants and the urban poor do not act in their own political best interest, migration exacerbates the conditions in cities while reinforcing rural electoral dominance.

Manipulation

Considerable geographic work focuses on the complexity of voting exclusion mechanisms for particular types of citizens, including international migrants (Collyer 2013); yet there is little on the role of internal migrants in political competition. Taking advantage of looser populations in electoral contests is part of the available options for democratic manipulation (Schedler 2002). Three ways in which migrants and the urban poor contribute to corrupted democratic contests are particularly important: where they vote, who they vote for, and whether their vote is weighted the same as others.

Returning to Rural Areas to Vote. The rural–urban power divide is reinforced by where voters can register to vote: Whereas urban residents can vote in urban location, migrants are often not legally registered there and therefore—regardless of how much time is spent in urban areas—often return to rural areas for census taking, voting, holidays, and significant events (Gugler 2002; Desai 2010; Beguy, Bocquier, and Zulu 2010).

This is not simply an administrative issue: The close ties that most urban migrants retain with their sending communities results in the "shifting of political loyalties to maintain access to state controlled patronage sources" (Galvan 2001, 59–60). This is underscored by the strongly ethnicized experience of migrants through urban social networks and associations that provide access to services, support, and opportunities on an ethnic-club basis. Hence, even urban residents are more likely to resort to patronage voting behavior based on ethnic markers (Desai 2010), as they remain supported by and integrated with rural power structures. These rural political identities limit the abilities of the poor to create and engage in collective action (Platteau 2000; see Robson and Roque [2001] for a critique of Angola's urban community organization).[5] Further, ethnic clustering in urban areas conspires to make urban zones less attractive or accessible regions for competing parties to encourage alternative political leanings.

Vote Buying and Busing. Another way to manipulate democratic contests is by directing migrants and urban poor to register and vote in contested electoral areas. This is an active reallocation of movable votes, where voter registration lists are superficially inflated above the resident population totals. Although increased registration can be due to demographic growth, it could also result from inflated registration and busing, which allows candidates to win contested, pivotal districts through the temporary relocation of migrants and the purchase of urban poor votes (Jensen and Justesen 2012).

There are several recent examples of vote buying and busing in Benin (Wantchekon 2003), Ghana (Lindberg 2003), Sao Tome and Principe (Vicente 2010), Nigeria (Bratton 2008), Kenya (Kramon 2009), Malawi, and Mozambique (Birch 2011). Inflated active registration lists are not a smoking gun but provide the foundations for electoral fraud facilitated by migrants and the urban poor. Active registration lists are updated before an election. This allows for a real-time map of where voters will cluster. Passive lists are less frequently updated. For example, in Kenya's 2013 election, both Central and Nairobi districts had registered voters over 100 percent ("Holding Their Breath" 2013, 1). Nairobi enrollment was 121.6 percent of the expected district electorate and the Central region had 107.2 percent of its electorate. Other areas of the state recorded totals from 60 percent upward, and the national registration total was 78 percent (see Nation Map 21 December 2012). Areas that were strongholds of candidate Uhuru Kenyatta from the Jubilee alliance (now president) had higher than average registration across several territories,[6] but the CORD alliance (in opposition) took Nairobi, as expected. Surpassed target percentages can be attributed to population transfers and settlement (Willis Otieno, Electoral Institute for Sustainable Democracy in Africa, personal communication, 18 February 2013). The center was most affected by distorted voting: Due to constraints and a March election, most of those who traditionally voted in their home districts were forced to register in the city.

Demographic Gerrymandering. The ways in which incumbents stay in power are sophisticated and catered to the local and external pressures experienced by regimes. Governments can manipulate the political geography and demography of states to make migrants and urban growth disappear, thereby reinforcing their largely rural democratic strategy. One simple way to counter urban opposition is to erase its evidence through embellishing favorable demographic trends in rural areas and reducing the impact of urban areas through a form of gerrymandering. This strategy is common in Zimbabwe, as described by O'Donovan (2003) using Electoral Institute of Southern Africa data. By manipulating voting rolls, the ZANU-PF regime in 1994 and 2000 was able to ensure that voters in potentially opposition (often urban) constituencies counted less than voters in rural ZANU-PF constituencies. This was done through the creation of voting profiles in which the demographic growth of Harare, Bulawayo, and other constituencies supportive to opposition candidates experience a stagnating or decreasing voting-age population from 1994 to 2000 (despite a reported potential population increase of 35 percent). Hence, all became electorally less valuable. Yet, rural areas had, as described by independent analysts, voting lists that were inflated up to 30 percent (due to names of dead, migrants, etc.). Indeed, in a number of rural provinces, the "number of people on the register exceeded that estimated population of the district" (O'Donovan 2003, 5). The practical avenue for fraud is through ballot stuffing on election day.

O'Donovan (2003) noted that a cleaned voter list might indeed lead to voter decreases in opposition areas, but the same effect should have been seen across the state and irrespective of party support. This strategy of making urban residents and migrants politically disappear creates unequal citizenship across urban and rural areas. This practice is ongoing according to reports about Zimbabwe's 2013 election (and also in the Democratic Republic of the Congo, Guinea, Egypt, Sierra Leone, and Nigeria).

Consequences and Conclusions

Resnick (2012) suggested that, as migration and urban growth continue, the urban poor might respond by insisting on improved urban circumstances through populist, programmatic political parties. Despite research suggesting that opposition candidates are likely to engage in populist appeals in urban areas, however, Bratton and Van de Walle (1992, 440) criticized the formation of opposition parties as often a "spontaneous convergence of diverse urban interests" and directed toward appealing to wealthier, educated subsections of the population. Indeed, a class distinction is evident in urban voting: Whereas ethnic voting is negatively correlated with levels of wealth in urban locations (Conroy-Krutz 2009), the urban poor might not be active supporters of opposition candidates unless linked to accessible patronage (Desai 2010). This combined research casts doubt on the notion of a monolithic urban constituency and the notion that the urban poor might reorient African democracy.[7]

Hence, through administrative, fiscal, and representational constraints, migrants and urban poor are effectively disenfranchised, used, or absent from the political process. Political candidates do not often represent their constituency; they remain outside of active and funded patronage networks, and through the undocumented residence conditions for migrants, their voting practices uphold rural dominance. Politicians respond to a political calculus that maximizes voter support for minimal efforts and costs; hence, demographically strong rural areas are far more attractive sites to position programs designed to increase allegiances. Despite growing urbanization, this political strategy is kept in place through constraints and obstacles for urban voters and potentially urban-focused parties.

This has serious consequences for urban development, as life in urban areas continues to grow more difficult than that in rural areas across African states. Sub-Saharan states are urbanizing without parallel national economic growth. UN-HABITAT projects that urban poverty will account for over 40 percent of total poverty in several African countries by 2020 (Ravallion, Chen, and Sangraula 2007; United Nations 2011). Consequently, the urban poor are excluded from many of the advantages of cities and, in a reversal of trends, are now worse off than the rural poor in some places (Mabogunje 2007; Potts 2010). The weak economic performance is due, in part, to the proliferation of informal work. The best estimates suggest that over 50 percent of the urban laborers are employed in informal livelihoods (Mabala 2011).

The urban poor have limited access to municipal services and are disproportionately affected by disease, violence, pollution, and insecure living conditions. Millions of urban dwellers have unsafe, unreliable, difficult, and possibly privatized access to water (McGranahan et al. 2009) and less than 10 percent of the population is connected to sewers (Satterthwaite 2010). In poorly managed cities, the rate of child mortality (under age five) is 25 percent, with most deaths related to the lack of infrastructure and services (Bolnick et al. 2006). These populations are living on the margins of security and safety.

To sustain a practice that benefits political elites, governments misrepresent and ignore problems by branding needy urban constituencies as largely illegal. In particular, regimes contend that informal urban livelihoods and settlements are illegal, and the lack of authorized resident identities prevents migrants from accessing justice, services, and representation (Desai 2010). Incidents of urban cleaning, where military and police forces destroy illegal slum settlements, are disturbingly common, as evidenced by recent raids in Zimbabwe (2005, 2007), Angola (2007), Kenya (2008, 2009, 2010) Nigeria (2000, 2009, 2013), Sudan (2005), South Africa (2010), Ghana (2002, threatened in 2011), Ethiopia (2011), and Uganda (2011).

By 2025, Africa's urban population is estimated to surpass its rural sector. Will the urban poor then be a key constituency? Perhaps, but presently, this population finds itself subject to a host of new ecological, security, and economic risks, higher rates of poverty, and political structures that limit their agency as citizens. Despite these challenges, migrants still engage in ever increasing rates of circular migration, while being forced to conform to rural and ethnicized political associations.

In conclusion, much of the research on both migration and urbanization does not consider how political institutions direct movement, create risks, and shape community fortunes. Although a debate continues on which factors motivate rural–urban movements, there is little question that rural development agendas are favored over needed urban programs and planning. These political and economic calculations have drastic effects on the lives of migrants, urban residents, and rural residents, whose livelihoods are closely tied to community members in urban locations. The focus on decentralization and democratization in this article, however, reveals how the institutional changes across sub-Saharan African have motivated governments to downplay risk in urban areas, conflate appeals to rural areas, and actively work against an urban political identity. Although poverty and state capacity are issues within developing states, political calculations—as shaped by changing institutional incentives—are a more compelling and thorough explanation for the increasing dire state of urban areas.

Notes

1. Circular migration refers to high rates of return and repeat movement, into either urban or other rural areas.
2. A counter to this is noted by Potts (2010).
3. Bolnick et al. (2006) reported that at least 60 percent of Africa's urban population lives in urban centers of fewer than half a million inhabitants and a large proportion in significantly smaller towns.
4. Other modes to deal with threats to the stability of a central regime are generally removed, delegitimized, or diffused by fragmenting rival political bases (Rakodi and Lloyd-Jones 2002).
5. Exceptions to this are states where unions and urban protest movements in the 1980s encouraged the transition to multiparty politics (Posner and Simon 2002).
6. This is a pivotal point: Kenya's 2010 constitution demanded a national distribution of votes to win national office.
7. The few countries where urban areas have produced competitive parties are characterized by strong union systems present before multiparty democracy (e.g., South Africa, Zambia). In those cases, limited fiscal support has had the greatest effect on the urban poor due to the urban support of opposition (see Resnick 2011; Conroy-Krutz 2012).

References

Bates, R. 1981. *Markets and states in tropical Africa: The political basis of agricultural policies*. Berkeley: University of California Press.

———. 1993. Urban bias: A fresh look. *Journal of Development Studies* 29 (1): 219–28.

Beauchemin, C. 2011. Rural–urban migration in West Africa. *Population, Space and Place* 17 (1): 47–72.

Beguy, D., P. Bocquier, and E. M. Zulu. 2010. Circular migration patterns and determinants in Nairobi slum settlements. *Demographic Research* 23 (20): 549–86.

Birch, S. 2011. *Electoral malpractice*. Oxford, UK: Oxford University Press.

Bolnick, J., H. Kanuyi, R. Mabala, G. McGranahan, D. Mitlin, S. Nkhoma, J. Oucho, et al. 2006. A pro-poor urban agenda for Africa: Clarifying ecological and development issues for poor and vulnerable populations. Human Settlements Discussion Paper Series 2 on Urban Change. London: International Institute for Environment and Development.

Boone, C. 1998. State building in the African countryside: Structure and process at the grassroots. *Journal of Development Studies* 34 (4): 1–31.

———. 2003. Decentralization as political strategy in West Africa. *Comparative Political Studies* 36 (4): 355–80.

Bratton, M. 2008. Vote buying and violence in Nigerian election campaigns. *Electoral Studies* 27 (4): 621–32.

Bratton, M., and N. Van de Walle. 1992. Popular protest and political reform in Africa. *Comparative Politics* 24 (4): 419–42.

Brettell, C. 2008. Theorizing migration in anthropology: The social construction of networks, identities, communities and globalscapes. In *Migration theory: Talking across disciplines*, ed. C. Brettell and J. Hollifield, 113–59. London and New York: Routledge.

Brooks, R., and J. Waters. 2010. Social networks and educational mobility. *Globalisation, Societies and Education* 8 (1): 143–57.

Collinson, M., S. Tollman, K. Kahn, and S. Clark. 2003. *Highly prevalent circular migration: Households, mobility and economic status in rural South Africa*. Johannesburg, South Africa: University of the Witwatersrand.

Collyer, M. 2013. A geography of extra-territorial citizenship: Explanations of external voting. *Migration Studies* 1 (1): 1–18.

Conroy-Krutz, J. 2009. Who are Africa's (non) ethnic voters? Evaluating theories on the salience of ethnicity in African electoral politics. Paper presented at the conference of the American Political Science Association, Toronto, Canada.

———. 2012. What determines the price of a vote? Motive, means, and opportunity in vote-buying practices. Paper presented at the Midwest Working Group Meeting in African Political Economy, December 7–8, Duke University, Durham, NC.

Conroy-Krutz, J., and C. Logan. 2012. Museveni and the 2011 Ugandan election: Did the money matter? *Journal of Modern African Studies* 50 (4): 625–55.

Crook, R. 2003. Decentralization and poverty reduction in Africa: The politics of local-central relations. *Public Administration and Development* 23 (1): 77–88.

De Haan, A., K. Brock, and N. Coulibaly. 2002. Migration, livelihoods and institutions: Contrasting patterns of migration in Mali. *Journal of Development Studies* 38 (6): 37–58.

Desai, R. 2010. The political economy of urban poverty in developing countries: Theories, issues, and an agenda for research. Wolfensohn Center for Development Working Paper 20. Washington, DC: Brookings Institute.

Dustmann, C. 2003. Return migration and the optimal migration duration. *European Economic Review* 47 (2): 353–67.

Foresight. 2011. Migration and global environmental change. Final project report. London: The Government Office for Science.

Galvan, D. 2001. Political turnover and social change in Senegal. *Journal of Democracy* 12 (3): 51–62.

Gould, W. T. S. 1995. Migration and recent economic and environmental change in East Africa. In *The migration experience in Africa*, ed. J. Baker and T. Akin Aida, 122–48. Uppsala, Sweden: Nordiska Afrikainstitutet.

Gugler, J. 2002. The son of the hawk does not remain abroad: The urban–rural connection in Africa. *African Studies Review* 45 (1): 21–41.

Harding, R. 2011. One for the road: Voting for public goods in Ghana. Working paper. New York: New York University.

———. 2012. Urban–rural differences in support for incumbents across Africa. Working Paper No. 120, Afrobarometer.

Holding their breath. 2013. *Africa Confidential*. 1 February: 54, 3, p. 1.

Homewood, K., E. Coast, and M. Thompson. 2004. Inmigrants and exclusion in east African rangelands: Access, tenure and conflict. *Africa* 74 (4): 567–610.

Ichino, N., and N. L. Nathan. 2013. Crossing the line: Local ethnic geography and voting in Ghana. *American Political Science Review* 107 (2): 344–61.

Jensen, P., and M. Justesen. 2012. The economic origins of vote buying in Africa. Paper presented at the meeting of the European Political Science Association.

Kasara, K. 2007. Tax me if you can: Ethnic geography, democracy, and the taxation of agriculture in Africa. *American Political Science Review* 101 (1): 159–72.

Kramon, E. 2009. Vote-buying and political behavior: Estimating and explaining vote-buying's effect on turnout in Kenya. Working Paper No. 114, Afrobarometer.

Laclau, E., and C. Mouffe. 2001. *Hegemony and socialist strategy: Towards a radical democratic politics*. 2nd ed. London: Verso.

Lilleør, H., and K. Van den Broeck. 2012. Economic drivers on migration and climate change in LDCs. *Global Environmental Change* 21:S70–S81.

Lindberg, S. I. 2003. "It's our time to 'chop' ": Do elections in Africa feed neo-patrimonialism rather than counter-act it? *Democratization* 10 (2): 121–40.

Mabala, R. 2011. Youth and "the hood"—Livelihoods and neighborhoods. *Environment and Urbanization* 23 (1): 157–81.

Mabogunje, A. 2007. Tackling the African "poverty trap": The Ijebu-Ode experiment. *Proceedings of the National Academy of Science* 104 (43): 16781–86.

McGranahan, G., D. Mitlin, D. Satterthwaite, C. Tacoli, and I. Turok. 2009. *Africa's urban transition and the role of regional collaboration*. Working paper 5, Human Settlements Working Paper Series, International Institute of Environment and Development.

Ndegwa, S. 2002. Decentralization in Africa: A stocktaking survey. Africa Region Working Paper Series No. 40, Washington, DC: World Bank.

Norris, P., and R. Mattes. 2003. Does ethnicity determine support for the governing party: Structural and attitudinal basis of partisan identification in 12 African nations. CSSR Working Paper No. 36. Cape Town, South Africa: Center for Social Science Research, University of Cape Town.

Nugent, P. 1999. Living in the past: Urban, rural and ethnic themes in the 1992 and 1996 elections in Ghana. *Journal of Modern African Studies* 37 (2): 287–319.

O'Donovan, M. 2003. Zimbabwe's democracy: quantifying the impact of electoral maladministration. Paper presented at the Electoral Institute of Southern Africa, Johannesburg, South Africa.

Perch-Nielsen, S., M. Bättig, and D. Imboden. 2008. Exploring the link between climate change and migration. *Climatic Change* 91 (1): 375–93.

Platteau, J.-P. 2000. *Institutions, social norms, and economic development*. Amsterdam: Harwood Academic.

Posner, D. 2005. *Institutions and ethnic politics in Africa*. New York: Cambridge University Press.

Posner, D., and D. Simon. 2002. Economic conditions and incumbent support in Africa's new democracies: Evidence from Zambia. *Comparative Political Studies* 35 (3): 313–36.

Potts, D. 2005. All my hopes and dreams are shattered: Urbanization and migrancy in an imploding African economy—The case of Zimbabwe. *Geoforum* 37 (4): 536–51.

———. 2010. *Circular migration in Zimbabwe & contemporary sub-Saharan Africa*. Rochester, NY: Boydell and Brewer.

Rakodi, C., and T. Lloyd-Jones, eds. 2002. *Urban livelihoods: A people-centered approach to reducing poverty*. London: Earthscan.

Raleigh, C. 2013. Migration, urbanization and political power in Africa. Brief for the Climate Change and African Political Stability Program, Austin, TX.

Raleigh, C., and L. Jordan. 2010. Climate change and migration: Emerging patterns in the developing world. In *Social dimensions of climate change: Equity and vulnerability in a warming world*, ed. R. Mearns and A. Norton, 103–23. Washington, DC: World Bank.

Ravallion, M., S. Chen, and P. Sangraula. 2007. New evidence on the urbanization of poverty. *Population and Development Review* 33 (4): 667–701.

Resnick, D. 2011. In the shadow of the city: Africa's urban poor in opposition strongholds. *Journal of Modern African Studies* 49 (1): 141–66.

———. 2012. Opposition parties and the urban poor in African democracies. *Comparative Political Studies* 45 (11): 1351–78.

Robson, P., and S. Roque. 2001. "Here in the city, everything has to be paid for": Locating the community in peri-urban Angola. *Review of African Political Economy* 28 (90): 619–28.

Roncoli, C., and K. Ingram. 2001. The costs and risks of coping with drought: Livelihood impacts and farmers' responses in Burkina Faso. *Climate Research* 19 (1): 119–32.

Salih, M., and A. Mohamed, eds. 2003. *African political parties: Evolution, institutionalization, and governance*. London: Pluto Press.

Satterthwaite, D. 2010. The scale of urban change worldwide 1950–2000 and its underpinnings. Human Settlements Discussion Paper Series on Urban Change No. 1, International Institute for Environment and Development, London.

Scarritt, J., and S. McMillan. 1995. Protest and rebellion in Africa: Explaining conflicts between ethnic minorities and the state in the 1980s. *Comparative Political Studies* 28 (3): 323–49.

Scarritt, J., and S. Mozaffer. 1999. The specification of ethnic cleavages and ethno political groups for the analysis of democratic competition in Africa. *Nationalism and Ethnic Politics* 5 (1): 82–117.

Schedler, A. 2002. The menu of manipulation. *Journal of Democracy* 13 (2): 36–50.

Skeldon, R. 2012. Going round in circles: Circular migration, poverty alleviation and marginality. *International Migration* 50 (3): 43–60.

Smit, W. 1998. The rural linkages of urban households in Durban, South Africa. *Environment and Development* 10 (1): 77–88.

Tacoli, C. 2002. Changing rural–urban interactions in sub-Saharan Africa and their impact on livelihoods: A summary. Human Settlements Working Paper 7, Rural-Urban Interactions and Livelihoods Strategies. London: International Institute for Environment and Development.

Therkildsen O. 1994. *Economic decline and democratic decentralization in sub-Saharan Africa: Conference on democratic decentralization in Asia and Africa.* London: Institute of Commonwealth Studies, University of London.

Todaro, M. 1969. A model of labor migration and urban unemployment in less developed countries. *American Economic Review* 59 (1): 138–48.

Tripp, A. 2010. *Museveni's Uganda: Paradoxes of power in a hybrid regime*. Boulder, CO: Lynne Reinner.

United Nations. 2011. *World urbanization prospects: The 2011 revision*. New York: United Nations.

Van de Walle, N. 1989. Rice politics in Cameroon: State commitment, capability, and urban bias. *Journal of Modern African Studies* 27 (4): 579–99.

———. 2001. *African economies and the politics of permanent crisis: 1979–1999*. Cambridge, UK: Cambridge University Press.

Van De Walle, N., and K. Butler. 1999. Political parties and party systems in Africa's illiberal democracies. *Cambridge Review of International Affairs* 13 (1): 14–28.

Vertovec, S. 2007. New directions in the anthropology of migration and multiculturalism. *Ethnic and Racial Studies* 30 (6): 961–78.

Vicente, P. C. 2010. Is vote-buying effective? Evidence from a field experiment in West Africa. http://www.pedrovicente.org/ (last accessed 1 May 2013).

Wantchekon, L. 2003. Clientelism and voting behavior: Evidence from a field experiment in Benin. *World Politics* 55 (3): 399–422.

World Bank. 2000. *The world development report 2000*. Washington, DC: World Bank.

———. 2008. *World development report 2008: Reshaping economic geography*. Washington, DC: World Bank.

Wunsch, J. 2001. Decentralization, local governance and "recentralization" in Africa. *Public Administration and Development* 21 (1): 277–88.

Following Migrant Trajectories: The Im/Mobility of Sub-Saharan Africans en Route to the European Union

Joris Schapendonk* and Griet Steel†

*Department of Geography, Planning, and Environment, Radboud University
†Institute of Development Policy and Management, University of Antwerp

We use two trajectory ethnographies that follow the migration processes of Sudanese and Nigerians heading for the European Union across space and time, to explore how the main theoretical principles of the mobilities debate add value to transnational migration research. We thereby particularly appreciate the relational ontology of the mobilities approach and its analytical focus on the differentiation of power and experiences in mobility processes. By providing in-depth insights into how migration trajectories evolve—that is, how they are produced, facilitated, slowed down, and blocked—we argue that a thorough analysis of migrants' im/mobilities helps us to reveal the spatial frictions, embodied efforts, and emotions that are inherent aspects of transnational engagements.

我们运用跟随苏丹与尼日利亚人前往欧盟迁徙过程的跨时空轨迹民族志，探讨流动辩论的主要理论原则，如何对跨国迁徙研究带来价值。据此，我们特别赏识流动取径的关係性主体论，及其对流动过程中的权力与经验差异的聚焦分析。透过提供洞见解析迁徙轨迹如何演变——亦即这些轨迹如何被生产、促成、减缓，并受到阻碍——我们主张，对于迁徙者的 (不) 流动之完整分析，协助我们揭露做为跨国参与内在面向的空间摩擦、身体化的努力及情绪。

Utilizamos dos etnografías de trayectoria que siguen los procesos de migración de sudaneses y nigerianos en ruta hacia la Unión Europea, a través del espacio y el tiempo, para explorar de qué maneras los principios teóricos del debate de movilidades añaden valor a la investigación sobre migración transnacional. De este modo particularmente apreciamos la ontología relacional del enfoque de movilidades y su concentración analítica sobre la diferenciación del poder y las experiencias en los procesos de movilidad. Al proporcionar una visión a profundidad sobre cómo evolucionan las trayectorias de migración—esto es, cómo son producidas, facilitadas, reducidas en intensidad y bloqueadas—argüimos que un sesudo análisis de las in/movilidades de los migrantes nos ayuda a revelar las fricciones espaciales, los esfuerzos encarnados y las emociones que son aspectos inherentes de los compromisos transnacionales.

The main argument of the mobilities turn—namely, that social science has not thoroughly dealt with mobility (Sheller and Urry 2006; Cresswell 2010)—also applies to migration studies. Although migration scholars have always focused on questions concerning human movement, their analytical principles have long been based on "sedentarist metaphysics" (Malkki 1992; Cresswell 2006). In other words, migration has mainly been studied from the position of fixed locations: the A and the B, the push and the pull, the place of origin and the destination. As a result, mobility processes have generally been reduced to residual in-between phases between two locations, and the migrants in question have been approached as moving from one state of fixity (in the place of origin) to another (in the destination).

The transnational turn (e.g., Glick Schiller, Basch, and Blanc-Szanton 1995; Portes 1997), however, debunked this notion. By focusing on migrants' multiple relationships that cross geographical, cultural, and political boundaries, transnationalism positions migrants in two places simultaneously (the sending and the receiving country). We might question, however, whether transnational studies overcome the bipolar dominance of migration conceptualizations, as their focus is mainly on linkages between only two places (Van Bochove 2012). Moreover, whereas much emphasis is put on transnational engagements and networks, less attention is paid to experiences of mobility and spatial frictions in mobility processes, let alone to the journeys between places of departure and arrival (Burell 2008; Schapendonk 2011).[1]

This article explores how the main principles of the mobilities paradigm contribute to a better understanding of transnational migration. To do so, we follow the empirical agenda of the mobilities turn by investigating the im/mobility dynamics of migration trajectories of Sudanese and Nigerian migrants heading for the European Union (EU). Although the trajectories of these two migrant groups are situated in different sociopolitical contexts, both exemplify the increasing connectivity of sub-Saharan African societies in terms of globalization, triggering people's aspirations to reach other places (Appadurai 2001), as well as the increasingly restrictive European migration policies that aim to immobilize potential immigrants (Van Houtum 2010). By providing in-depth insights into how these trajectories evolve (i.e., how they are produced, facilitated, slowed down, and blocked) and how they are experienced in differentiated ways, the mobilities perspective helps us to move away from bipolar and frictionless conceptualizations of transnational migration.

Our empirical analysis is based on two trajectory ethnographies that go beyond the idea of migration as a one-off relocation. In addition to reconstructing migration histories on the basis of in-depth interviews, we followed the actual twists and turns of migration trajectories by way of translocal engagements (e.g., telephone calls and Internet conversations) and follow-up visits to our respondents in different places and during different times of their trajectories. Although the two research projects share this mobile methodological approach (Büscher and Urry 2009), they started from different geographical entry points and thus resulted in complementary insights.

The first project started in the EU with active engagement in one of the Sudanese communities in Belgium. The project sheds light on the migration experiences of fifty Sudanese migrants, the majority of whom are men. We followed eight of these migrants through space and time, through The Netherlands, France, and Sudan. This "following strategy" reveals how the trajectories of Sudanese migrants are inherently related to the im/mobility of their relatives in Sudan and elsewhere.

The second project, which focuses on Nigerian migration, started to investigate migrants' trajectories from two important transit spaces in the context of sub-Saharan African migration to Europe: Morocco and Turkey. We interviewed twenty Nigerian migrants (fifteen men and five women) in these countries, and seven of them (all male) were followed in their attempts to move onward to the EU. In this regard, respondents have been revisited in Greece and Italy. This project has garnered particularly rich insights into migrants' waiting statuses and unexpected opportunities prior to their reaching the EU, as well as into their uncertainty and changing aspirations during their first period in the EU.

Before analyzing the dynamics of the migration trajectories under study, we explore how the main theoretical principles of mobilities research add value to transnational migration studies. We thereby particularly appreciate its relational ontology and analytical focus on the differentiation of power and experiences in mobility processes.

An Im/Mobilities Approach to Migration

Transnational migration research has been a main source of inspiration for the mobilities debate in terms of the theorization of interconnectivity and deterritorialized communities (e.g., Blunt 2007), but the mobilities debate has the potential to enrich our understanding of the dynamics that constitute migrants' transworlds. With their analytical emphasis on the relationality of mobility and stasis (e.g., Hannam, Sheller, and Urry 2006; Urry 2007; Cresswell and Merriman 2011), mobilities scholars argue that mobility and immobility are not unconnected opposites but are instead interrelated effects. This implies that they pay attention to the role of fixed infrastructures, networks, and places in mobility processes, as well as to intersecting periods of rest and blocked mobilities. This relational notion of im/mobility inevitably involves questions of justice and power, which is generally referred to as the "politics of mobilities" (Cresswell 2010; Ernste, Martens, and Schapendonk 2012). The political and power-ridden character of im/mobility reflects differences in terms of access and speed and articulates that the mobility potential of people depends very much on their positionality in time and space (Sheppard 2002; Massey 2005). So-called mobility regimes are responsible for the maintenance of high levels of inequality by privileging the movement of some while stigmatizing and hindering the movement of others (Shamir 2005; Glick Schiller and Salazar 2013). The facilitation of the free movement of EU citizens between EU member states, for instance, is an important factor that explains the fortification of borders to immobilize those who originate from outside the EU (Verstraete 2001). In this sense, mobility should be studied not only as connection "but also as an aspect of new confinements and modes of exploitation" (Glick Schiller and Salazar 2013, 190).

The politics of im/mobilities, however, is reflected not only in the attempts by states and suprastates to manage cross-border movements. From a micro perspective, the mobilities of some people could have the effect of immobilizing other people. In the context of migration, global care chain studies (Hochschild 2000), livelihood approaches (Kothari 2003), and transnational family analyses (Parreñas 2005) have underlined that the movements of migrants considerably affect the mobility opportunities of their household members in negative ways, basically due to changing responsibilities in terms of care-taking (Steel, Winters, and Sosa 2011). Other studies indicate that transnational engagements (e.g., remittances) might coincide with, and partly rely on, migrants' enduring immobility in their destination country (Bailey et al. 2002). The mobilities approach builds further on the diverse power geometries (Massey 2005) of migration processes by analyzing how mobility and immobility are constantly contingent on each other (Adey 2006).

Mobilities scholars also maintain that mobilities are always full of meaning. They perceive mobilities not as empty spaces and dead time but as experienced and value-laden practices that provoke differentiated emotions across lines of class, age, and gender (Kaplan 2003; Walton-Roberts and Pratt 2005; Cresswell 2006). In addition to everyday, routinized, joyful, and comfortable mobilities, critical studies investigate the daily realities of migrants who are confronted with mobility regimes that create specialized zones and territorialities of confinement (Mountz 2010). They shed light on the frustrations related to prolonged waiting (Jeffries 2008; Conlon 2011), enforced immobility (Hyndman and Giles 2011), and the time spent in transit (Schapendonk 2012; Suter 2012), which are the direct effects of bordering practices.

The mobilities approach thus challenges migration researchers to become increasingly sensitive to the power dynamics and differentiated meanings attached to migration processes. It does not take transnational interconnectivity for granted but carefully analyzes (1) the ways im/mobilities are produced in transnational processes, (2) the ways mobilities come with power differences in terms of access and speed, and (3) how these power differences are reflected in mobility experiences and mobility relations. These three aspects are at the core of our empirical analysis of the migration trajectories of Sudanese and Nigerian migrants. To better comprehend these trajectories, we first outline the sociopolitical background to Sudanese and Nigerian migration to the EU.

Contextualizing Sudanese and Nigerian Migration to the EU

With its diverse cultures of mobility, sub-Saharan Africa is often portrayed as the subcontinent where sedentarity is questioned in profound ways (De Bruijn, Van Dijk, and Foeken 2001). Although population movements have had different patterns across the African continent, extracontinental migration in sub-Saharan Africa has long been facilitated by postcolonial links with European countries (Bakewell and De Haas 2007). Today, however, sub-Saharan African migrants are increasingly confronted with the EU's strict visa regimes, lengthy asylum procedures, fortified borders, and externalized border controls. As a result, many African voyages to the north have become lengthy, risky, and fragmented undertakings (Collyer 2007; Schapendonk 2011).

Because of its central location at the crossroads of several trans-Saharan mobility corridors, Sudan has long been a country from which migrants, pastoralists, and caravan traders set off to move across the region (Abusharaf 1997). In the last four decades, Sudanese emigration involves mixed flows involving both refugee and labor mobility. Although the majority of Sudanese refugees are to be found in neighboring countries, scholars have particularly paid attention to the overseas migration movements of southern Sudanese[2] people toward the United States (Abusharaf 2002; Shandy 2007), Canada (Simich, Este, and Hamilton 2010), and Europe (Weissköpel 2009). In quantitative terms, however, labor migration from especially northern Sudan to Europe and the oil-producing countries in the Middle East is at least as important as southern Sudanese refugee movements. In this context, some estimates indicate that by the end of the twentieth century, Sudan had lost almost two thirds of its stock of university graduates (Elnur 2002). Although highly skilled migrants are often considered voluntary movers, this emigration cannot be understood in isolation from the authoritarian regimes and civil wars in Sudan. Furthermore, these migrants have paved the way for a much more heterogeneous population of regular and irregular migrants to reach European territory. Our respondents originate mainly from northern Sudan.

Nigeria's recent history is less characterized by violent conflict. Nevertheless, its emigration profile is also shaped by postindependence political and economic instability (Black et al. 2004). The oil boom in the 1970s, which drew to Nigeria many migrants from West African countries, was followed by a period of economic downturn, communal conflicts, and political repression that triggered significant emigration flows, especially of highly skilled migrants toward the United States (Reynolds 2004) and the United Kingdom (Black et al. 2004). In recent years, Nigerian emigration patterns have diversified in geographical terms, with Italy, Spain, and China emerging as important destinations. An estimated 329,000 Nigerian migrants reside in the EU, making them one of the largest sub-Saharan African communities in Europe (Adepoju and Van der Wiel 2010). European policymakers are particularly concerned about the involvement of Nigerians in irregular migration, including unauthorized boat migration and human trafficking (Carling 2006). It appears that the majority of the recently arrived Nigerian migrants in the EU originate from the country's southern regions (Adepoju and Van der Wiel 2010), which is also the case for our respondents.

From the perspective of EU member states, Sudanese and Nigerian migrations are generally classified as "unwanted migrations" (De Haas 2008). Sudanese migration is partly shaped by discourses of asylum, whereas Nigerian migration is predominantly linked to irregular migration. In reality, however, this categorical division is less clear-cut. Many Nigerians end up in European asylum procedures, and many Sudanese enter the EU in unauthorized ways or become irregular migrants when their asylum applications are rejected. As our empirical sections further illustrate, by following migrants' trajectories, one becomes more sensitive to the fact that migrants easily jump over static policy categories during their mobility processes (Schuster 2005; Schapendonk 2011).

Sudanese Trajectories

In December 2010, I (the Belgian female researcher of the team) went to Khartoum (Sudan) to attend the marriage party of Adam and Laila.[3] Adam is a Sudanese friend who now lives in The Netherlands. During the party, I sat at a table with various Sudanese migrants I had first met in Belgium. Among them was Mohammed, who introduced me to his new wife, Shiman. They had been married for just one week.

Adam and Mohammed have been very close friends since 1999, when they met at an asylum-seekers center in The Netherlands. After a lengthy procedure, both had their asylum applications rejected. For three years, the two undocumented men wandered the country under the constant risk of being detained by the Dutch police. Since this situation was hopeless, Mohammed and Adam decided to seek asylum in the neighboring country of Belgium. While in Belgium, Mohammed met, and soon married, his Belgian girlfriend. The marriage profoundly influenced his legal status, as he received a temporary residence permit in Belgium, which allowed him to travel to Sudan to visit his family after a period of more than ten years of physical separation. In the end, Mohammed's marriage did not work out. Nonetheless, the couple decided to finalize their divorce only after Mohammed's temporary permit had been upgraded to permanent in 2010.

Adam waited four years for a decision on his asylum application. He was highly frustrated about this, as he missed his family tremendously. In 2007, he decided to move back to The Netherlands to benefit from a one-time regularization program ("Generaal Pardon"). As he had gone to Belgium without authorization, the Dutch authorities had not noticed his temporary absence and granted him a permanent residence permit based on his long-term stay in the country.

A month after their marriage ceremonies in Sudan, Adam and Mohammed returned to Europe. From that moment, the trajectories of the two couples diverged. Laila arranged a tourist visa at the Dutch Embassy in Khartoum and arrived heavily pregnant in The Netherlands, eight months after their marriage. Adam and Laila immediately applied to the Dutch authorities for family reunification. The permanent resident status of Adam, in combination with Laila's pregnancy, led to a positive outcome.

Shiman and Mohammed, however, did not have the financial means or the social contacts to apply for a tourist visa. They therefore decided to start the family reunification procedure outside Europe. As Sudan does not have a Belgian embassy, they had to travel to the embassy in Cairo (Egypt) to start this procedure. After two months of living together in Cairo, Mohammed and Shiman (who appeared to be pregnant) separated again: Shiman went back to Sudan, and Mohammed returned to Belgium.

At the time of writing, Adam and Laila are living with their two daughters in The Netherlands. Due to the negative response of the Belgian immigration authorities, Mohammed, Shiman, and their recently born daughter are still separated by the EU's external border. Every time I meet Mohammed, he tells me how sad he is about being unable to watch his little child growing up.

These empirical reconstructions, which highlight just a few snapshots in the long and fragmented migration trajectories of the informants, sheds light on certain recurrent mobility patterns that have been echoed by several other Sudanese migrants. The continuities and discontinuities of their migration processes exemplify the "politics of differentiation" (Mountz 2011) in terms of access and speed. The four informants experienced noteworthy spatial frictions, but in different forms and at different times. Mohammed is currently facing a significant hindrance in his life, as he is geographically separated from his wife and recently born child. Through his previous marriage to a Belgian woman, however, he had increased access to mobility (which was manifested in the possibility of visiting his family in Sudan), whereas Adam was still in limbo, waiting for the final decision on his asylum application. Although Shiman and Laila married around the same time, their trajectories toward Europe took different courses: Laila soon managed to be united with her husband in The Netherlands, whereas Shiman's migration toward Europe remains blocked in Sudan. She is one of the many Africans aspiring to reach other places while lacking the ability to do so (Carling 2002).

Shiman's immobility with respect to the EU, however, does not mean that she has been immobile in absolute terms. The fact that she and Mohammed both traveled to Egypt to apply for family reunification indicates that migration processes may already require international mobility before the actual departure to a specific destination. As such, the movements of Shiman and Mohammed put into perspective the general notion that the predeparture phase is a moment of nonmovement. These preparatory efforts, however, do not guarantee a successful migration project. In fact, as their synchronized travel to Egypt was a costly event, it actually harmed Shiman's and Mohammed's mobility potential.

The four trajectories also emphasize the gendered im/mobility relations in Sudanese migration trajectories. As the case of Shiman and Laila exemplifies, a woman's mobility toward Europe could depend on the opportunities afforded by the fact that her husband lives in Europe. In line with this, almost all of the Sudanese female informants had obtained residency papers through marriage to Sudanese partners who were already living in Europe. In the same vein, Mohammed's claim to European citizenship also depended crucially on his former marriage to a European woman. Thus the most intimate spheres of migrants' relationships often shape mobility opportunities and bordering practices.

Nigerian Trajectories

In April 2008, I (the Dutch male researcher of the team) went to a small apartment in Tarlabaşi, a migrant neighborhood in Istanbul (Turkey). The apartment was shared by five young Nigerian men who found themselves "in transit," as they had entered Istanbul in the hope of reaching Europe from there. One of them—Destiny—had first left his country in 2002, when he set off for Johannesburg (South Africa). He was refused entry, however, and the South African immigration authorities repatriated him to Nigeria. Back in Lagos, it took Destiny some years to save enough money to leave again. This time, inspired by his Pentecostal pastor, he chose "destination Europe." As he could not obtain a visa for any European country, he arranged a direct flight to, and a three-month visa for, Turkey through a local migration broker.

I spent a month with Destiny in Istanbul, and when I left, I wondered whether his dream of a European life and, ultimately, his goal of leading a transnational life of traveling back and forth between Europe and Nigeria, would ever become reality.

In August 2008, Destiny e-mailed me to say that he had reached Greece. When I visited him there in 2010, he told me all about his border crossing. He had paid his "connection man" in Turkey $1,200 for a "guaranteed cross"; in other words, repeated attempts would be made to get him to Greece. After handing over his money, the connection man instructed him to check into a certain hotel in a Turkish coastal city.

Destiny waited impatiently in the hotel for seven days. Finally, in the middle of the night, some men knocked on his door to take him, together with some twenty other migrants, to a beach. When they got there, they boarded an inflatable boat and set off. Their passage was nerve-racking, and the boat started to take on water; Destiny feared that he would never reach Europe alive. In the end, he and the other passengers were picked up by a passing fishing boat and taken to the Greek island of Samos.

Destiny was detained on Samos for eight days before the immigration authorities transferred him to Athens. After an intermediate stay in the Greek capital to finalize his asylum application, he moved to Crete to make some money as a street vendor. According to Destiny, although Crete was a far better place than Istanbul, he had not yet reached the Europe he had in mind. He wanted to move onward. In October 2011, Destiny informed me during an Internet conversation that he had reached Switzerland via Italy and had applied for asylum. Two months later, his application had been rejected and the Swiss authorities had started the repatriation procedure. Destiny decided to return to Italy, where he is now hoping to settle.

Over the years, I also have remained in contact with three of the other Nigerian men I first met in the apartment in Istanbul. Apart from Destiny, only one reached the EU. He had a similar boat passage to Greece, but unlike Destiny, he is still living there. One of the other men returned to Nigeria, mainly because he found trying to reach Europe too risky. He referred to Istanbul as "Istanblock."

The third man—Vincent—never left Istanbul. Instead of moving to Europe, he became involved in the informal transnational textile trade between Turkey and Nigeria. His new role had led to a significant improvement in his socioeconomic position. His frustrations regarding his inability to reach the EU had been replaced by feelings of excitement about life in Istanbul. His aspiration to reach Europe had disappeared, as he stated on the telephone in early 2010: "Europe is far now. I will not travel anymore."

The presence of Nigerian migrants in Istanbul is often discussed in the context of transit migration, referring to migrants' waiting periods in third countries before they move onward to their destinations (Collyer, Düvell, and De Haas 2012). The im/mobility experiences of these Nigerian men, however, complicate the linear notion of transit migration, as though migrants depart, go through a transit phase, and then settle. Destiny's case illustrates that migrants might experience different phases of immobility in their processes of movement. Consequently, trajectories could have multiple moments of departure, and destinations are possibly in between and might soon turn into transit places (Grillo 2007). At the same time, Vincent's case suggests that migrants can move out of transit situations without physically moving onward. Because of an unforeseen opportunity within the transnational trade, Vincent's transit place has turned into a de facto destination.

The contrasting trajectories of Destiny and Vincent underline the argument that neither mobility nor immobility should be considered inherently positive or negative. Whereas nomadic discourses usually associate mobility with freedom, agency, transgression, and cosmopolitanism (Cresswell 2006), Destiny's series of movements is characterized by a lack of freedom, restlessness, and boundaries. Migrants might become "stuck in the process of movement," denoting that mobility and immobility experiences could overlap (Adey 2006) and that mobility can indeed produce entrapments (Glick Schiller and Salazar 2013). In a similar context, one of our respondents used the metaphor of "being stuck on a carousel" to express his experience of being forced to move from place to place. At the same time, Destiny's periods of immobility in Turkey and Greece had little to do with a positive form of emplacement; rather, these periods were unwelcome and enforced interruptions in a mobility process. Vincent's trajectory shows that such unwelcome immobilities can give rise to new opportunities and enhanced freedoms. This observation clearly illustrates the ambiguity of transit situations (Suter 2012). Transit statuses are commonly associated with punctuated periods of getting stuck, waiting, and limbo. Migrants actively experience these statuses and, as such, migrants might strategically incorporate them in their future life projects (Conlon 2011).

These Nigerian migration trajectories indicate that migrants' freedom to move is inherently linked to their freedom to stay somewhere (Sager 2008). In other words, increasing mobility is only desirable if there is room to stay at a preferred location (De Haas and Rodríquez 2010). Conversely, settlement in a place is most comfortable for the migrant when incoming and outgoing movement is possible. Such im/mobility dialectics touch on the widely debated argument of transnational scholars that migrants' integrations can go hand in hand with transnational orientations (Portes, Escobar, and Radford 2007). Put differently, when migrants are able to settle, new forms of "grounded mobilities" (Blunt 2007) are likely to emerge.

Conclusions

Although transnational migration studies have successfully debunked static notions of migration, they have dealt only marginally with mobility processes. A thorough analysis of im/mobility relations and power differentiations in mobility experiences contributes to a better understanding of migrants' transworlds for at least three reasons. First, it helps us to further question the linear and bipolar logics on which many migration conceptualizations are built. By carefully following migrants' trajectories, we observe that migration processes do not follow the conventional order of uprooting–movement–regrounding. International travel might occur in predeparture phases, and migrants' destinations might turn into transit places during the process of movement. As a consequence of these dynamics, migrants' trajectories often involve more than two locations. Although a nonlinear and multipolar perspective on migrants' transworlds will strengthen the emphasis on process, interdependency, and flows, the constant search for im/mobility dialectics prevents us from lapsing into nomadic conceptualizations of transnationalism, as it creates room for fixity and emplacement in the analysis of mobilities.

Second, whereas transnational migration research has garnered rich qualitative insights into migrants' transnational engagements, the im/mobilities approach is particularly helpful in exposing the embodied efforts, emotions, and frictions that are inherent parts of these engagements. In that sense, we do not take transnational interconnectivity for granted but instead emphasize that it is something that requires performances, investments, and maintenance from the people involved. The mobility efforts of Shiman and Mohammed, for instance, produced more involuntary immobility and led to transnationally shared frustrations. Similarly, the im/mobility experiences of Destiny point at the many en route bordering processes. His trajectory exemplifies that some migration regulations are designed to immobilize migrants, whereas others prevent migrants from staying put and, therefore, produce restless mobilities.

These two arguments are strongly related to a sense of process geography (Appadurai 2001). As study objects become mobile, researchers have to think of mobile and process-like methods, such as traveling with migrants (Büscher and Urry 2009). The third and final argument in favor of a mobilities approach to investigate transnational migration, however, points to the importance of a combined methodology of place-based and process-like research. Analyzing migrants' trajectories across time and space has resulted in an in-depth notion of how mobility is embodied by migrants during different phases of their life projects. At the same time, it has enabled us to identify specific places of immobilization and mobilization. European embassies on the African continent, for instance, are significant thresholds where the prolonged waiting of would-be movers manifests in space. As a response to the immobilization practices of the EU's visa regime, numerous brokering services have emerged in Africa to assist the would-be movers to overcome their involuntary immobility. These are places of mobility productions, specialized in finding new entry points and inventing alternative travel modes. Further along the road to the EU, there are various kinds of mobility hubs, relay places, and places of confinement where the production, politics, and experiences of im/mobilities express themselves. Thus, to advance our understanding of im/mobility dynamics and transnational migration, researchers sometimes also have to stay put.

Notes

1. An important exception is the literature on border crossings in the context of Mexican migration to the United States (e.g., Singer and Massey 1998).
2. In July 2011, the Republic of South Sudan became independent from Sudan following a referendum in which the Sudanese voted almost unanimously for secession. This secession was intended to put an end to a long and violent conflict in the region.
3. All names have been changed to pseudonyms.

References

Abusharaf, R. M. 1997. Sudanese migration to the new world: Socio-economic characteristics. *International Migration* 35 (4): 513–36.

———. 2002. *Wandering Sudanese migrants and exiles in North America*. New York: Cornell University Press.

Adepoju, A., and A. Van der Wiel. 2010. *Seeking greener pastures abroad. A migration profile of Nigeria*. Ibadan, Nigeria: Safari Books.

Adey, P. 2006. If mobility is everything then it is nothing: Towards a relational politics of (im)mobilities. *Mobilities* 1 (1): 75–94.

Appadurai, A. 2001. Grassroots globalization and the research imagination. In *Globalization*, ed. A. Appadurai, 1–21. Durham, NC: Duke University Press.

Bailey, A. J., R. A. Wright, A. Mountz, and I. M. Miyares. 2002. (Re)producing Salvadoran transnational geographies. *Annals of the Association of American Geographers* 92 (1): 125–44.

Bakewell, O., and H. De Haas. 2007. African migrations: Continuities, discontinuities and recent transformations. In *African alternatives*, ed. L. De Haan, U. Engel, and P. Chabal, 95–117. Leiden, The Netherlands: Brill.

Black, R., S. Ammassari, S. Mouillessaux, and R. Rajkotia. 2004. *Migration and pro-poor policy in West Africa*. Brighton, UK: Development Research Centre on Migration, Globalisation and Poverty, University of Sussex.

Blunt, A. 2007. Cultural geographies of migration: Mobility, transnationality and diaspora. *Progress in Human Geography* 31 (5): 684–94.

Burell, K. 2008. Materialising the border: Spaces of mobility and material culture in migration from post-socialist Poland. *Mobilities* 3 (3): 353–73.

Büscher, M., and J. Urry. 2009. Mobile methods and the empirical. *European Journal of Social Theory* 12 (1): 99–116.

Carling, J. 2002. Migration in the age of involuntary immobility: Theoretical reflections and Cape Verdean experiences. *Journal of Ethnic and Migration Studies* 18 (1): 5–42.

———. 2006. *Migration, human smuggling and trafficking from Nigeria to Europe*. Geneva: International Organization for Migration.

Collyer, M. 2007. In-between places: Trans-Saharan transit migrants in Morocco and the fragmented journey to Europe. *Antipode* 39 (4): 668–90.

Collyer, M., F. Düvell, and H. De Haas. 2012. Critical approaches to transit migration. *Population, Space and Place* 18 (4): 407–14.

Conlon, D. 2011. Waiting: Feminist perspectives on the spacings/timings of migrant im/mobility. *Gender, Place & Culture*, 18 (3): 353–60.

Cresswell, T. 2006. *On the move: Mobility in the modern western world*. London and New York: Routledge.

———. 2010. Towards a politics of mobility. *Environment and Planning D: Society and Space* 28 (1): 17–31.

Cresswell, T., and P. Merriman, eds. 2011. *Geographies of mobilities: Practices, spaces, subjects*. Farnham, UK: Ashgate.

De Bruijn, M., R. Van Dijk, and D. Foeken, eds. 2001. *Mobile Africa: Changing patterns of movement in Africa and beyond*. Leiden, The Netherlands: Brill.

De Haas, H. 2008. The myth of invasion. The inconvenient realities of migration from Africa to the European Union. *Third World Quarterly* 29 (7): 1305–22.

De Haas, H., and F. Rodríquez. 2010. Mobility and human development: Introduction. *Journal of Human Development and Capabilities* 11 (2): 177–84.

Elnur, I. 2002. The second boat of Africa's new diaspora: Looking at the other side of the global divide with an emphasis on Sudan. *African Issues* 30 (1): 37–41.

Ernste, H., K. Martens, and J. Schapendonk. 2012. The design, experience and justice of mobility. *Tijdschrift voor Economische en Sociale Geografie* 103 (5): 509–15.

Glick Schiller, N., L. Basch, and C. Blanc-Szanton. 1995. From immigrant to transmigrant: Theorizing transnational migration. *Anthropological Quarterly* 68 (1): 48–63.

Glick Schiller, N., and N. B. Salazar. 2013. Regimes of mobility across the globe. *Journal of Ethnic and Migration Studies* 39 (2): 183–200.

Grillo, R. 2007. Betwixt and between: Trajectories and projects of transmigration. *Journal of Ethnic and Migration Studies* 33 (2): 199–217.

Hannam, K., M. Sheller, and J. Urry. 2006. Mobilities, immobilities and moorings. *Mobilities* 1 (1): 1–22.

Hochschild, A. R. 2000. The nanny chain. *The American Prospect* 11 (3): 1–4.

Hyndman, J., and W. Giles. 2011. Waiting for what? The feminization of asylum in protracted situations. *Gender, Place and Culture* 18 (3): 361–79.

Jeffries, C. 2008. Waiting. *Environment and Planning D: Society and Space* 26:954–58.

Kaplan, C. 2003. Transporting the subject: Technologies of mobility and location in an era of globalization. In *Uprooting/regroundings: Questions of home and migration*, ed. S. Ahmed, C. Castaneda, A. M. Fortier, and M. Sheller, 207–24. Oxford, UK: Berg.

Kothari, U. 2003. Staying put and staying poor. *Journal of International Development* 15 (5): 645–57.

Malkki, L. 1992. National geographic: The rooting of peoples and the territorialization of national identity among scholars and refugees. *Cultural Anthropology* 7 (1): 24–44.

Massey, D. 2005. *For space*. London: Sage.

Mountz, A. 2010. *Seeking asylum. Human smuggling and bureaucracy at the border*. Minneapolis: University of Minnesota Press.

———. 2011. Where asylum-seekers wait: Feminist countertopographies of sites between states. *Gender, Place & Culture* 18 (3): 381–99.

Parreñas, R. S. 2005. *Children of global migration: Transnational families and gendered woes*. Stanford, CA: Stanford University Press.

Portes, A. 1997. Immigration theory for a new century: Some problems and opportunities. *International Migration Review* 32 (4): 799–825.

Portes, A., C. Escobar, and A. W. Radford. 2007. Immigrant transnational organizations and development: A comparative study. *International Migration Review* 41 (1): 242–81.

Reynolds, R. R. 2004. "We are not surviving, we are managing": The constitution of a Nigerian diaspora along the contours of the global economy. *City & Society* 16 (1): 15–37.

Sager, T. 2008. Freedom as mobility: Implications of the distinction between actual and potential travelling. In *Spaces of mobility. The planning, ethics, engineering and religion of human motion*, ed. S. Bergmann, T. Hoff, and T. Sager, 243–67. London: Equinox.

Schapendonk, J. 2011. *Turbulent trajectories: Sub-Saharan African migrants heading north*. PhD thesis, Radboud University, Nijmegen, The Netherlands.

———. 2012. Migrants' im/mobilities on their way to the EU: Lost in transit? *Tijdschrift voor Economische en Sociale Geografie* 103 (5): 577–83.

Schuster, L. 2005. The continuing mobility of migrants in Italy: Shifting between places and statuses. *Journal of Ethnic and Migration Studies* 31 (4): 757–74.

Shamir, R. 2005. Without borders? Notes on globalization as a mobility regime. *Sociological Theory* 23 (2): 197–217.

Shandy, D. J. 2007. *Nuer-American passage: Globalizing Sudanese migration*. Gainesville: University Press of Florida.

Sheller, M., and J. Urry. 2006. The new mobilities paradigm. *Environment and Planning A* 38:207–26.

Sheppard, E. 2002. The spaces and times of globalization: Place, scale, networks and positionality. *Economic Geography* 78 (3): 307–30.

Simich, L., D. Este, and H. Hamilton. 2010. Meanings of home and mental well-being among Sudanese refugees in Canada. *Ethnicity and Health* 15 (2): 199–212.

Singer, A., and D. S. Massey. 1998. The social process of undocumented border crossing among Mexican migrants. *International Migration Review* 32 (3): 561–92.

Steel, G., N. Winters, and C. Sosa. 2011. Mobility, translocal development and the shaping of development corridors in (semi-)rural Nicaragua. *International Development Planning Review* 33 (4): 409–28.

Suter, B. 2012. *Tales of transit: Sub-Saharan African migrants' experiences in Istanbul*. PhD thesis, Malmö University, Malmö, Sweden.

Urry, J. 2007. *Mobilities*. Cambridge, UK: Polity Press.

Van Bochove, M. 2012. Truly transnational: The political practices of middle-class migrants. *Journal of Ethnic and Migration Studies* 38 (10): 1551–68.

Van Houtum, H. 2010. Human blacklisting: The global apartheid of the EU's external border regime. *Environment and Planning D: Society and Space* 28 (6): 957–76.

Verstraete, G. 2001. Technological frontiers and the politics of mobility in the European Union. *New Formations* 43:26–43.

Walton-Roberts, M., and G. Pratt. 2005. Mobile modernities: A South Asian family negotiates immigration, gender and class in Canada. *Gender, Place & Culture* 12 (2): 173–95.

Weissköpel, C. 2009. Traversing cultural sites. Doing ethnography among Sudanese migrants in Germany. In *Multi-sited ethnography: Theory, praxis and locality in contemporary research*, ed. M. A. Falzon, 251–70. Farnham, UK: Ashgate.

North Korean Women's Narratives of Migration: Challenging Hegemonic Discourses of Trafficking and Geopolitics

Eunyoung Choi

Independent Scholar, Seoul, South Korea

This article recovers the subaltern stories of North Korean migrant women living and working underground in China. This work begins to unravel relations of power operating through hegemonic discourses of human trafficking. International human rights agencies and the U.S. government often make totalizing statements that categorize North Korean migrant women in China without authorization as trafficked: powerless victims without agency who need to be rescued with the assistance of international society. Based on in-depth interviews with North Korean migrants and extensive document analysis, however, I argue that the strategic spotlight on North Korean human trafficking has actually worsened the condition of North Koreans in China by leading to more crackdowns by both North Korean and Chinese authorities who consider those international efforts as political intervention and threats to their national security. In addition, current discourses on North Korean human trafficking are grounded in Western liberal discourses of the universality of human rights. As a result, trafficking discourses and policies ignore the complicated nature of global political and economic structures that shape women's basic economic security and their decision to migrate. Through critical examination of contemporary geopolitics, this article shows the urgency of democratic and decolonized ways of understanding the trafficking of North Korean women and of offering them support.

本文恢復在中国非法生活及工作的北韩移民女性的底层人民故事。此工作首先揭露透过人口贩运的霸权论述运作的权力关係。国际人权组织与美国政府, 经常对中国境内的北韩非法移民女性进行整体化的宣称, 将之描绘为受到人口贩运, 且不具能动性、需要透过国际社会的协助进行解救的受害者。然而, 我根据对北韩移民者的深度访谈与大规模的档案分析, 主张对北韩人口贩运的策略性关注, 实际上却使得北韩与中国政府视这些国际努力为政治介入及对国家安全的威胁, 因而同时引发北韩与中国政府更多的打压, 导致了中国境内北韩人民处境的恶化。此外, 当今对于北韩人口贩运的论述, 是以西方自由主义对于普世人权的论述为根据。人口贩运的论述与政策, 因而忽略了形构女性基本经济安全及其决定迁徙的全球政治经济结构的复杂本质。透过对当代地缘政治的批判性检视, 本文呈现以民主且去殖民化的方式理解北韩女性人口贩运、并给予其支持的迫切性。

Este artículo rescata las historias de las propias mujeres migrantes norcoreanas que viven y trabajan escondidas en China. Para empezar, este trabajo desenreda las relaciones de poder que obran en los discursos hegemónicos relacionados con tráfico humano. Las agencias internacionales de derechos humanos y el gobierno de los EE.UU. a menudo se pronuncian con fuertes declaraciones que categorizan a las migrantes norcoreanas ilegales en China como sujetos de tráfico humano: víctimas indefensas sin quien las represente, que merecen ser rescatadas con la ayuda de la sociedad internacional. Con todo, a partir de entrevistas a profundidad con migrantes norcoreanas, además de amplio análisis documental, arguyo que la estrategia de sacar a la luz pública el tráfico humano de norcoreanas en realidad ha empeorado la condición de estas inmigrantes en China, al generar una represión más severa por parte de las autoridades norcoreanas y chinas que califican aquellos esfuerzos internacionales como un tipo de intervención política y amenazas para la seguridad nacional de sus países. Además, la apreciación actual sobre el tráfico humano de norcoreanos se basa en las opiniones liberales occidentales sobre la universalidad de los derechos humanos. El resultado es que los discursos y políticas sobre este tipo de tráfico ignoran la naturaleza compleja de las estructuras políticas y económicas globales que configuran la seguridad económica básica de las mujeres y sus decisiones sobre migrar. Mediante un examen crítico de la geopolítica contemporánea, este artículo pone de presente la urgencia de enfoques democráticos y descolonizados para comprender el tráfico al que son sometidas mujeres norcoreanas y para brindarles apoyo.

Discourses surrounding and practices against the trafficking of North Korean women are fraught with geopolitical tensions. Since early 2000, governments, international agencies, nongovernmental organizations (NGOs), activist groups, and international media have expressed enormous concern over the human trafficking of North Korean women. Most reports on the topic focus on raising international awareness of the human rights abuses that trafficked North Korean women face in China (Human Rights Watch 2002; Kang 2004; Muico 2005; Amnesty International 2009; U.S. State Department 2002–2012). They emphasize the need for international cooperation to put pressure on the governments of North Korea and China and to rescue and protect North Korean women.

According to the former director of the U.S. State Department's Office to Monitor and Combat Trafficking in Persons, between 80 and 90 percent of North Korean migrants in China are actually trafficked persons, and a large proportion of those victims end up in sexual slavery (Bullock 2005). This frightening statistic is frequently presented as evidence of the brutality of the North Korean regime. It is also used to justify the earlier 2004 North Korean Human Rights Act, a U.S. law aimed at providing humanitarian assistance to North Koreans both inside and outside the country, that grants financial assistance to nonprofit organizations seeking to promote democracy, human rights, and the development of a market economy in North Korea. Lawmakers also sought to pressure the Chinese and North Korean governments to end their campaigns of locating and forcibly returning North Koreans entering China without permission. Although the law brought much-needed attention to the plight of trafficked North Korean women, the emphasis placed on the irresponsible and brutal policies of North Korea and China (U.S. State Department 2006) served to further obfuscate the broader conditions that shape trafficking of North Korean women and the women's own migration experiences.

This article aims to foreground the voices and migration experiences of North Korean women against a backdrop of contemporary geopolitics and the antitrafficking efforts developed in relation to them. Current discourses on North Korean human trafficking are often framed by Western liberal ideals of universal human rights. As a result, antitrafficking discourses and policies ignore the complicated nature of the global political and economic structures that shape North Korean women's migration and efforts for survival. Furthermore, popular depictions construct North Korean women as powerless victims of sexual violence, in need of protection from the North Korean and Chinese authorities. Both are problematic because such discourses serve as justification for foreign policy interventions that ultimately harm rather than help North Korean women. Taking a feminist geopolitical approach, I seek to challenge totalizing discourses that categorize North Korean women as trafficked, powerless victims in need of rescue by enlightened nations.

This research is based on a critical review of the *Trafficking in Persons Reports* produced by the U.S. government, as well as in-depth interviews conducted with North Korean women in the North Korean Chinese borderland of Yanbian between 2003 and 2007 and in Seoul in summer 2012. I examine the policies and dialogues regarding the migration of North Koreans among and within states and international organizations and their impact on North Korean women's daily experiences of human trafficking. As undocumented migrants in China are highly vulnerable and publicly invisible, I conducted research there with the support of an international NGO that provided humanitarian aid to North Korean migrants. Fourteen key informants gave multiple interviews. Each had lived during different time periods in Yanbian including Yanji city, Tumen, Loingjing, Helong, Dunhua, Huchun, Wanquing, and Antu. Most participants were either North Korean women who had married Chinese men and settled in rural areas or stayed in shelters provided by humanitarian organizations.

Drawing on these in-depth interviews with trafficked women in the North Korean–Chinese borderland and other data, I explore how North Korean women understand their own experiences of migration and why these understandings differ from public discourses. In doing so, I demonstrate how the imperial geopolitics of the region are implicated in the international community's concern for North Koreans and the antitrafficking policies that have been created. In using the term *imperial geopolitics* I seek to draw attention to the ongoing practices of political domination exercised by powerful states and the importance that discourses of cultural and political supremacy play in contemporary strategies to leverage control over the human and natural resources, as well as capital, of subordinated states. Many scholars view the United States as an imperializing state that relies on geographical tropes to produce a complex taxonomy of places categorized by the level of "danger," "threat," or "safety" they present (O'Tuathail and Agnew 1992; Dalby 1994). In this U.S.-produced geopolitical imagination, North Korea has been labeled an "axis of evil" and major barrier to global expansion of liberal democracy and market economy. By unraveling the power relations that shape discourses and migrations, I argue that North Korean women's own narratives about migration challenge the discourses and policies that cement their subaltern status as powerless victims.

Feminist Geopolitics, Migration, and Human Trafficking

By shifting from a rigidly territorialized view of politics toward more nuanced understandings of the spatiality of power, scholars have drawn attention not only to the geographical assumptions and designations that shape political debates and practice but also to the widely differing spheres and subjects entangled in the spatial politics of states (Megoran 2008). The field of critical geopolitics in general and feminist contributions therein have illuminated both the role of geographical knowledge in the discursive construction of space, as well as the wide range of nonstate actors and places outside formal state institutions implicated in the exercise of political practice (Gilmartin and Kofman 2004). Feminist geopolitics scholars, in particular, try to recover untold stories and unrecognized subjects of conventional geopolitics (Hyndman 2004).

Over the last decade, scholars have emphasized the role of the state in international migration and recognized migration policies as fundamentally geopolitical practices (Coleman 2007; Hyndman 2012). In particular, recent literature on the geopolitics of migration has focused on the aims and effects of exclusionary sovereign power to regulate immigration along national borders (Mountz 2010). Contestations between economic and asylum-seeking undocumented migration, often executed by human smugglers, and receiving governments' efforts at border controls have received close academic scrutiny (e.g., Geiger and Pecoud 2010). Although the field of critical geopolitics has moved us away from accounts of world politics predicated on states and their territorializing impulses, there remains much to learn about the relationship between the changing spatial politics of states, geopolitical discourses that support them, and their influence on the everyday political practices of a wide range of political subjects, especially beyond Europe and North America. Nowhere is the need for greater knowledge more aptly demonstrated than in the literature exploring the geopolitics of human trafficking. There, the geographical discourses of nonstate actors and imperial impulses of states have combined to produce political practices that threaten, rather than protect, the security of undocumented migrants.

Research on human trafficking since 2000 has taken different tacks. Writers who adopt a human rights perspective portray trafficked women as powerless victims of "transnational organized crime" in need of protection and assistance (Kristof 2004; Landesman 2004). Those supporting a criminal justice approach (United Nations Convention against Transnational Organized Crime 2000; Council of Europe 2005) believe that trafficked women need to be controlled and regulated to protect national security. From this perspective, trafficked women are considered illegal and their bodies deemed vectors for the spread of HIV and other diseases (Sharma 2005). Much of the focus has also been on sex trafficking only, a fact that feminist scholars argue has obscured other forms of trafficking and, crucially, the relationship among labor, migration, and international relations (Chuang 2006; Zheng 2010; Bernstein 2012; Brennan 2013). Furthermore, heightened emphasis on sex trafficking reinforces gendered and racialized constructions of the powerlessness of "victims."

In addition, international organizations like the United Nations explicitly link human trafficking to efforts to tackle all types of organized criminal activities that cross national borders, including the movement of persons, weapons, or drugs (Kempadoo 2005, 39). Here, trafficking works as master metaphor for the undesirable underside of globalization that erodes the nation-state's boundaries and threatens its sovereignty. Preventing trafficking through immigration control has therefore become a great concern for receiving states. Antitrafficking has become almost synonymous with a war on international criminal activities (Kempadoo 2005), an observation also made by Bernstein (2010), who argues that feminist antitrafficking activism is committed to carceral paradigms of social justice and supportive of modes of state and NGO engagement that use women's human rights to justify military interventions.

Postcolonial and transnational feminist scholars find these approaches to human trafficking problematic (Doezema 2000, 2010; Kempadoo 2005, 2012; Sharma

2005; Sudbury 2005). Kempadoo (2005), for example, argues that efforts by countries in the Global North to control the international migration of women in the Global South are sexist and racist. She shows that the antitrafficking movement is firmly linked to the "criminalization of migrant women from the global south and to greater policing and control of their mobility, bodies, and sexuality" (39). Similarly, Sudbury (2005) shows the structural conditions such as asymmetrical economies that prompt Jamaican women to work as carriers of drugs across international borders, decisions that bring them into the punitive carceral space of the UK prison system.

Transnational feminists who study human trafficking problematize the images of trafficked women as powerless and innocent as portrayed by international human rights organizations and by Western feminists (Sudbury 2005; Cheng 2010; Zheng 2010; Kempadoo 2012). These scholars are especially critical of the work of abolitionist feminists who view trafficking as a form of sexual exploitation caused by patriarchy and capitalist exploitation; that is, as harmful to women and a violation of women's rights (Limoncelli 2009, 261–62). Transnational feminists argue that this approach generally avoids addressing the basic economic insecurity faced by trafficked women and its relationship to the patterns of uneven development generated within the globalizing capitalist system. To counter these hegemonic and geopoliticized discourses of human trafficking, transnational feminists try to include the views, experiences, and perspectives of women of the Global South in current debates and representations of human trafficking (Doezema 2010; Kempadoo 2012).

This article builds on these approaches. The critiques posed by transnational feminists are particularly important in the case of North Korea because of the impact of Cold War tensions on the lives of migrants. Those who cross the North Korean–Chinese border move from a relatively closed socialist system in a totalitarian regime to a global capitalist economy where they are subject to the influence and interests of powerful state and nonstate organizations, most notably, the United States, China, South Korea, and the United Nations. North Korea is isolated within this globalizing capitalist system and increasingly targeted by international criticism for its nondemocratic political system and its stated ambition to develop nuclear weapons.

Although much research on the implications of these shifts has focused on the changing geopolitical power held by states and regional levels of security (Hogan 1992; Buchanan 1999; Christensen 1999; Ross 1999; S.-H. Lee 2003; Shambaugh 2005; Beeson 2007), few scholars have examined how the changing geopolitics of the region differently affect the daily lives of North Koreans (Park 2003). The current debates and practices around the migration of North Korean women should therefore be understood as crucial elements of an emerging imperial geopolitics of trafficking. This geopolitical field promotes the humanitarian concerns and human rights discourses of Western NGOs and international aid agencies to effect political practices that ultimately support the imperial ambitions of powerful states. In the context of North Korea, the geopolitics of trafficking can be observed in the geographical designations that the U.S. government uses to guide foreign policy in the region, humanitarian discourses used by Western NGOs to influence this process, and economic implications of these policies for the relationship between the United States, North Korea, and China.

Humanitarianism, Human Rights, and the Imperial Response to Trafficking

The line between trafficking and smuggling is blurry, yet those anxious to protect exploited women often construct trafficked women as innocent and powerless victims in need of rescue (Sharma 2005). Such depictions of North Korean women have been used by human rights agencies such as Human Rights Watch (2002), Amnesty International (2009), and Anti-Slavery International (2005) and are evident in the North Korean Human Rights Act (2004) and the U.S. Department Trafficking in Persons reports (2003–2013). Agustín (2006) claims that this tendency is the result of moral panic about prostitution in Western society and the avoidance of uncomfortable truths about the willingness of many women to be trafficked. Women categorized as powerless slaves often receive public sympathy and support, but smuggled women willing to be trafficked garner no such sympathy and are, instead, seen as agents of sexual immorality (Kempadoo and Doezema 1998).

An imperialist understanding of human trafficking can be found in the U.S. State Department's annual Trafficking in Persons Report (U.S. State Department 2003–2013). The U.S. government understands human trafficking as a multidimensional threat to global health, national security, and human security. To combat this threat with diplomatic tools, the U.S. government compiles information on more than 180 countries. Based on these data, the United States categorizes other nation-states into four groups (Tier 1, Tier 2, Tier 2 Watch List, or Tier 3) and announces that the third-tier countries—the least favorable rating among the categories—might be subject to certain sanctions by the U.S. government, including withholding of nonhumanitarian, non-trade-related assistance funds.

North Korea has been included in this annual report since 2003, rated annually as one of the worst offenders (Tier 3). In June 2013, China was also downgraded from Tier 2 to Tier 3 and consequently could be subject to sanctions including the cancellation of cooperative programs and economic assistance. Although economic sanctions such as the withdrawal of aid are unlikely to have a major impact on the Chinese economy (between 2001 and 2012 the United States provided only $338 million in economic assistance), the new rating could negatively affect bilateral U.S.–China relations.

The Trafficking in Persons reports have been highly contentious, not only because of the position of authority assumed by the United States in categorizing states but also because of ongoing concerns about methodologies used to determine tier placement and compliance (Moon 2008; Gallagher 2010; Government Accounting Office 2010). By categorizing nation-states based on government efforts to combat trafficking, the United States is producing new spatial imaginaries of the world. Critics argue that this taxonomy mirrors binary understandings of power and spatiality—East and West, security and danger, freedom and oppression—reminiscent of the war on terror (Gregory and Pred 2006). Indeed, as scholars have observed (Kempadoo 2005; Gallagher 2010), in 2009 Iran and a number of other Muslim countries were categorized as Tier 3, alongside Burma and Cuba. Meanwhile, Australia's position in Tier 1 never changed, even though it took several years to meet minimum standards. These discrepancies confirm suspicion about the geopolitics surrounding categorization on the issue of trafficking; categorization is ideologically and politically motivated rather than rooted in facts.

The credibility of the data published in the Trafficking in Persons reports was tested in the 2007 report, stating that the North Korean government did not acknowledge human trafficking along the North Korea–China border. Yet, as my interviews with North Korean women in China revealed, the North Korean government actively searches for human traffickers in the name of eliminating nonsocialist activities. My informants even claimed that the North Korean government had conducted open executions in various sites in the Northern Hamkyung province. Far from failing to make an effort to deter the trafficking of persons, secret video recordings of the public execution of three accused traffickers in Hoeryeong City in 2005 indicated that the North Korean government had indeed taken steps to combat trafficking, albeit in ways that remain highly problematic.[1]

Growing international awareness of the human rights of North Koreans has not led to positive changes in the political and juridical practices of North Korea and China (Chang, Haggard, and Noland 2008). As Sharma (2005) insists, antitrafficking actions in the name of rescuing trafficked victims have legitimated repressive states' immigration controls. Specifically, the U.S. government's efforts to expose human rights abuses faced by North Korean migrants is regarded as an unwelcomed political intervention by North Korea and China, who assert that the United States exploits the moral vocabulary of universal human rights for its own imperial geopolitical ends (Yoon 2004; Kwak 2005). Since 2000, China has controlled its borders more tightly and repatriated North Koreans more actively, even though it knows that deported North Koreans receive harsh punishment from their home country (Margesson, Chanlett-Avery, and Bruno 2007). Militarization of the North Korea–China border pushed migration broker fees much higher and undocumented migrations further underground. Thus, the geopolitics surrounding North Korean human trafficking has made North Korean migrations more difficult, dangerous, and clandestine.

The Human Face of Human Trafficking

Since the mid-1990s, North Koreans have crossed the North Korea–China border unofficially for their own survival. Extreme food shortages and severe economic deprivation produce these flows (Lee et al. 2009). Even though the North Korea–China border is highly militarized on both sides, professional brokers guide migrants to China through bribing or avoiding soldiers. Poor young women with no money either to bribe soldiers or to obtain assistance from smugglers could get trafficked. Increased demands for young single women in rural China and expanding global markets for brides have commodified North Korean women's bodies, making them targets for human trafficking and forced marriages. Some are deceived in North Korea and sold after being transported into China, but others migrate more voluntarily via human trafficking. The following story is quite typical. Traffickers recruit North Korean women by deception, coercion, or abduction. They often seek out impoverished young women who are starving and deceive them with tales of the "Chinese dream."

> I met one old lady at the Musan railway station while I was returning back home with my mother. I was really hungry and exhausted. My mom was sick. While we sat in the railway station, the old lady came to us. After she saw my mom, she felt sympathy for us and gave us some potato cake. . . . While we ate the potato cake, she suggested that I go to China to get a job. The old woman said that her daughter went to China and sent some money to North Korea every month for her family. Owing to her daughter, no one died from hunger in her family. . . . I asked my mom to allow me to go to China with her daughter. . . . Once we [the old woman's daughter and I] came to China, we went to one house. While I was having dinner in that house, the woman who crossed the river with me disappeared. I've never met her again. A couple in the house told me that it is really dangerous to live in China as an illegal North Korean and suggested that I get married. I was only eighteen at that time. I was so afraid and did not know what to do. Therefore, I followed them and met a Chinese man who is now my husband. He does not allow me to work. He does not even allow me to go outside of the house. His family is always watching me, and I always stay at home. Later, I heard that my husband paid 4,000 yuan in 1998. (Interview with a twenty-six-year-old North Korean migrant in Antu, China, 2005)

This type of deception and sale of women into marriages happened mostly in the 1990s. Few North Koreans had heard about human trafficking at that time and, as a result, women became targets. Both men and women work as recruiters for trafficking in North Korea. Interestingly, many trafficked women mentioned older women, young girls, or disabled people as recruiters. The images of the weak are utilized to deceive. Some interviewees told me that they followed traffickers to get food without realizing that they crossed the border.

The majority of trafficked North Korean women find themselves confined in China while they await sale into marriage.[2] Although not the most common form of trafficking, some women are trafficked into the sex industry (Muico 2005, 5). Because North Korea treats returnees harshly, female migrants often agree to arranged marriage or work in the sex industry. The criminalization of migrants as illegal tends to push North Korean women into invisible, private spaces, mostly through forced marriage. The following interview shows how an undocumented North Korean woman had to endure extreme violence in the process of trafficking. The interviewee's father was a professor at a university, and she studied ceramic art at a college in a city of northern North Korea. Nonetheless, she reported that her college degree did not mean anything in North Korea's difficult economic conditions, so she decided to go to China to avoid dying of hunger.

> I heard about China from a person. She said that she had an aunt who was quite old and lived in China by herself without children. If I came to China and took care of her, her aunt would treat me like her daughter. I believed only half of those words, but I decided to leave the land that could not provide me a meal even if I starved to die. . . . I knew that I should not believe anyone in that fiendish ear. However, her aunt in China sounded as the only hope for me. She wanted to cross the border that night to visit her aunt. I asked her to give me some time to go to my house and tell my parents. But she said we did not have enough time. She paid some money to North Korean soldiers, and we safely came to China. Once we arrived in China, a car was waiting for us and brought us to a house in a city. In the house, just in front of me, she was bargaining over the price of me and then left soon. I realized that I was deceived and sold. I tried to run away, but they took my shoes and, instead, gave me a pair of slippers. They even followed me when I went to the toilet. That night I was raped. I resisted, but when they told me that they could ask the police to deport me, I gave up further resistance. From the next day on, the human traffickers treated me like their wife at night and, during the daytime, they showed me to the people who visited the house to buy me. A person said I was too short. The other person said I was too expensive. I hoped to be a doll. I was sold to a Chinese for the price of 7,000 yuan (around US$1,000). I was thirty at that time. It happened in 1997.

Sexual violation, including rape, is often used by traffickers to obtain "consent" to marry. Marriages between undocumented North Koreans and Chinese, however, are not officially accepted, so North Korean women in de facto marriage relations with Chinese men continuously struggle not to be detected by Chinese authorities and to avoid deportation.

Some women voluntarily involve themselves in trafficking. Since about 2000, the North Korean government has publicly announced the dangers of human trafficking, but still some women choose to use trafficking as an escape route. Most North Korean women I met said that they simply found it preferable to marry Chinese men rather than see their families die or live in poverty. Considering the political and economic situation that pushed North Korean women from their home countries and the extraordinary circumstances that North Korean women face in China, however, fully consensual involvement in human trafficking is rare. Additionally, even North Korean women who consciously and willingly move abroad to improve their lives and that of their families cannot fully know the dangers of the clandestine routes they must take to cross the border and the vulnerable status awaiting them in China. Even when North Korean women voluntarily contact marriage brokers or human traffickers to cross the North Korea–China border, the money paid by marriage partners in China goes to the intermediaries or kidnappers. Neither the North Korean woman nor her family receives anything (Lankov 2004, 861).

In spite of such difficult circumstances, North Korean women are not powerless. They negotiate and choose even with limited options. The following quote is testimony from North Korean poet Jini Choi (2005, 203):

> My friend told me that the quickest way to settle down in China is a marriage. He also told me to marry first and then run away if I cannot endure that marriage. . . . I agreed to come to China and then be sold into marriage.

In interviews with Choi in South Korea in 2006, she also criticized the moralizing discourses assigned to trafficked women. She told me that she hopes more women can escape North Korea, even through trafficking, rather than staying and waiting to die.

> I was sold into marriages three times and severely beaten and sexually abused almost every night while I lived with the third husband in China. As I have experienced those beatings, I know the realities of trafficking in North Korean women. Nonetheless, I personally hope that more and more North Korean women can cross the border to China and then come to safe third countries, even through . . . human trafficking and providing sex. Don't think that we are sexually immoral. If you have never starved more than three days in your life and seen your family die because of hunger in front of your eyes, you do not have a right to judge us.

Choi understands commercial marriage, and potential abuse in that marriage, as a strategy for survival and mobility. Her argument is supported by evidence that most women deported to North Korea subsequently returned to China, even if trafficked as a means to cross the border (Kim 2010). Anti-Slavery International also observed that despite miserable circumstances in China, many trafficked North Korean women find their life there better than starving to death in North Korea (Muico 2005, 5–6).

Conclusions

Stricter border controls by both North Korea and China, expanding global markets for brides and reproductive and sex workers, and extreme poverty in North Korea have all increased the trafficking of North Korean women. Trafficking is very risky and North Korean women experience sexual violence. Nonetheless, these women participate in trafficking to escape from life-threatening food shortages in North Korea. Even with limited choices and lack of information, North Korean women use their femininity as a tool to cross borders.

Drawing on empirical research conducted in North Korea–China borderlands, I showed that the strategic spotlight on North Korean human trafficking has worsened conditions of undocumented North Korean migrants in China by leading to more crackdowns by both North Korean and Chinese authorities that consider international attention as political intervention and a threat to national security. Based on these findings, I have argued that North Korean women's own narratives about migration can challenge Western depictions of them as powerless victims of human trafficking in need of human rights protection. Such discourses serve not only to deprive North Korean women of agency but also play a role in the strategic geopolitics that currently shape international critiques of North Korea and China.

There is an urgent need for more democratic and decolonized ways of understanding the trafficking of North Korean women and offering support. Deconstructing the colonizing discourses that portray women as hapless victims and in so doing reinforcing Western notions of cultural and political supremacy is an important step in the process. But equally important is the process of "mapping," producing cartographies that link North Korean women's migration and their struggles to survive to the strategies deployed by the United States, Europe, China, and North Korea. These feminist countertopographies (Katz 2001) reveal the efforts of nation-states to leverage economic and political control over each other with antitrafficking agendas, discourses, and policies. Outlining the topography and countertopography of trafficking brings into relief neoliberal state geopolitics at work and the importance of confronting these imperial impulses to the end of trafficking itself. Drawing on North Korean women's voices to detail their own migration experiences is the most important element in the process of tracking and revealing these contours.

Acknowledgments

This article was written by Dr. Eunyoung Choi and embodies research that she conducted for her doctoral dissertation (completed in 2010). Ill health prevented Eunyoung from carrying out all of the revisions requested in early reviews of the article. These revisions were completed by Beverley Mullings and Alison Mountz, both of whom were members of Eunyoung's doctoral committee and who drew on Eunyoung's dissertation to make revisions. The ideas expressed in this article are a reflection of Eunyoung's intellectual and emotional labor, which would not have been possible without the courageous participation of the North Koreans who so generously shared their experiences despite difficult circumstances. As Eunyoung states in her dissertation, "The North Korean migrants in China whom I owe the largest debt of gratitude remain anonymous in order to protect them." This article was significantly strengthened by the very constructive comments offered by three anonymous reviewers and the guidance and encouragement of Dr. Richard Wright, editor of the *Annals*, throughout this process. We thank them profusely on Eunyoung's behalf.

Notes

1. The execution was secretly recorded on video, a copy of which was smuggled outside the country and appeared first on Japanese television and later in a British Broadcasting Corporation (BBC) documentary (BBC 2004). It was reported that a widespread investigation was launched inside of North Korea to find the people responsible for making and exporting the video and who themselves were involved in trafficking.
2. The significant imbalance in the ratio of women to men in China, an outcome of its one-child policy, continues to fuel a demand for North Korean women, who are often sold by marriage brokers as "brides" to Chinese men (see Kim 2010).

References

Agustín, L. M. 2006. The disappearing of a migration category: Migrants who sell sex. *Journal of Ethnic and Migration Studies* 32 (1): 29–47.

Amnesty International. 2009. Amnesty International report 2009—State of the world's human rights. London: Amnesty International Publications. http://news.bbc.co.uk/2/shared/bsp/hi/pdfs/28_05_09_amrep.pdf (last accessed 4 December 2013).

Anti-Slavery International. 2005. Seoul. London: Anti-Slavery International. http://www.antislavery.org/includes/ documents / cm _ docs /2009/f/full_korea_ report_ 2005.pdf (last accessed 8 July 2013).

Beeson, M. 2007. *Regionalism and globalization in East Asia: Politics, security, and economic development.* New York: Palgrave Macmillan.

Bernstein, E. 2010. Militarized humanitarianism meets carceral feminism: The politics of sex, rights, and freedom in contemporary anti-trafficking campaigns. *Signs: Journal of Women in Culture and Society, special issue on Feminists Theorize International Political Economy* 36 (1): 45–71.

———. 2012. Carceral politics as gender justice? The "traffic in women" and neoliberal circuits of crime, sex, and rights. *Theory and Society* 41:33–259.

Brennan, D. 2013. *Life interrupted: Trafficking into forced labor in the United States.* Durham, NC: Duke University Press.

British Broadcasting Company (BBC). 2004. *Access to evil.* Documentary. London: BBC.

Buchanan, T. 1999. The coming decade of change on the Korean peninsula: Implications for Northeast Asia and the United States. *East Asia* 17 (4): 7–29.

Bullock, T. 2005. North Korean refugees frequent victims of human trafficking: State's Miller cites rampant trafficking from North Korea into China. http://iipdigital.usembassy.gov/st/english/article/2005/07/20050720153430tjkcollub0.4683344.html#axzz2YQ5FpzP4 (last accessed 8 July 2013).

Chang, Y., S. Haggard, and M. Noland. 2008. Migration experiences of North Korean refugees: Survey evidence from China. Working Paper series WP 08-4, Peterson Institute for International Economics, Washington, DC.

Cheng, S., ed. 2010. *On the move for love: Migrant entertainers and the U.S. military in South Korea.* Philadelphia, PA: University of Pennsylvania Press.

Choi, J. 2005. *Kukgyŏngŭl se pŏn kŏnnŏn yŏja ch'oe chini* [A woman who crossed the border three times]. Kyŏnggido, South Korea: Book House.

Christensen, T. 1999. China, the U.S.–Japan alliance, and the security dilemma in East Asia. *International Security* 23 (4): 49–80.

Chuang, J. 2006. Beyond a snapshot: Preventing human trafficking in the global economy. *Indiana Journal of Global Legal Studies* 13 (1): 137–63.

Coleman, M. 2007. A geopolitics of engagement: Neoliberalism, the war on terrorism, and the reconfiguration of US immigration enforcement. *Geopolitics* 12 (4): 607–34.

Council of Europe. 2005. *Council of Europe Convention on action against trafficking in human beings.* 16 May, CETS 197. http://www.refworld.org/docid/43fded544.html (last accessed 4 December 2013).

Dalby, S. 1994. Gender and critical geopolitics: Reading security discourse in the new world order. *Environment and Planning D: Society and Space* 12:595–612.

Doezema, J. 2000. Loose women or lost women? The reemergence of the myth of white slavery in contemporary discourses of trafficking in women. *Gender Issues* 18 (2): 23–50.

———. 2010. *Sex slaves and discourse masters: The construction of trafficking.* London: Zed Books/Palgrave Macmillan.

Gallagher, A. 2010. Improving the effectiveness of the international law of human trafficking: A vision for the future of the US Trafficking in Persons reports. *Human Rights Review* 12 (1): 381–400.

Geiger, M., and A. Pecoud, eds. 2010. *The politics of international migration management.* New York: Palgrave Macmillan.

Gilmartin, M., and E. Kofman. 2004. Critically feminist geopolitics. In *Mapping women, making politics: Feminist perspectives on political geography*, ed. L. A. Staeheli, E. Kofman, and L. Peake, 113–26. London and New York: Routledge.

Government Accounting Office. 2006. *Human trafficking: Better data, strategy, and reporting needed to enhance U.S. anti-trafficking efforts abroad.* Report to the Chairman, Committee on the Judiciary and the Chairman, Committee on International Relations, House of Representatives, GAO-06-825. Washington, DC: Government Accounting Office.

Gregory, D., and A. Pred, eds. 2006. *Violent geographies: Fear, terror and political violence*. London and New York: Routledge

Hogan, M. 1992. *The end of the cold war: Its meaning and implications*. Cambridge: Cambridge University Press.

Human Rights Watch. 2002. The invisible exodus: North Koreans in the People's Republic of China. *Human Rights Watch Report* 14 (8): 1–38. http://www.hrw.org/legacy/reports/2002/northkorea/ (last accessed 8 July 2013).

Hyndman, J. 2004. The (geo)politics of gendered mobility. In *Mapping women, making politics: Feminist perspectives on political geography*, ed. L. A. Staeheli, E. Kofman, and L. Peake, 169–84. London and New York: Routledge.

———. 2012. The geopolitics of migration and mobility. *Geopolitics* 17 (2): 243–55.

Kang, C. Y. 2004. Chaechung Talbukyŏsŏngdlui Saenghwal Silt'ae [A study on the life situation of North Korean female refugees in China]. *Yŏsŏngyŏngu Nonchong* [Journal of Women's Studies] 19:59–77.

Katz, C. 2001. On the grounds of engagement: A topography for feminist political engagement. *Signs* 26 (4): 1213–34.

Kempadoo, K. 2005. Victims and agents of crime: The new crusade against trafficking. In *Global lockdown: Race, gender, and the prison-industrial complexes*, ed. J. Sudbury, 35–55. London and New York: Routledge.

———. 2012. Introduction: Abolitionism, criminal justice, and transnational feminism: Twenty first-century perspective on human trafficking. In *Trafficking and prostitution reconsidered—New perspectives on migration, sex work, and human rights*, ed. K. Kempadoo, J. Sanghera, and B. Pattanaik, vii–xlii. London: Paradigm.

Kempadoo, K., and J. Doezema, eds. 1998. *Global sex workers: Rights, resistance and redefinition*. London and New York: Routledge.

Kim, J. 2010. Trafficked: Domestic violence, exploitation in marriage, and the foreign-bride industry. *Virginia Journal of International Law* 51 (2): 443–505.

Kristof, N. D. 2004. Inviting all democrats. *New York Times* 14 January. http://www.nytimes.com/2004/01/14/opinion/14KRIS.html (last accessed 8 July 2013).

Kwak, H. R. 2005. *Pukhanit'aljumin hyŏnhwanggwa munje: Saryewa chiwŏnbangan* [The current situation of North Korean migrants and their problems: Some examples and supporting plans]. P'aju, South Korea: Hangukhaksuljŏngbo [Korean Academic Information].

Landesman, P. 2004. The girls next door. *New York Times Magazine* 25 January. http://www.nytimes.com/2004/01/25/magazine/the-girls-next-door.html (last accessed 8 July 2013).

Lankov, A. 2004. North Korean refugees in northeast China. *Asian Survey* 44 (6): 856–73.

Lee, K., S. Choi, S. Kim, K. Lee, and S. Lim, 2009. *White paper on human rights in North Korea*. Seoul, South Korea: Korea Institute for National Unification.

Lee, S.-H. 2003. Terrorism and asymmetric war: Is North Korea a threat? *East Asia* 20 (2): 21–47.

Limoncelli, S. A. 2009. Human trafficking: Globalization, exploitation, and transnational sociology. *Sociology Compass* 3 (1): 72–91.

Margesson, R., M. Chanlett-Avery, and A. Bruno. 2007. North Korean refugees in China and human rights issues: International response and U.S. policy options. CRS Report for Congress RL34189, Congressional Research Service, Washington, DC.

Megoran, N. 2008, Militarism, realism, just war, or nonviolence? Critical geopolitics and the problem of normativity. *Geopolitics* 13 (3): 473–97.

Moon, K. H. S. 2008. Beyond demonization: A new strategy for human rights in North Korea. *Current History* 107:263–68.

Mountz, A. 2010. *Seeking asylum: Human smuggling and bureaucracy at the border*. Minneapolis: University of Minnesota Press.

Muico, N. K. 2005. *An absence of choice: Sexual exploitation of North Korean women in China*. London: Anti-Slavery International. http://www.antislavery.org/includes/documents/cm_docs/2009/f/full_korea_report_2005.pdf (last accessed 8 July 2013).

ÓTuathail, G., and J. Agnew. 1992. Geopolitics and discourse: Practical geopolitical reasoning in American foreign policy. *Political Geography* 11 (2): 190–204.

Park, H.-S. 2003. *Hyundai Pukhan Sahwoewa Kajok* [Contemporary North Korean society and family]. Seoul: Hanul.

Ross, R. 1999. The geography of the peace: East Asia in the twenty-first century. *International Security* 23 (4): 81–118.

Shambaugh, D. 2005. China engages Asia: Reshaping the regional order. *International Security* 29 (3): 64–99.

Sharma, N. 2005. Anti-trafficking rhetoric and the making of a global apartheid. *NWSA Journal* 17 (3): 88–111.

Sudbury, J. 2005. "Mules," "Yardies," and other folk devils: Mapping cross-border imprisonment in Britain. In *Global lockdown: Race, gender, and the prison-industrial complex*, ed. J. Sudbury, 167–82. London and New York: Routledge.

United Nations Convention against Transnational Organized Crime (Organized Crime Convention). 2000. G.A. Res. 55/25, Annex I, 55 U.N. GAOR Supp. (No. 49) at 44, U.N. Doc. A/45/49 (Vol. I)(2001), adopted 15 November 2000, entered into force 29 September 2003. http://www.unodc.org/documents/treaties/UNTOC/Publications/TOC%20Convention/TOCebook-e.pdf (last accessed 4 December 2013).

U.S. Congress, House. *North Korean Human Rights Act of 2004*. HR 4011:108. http://www.govtrack.us/congress/bills/108/hr4011 (last accessed 8 July 2013).

U.S. State Department. 2006. *Trafficking in persons report June 2006*. Washington, DC: U.S. State Department. http://www.state.gov/j/tip/rls/tiprpt/ (last accessed 8 July 2013).

———. 2003–2013. Trafficking in persons report. Washington, DC: U.S. State Department. http://www.state.gov/j/tip/rls/tiprpt/ (last accessed 8 July 2013).

Yoon, I. J. 2004. *The Korean diaspora: Migration, adaptation, and identity of overseas Koreans*. Seoul, South Korea: Korea University Press.

Zheng, T., ed. 2010. *Sex trafficking, human rights, and social justice*. New York: Routledge.

Environmental Hazards as Disamenities: Selective Migration and Income Change in the United States from 2000–2010

J. Matthew Shumway,* Samuel Otterstrom,* and Sonya Glavac†

**Department of Geography, Brigham Young University*
†Division of Geography and Planning, University of New England, Australia

An emerging area of migration research is the complex relationship between migration and environmental hazards, broadly defined. Environmental hazards are best viewed through a vulnerability lens, which has two components. The first is exposure—the frequency and duration of the hazardous events—and the second is adaptive capacity—the ability of communities to mitigate, deflect, or absorb the effects of exposure. Because migration is selective of individuals and places, it changes both the population's size and composition, thus affecting its exposure and adaptive capacity. In this article we examine how migration varies among sets of counties that experience significantly different exposures to all environmental hazards in the United States. We create an environmental hazards impact index in an attempt to measure the impacts of environmental hazards at the county level over a period of years. We found that counties that experience the greatest impacts from environmental hazards are losing income as a result of the migration. In counties with the highest impacts, income is lost through both net outmigration as well as income loss through out-migrants having higher incomes than in-migrants.

迁徙与广义定义的环境灾害之间的复杂关联, 是新兴的迁徙研究领域。环境灾害最适合透过脆弱性的视角检视之, 并具有两大组成元素: 首先是暴露——灾害事件的频率与期程, 再者是调适能力——社群减轻、避开或吸收暴露效应的能力。迁徙具有个人和地方的选择性, 因此会同时改变人口的规模与组成, 从而影响了人口的暴露及其调适能力。我们在本文中, 检视对所有的环境灾害经历不同暴露程度的美国各州中, 迁徙如何有所变异。我们创造了环境灾害影响指标, 试图衡量在数年之间, 州层级的环境灾害冲击。我们发现, 经历了最为剧烈的环境灾害冲击的各州, 正因人口迁徙的关係, 导致所得逐渐降低。受到最为严重冲击的各州, 同时由于人口的淨移出, 以及移出者较移入者拥有更高的收入, 因而丧失该州的所得。

Un área emergente en las investigaciones sobre migración es la compleja relación que existe entre migración y riesgos ambientales, en sentido amplio. Los riesgos ambientales se miran mejor a través de una lente de vulnerabilidad, que tiene dos componentes. El primero de ellos es la exposición—frecuencia y duración de los eventos calamitosos—y el segundo es la capacidad adaptativa— la habilidad de las comunidades para aliviar, reorientar o absorber los efectos de la exposición. Debido a su carácter selectivo en términos de individuos y lugares, la migración cambia tanto el tamaño de la población como su composición, afectando de ese modo su exposición y capacidad adaptativa. En este artículo examinamos la manera como la migración varía entre conjuntos de condados que experimentan exposiciones significativamente diferentes a todos los riesgos ambientales en los Estados Unidos. Creamos un índice de impacto de los riesgos ambientales a título de intento para medir estos impactos a nivel de condado en un determinado período de años. Hallamos que los condados que sufren los más altos impactos por riesgos ambientales están perdiendo ingresos como resultado de la migración. En los condados que tienen los más altos impactos, el ingreso se pierde a través de emigración neta lo mismo que por las pérdidas resultantes de las diferencias de ingreso de los emigrantes con ingresos más altos en comparación con los que ganan los inmigrantes.

Scholars writing on the relationship between migration and environmental hazards often take a hazards perspective (Hunter 2005; McLeman and Hunter 2010). That is, migration is perceived within the context of hazard vulnerability. This relationship is explained in terms of whether or not and to what extent migration increases or decreases the social vulnerability of preparation for or recovery from environmental hazards (Cutter et al. 2008; McLeman 2009; Johnson and Krishnamurthy 2010; Emrich and Cutter 2011).

Another principal dimension of research on environmental hazards and migration is how acute (e.g., Hurricane Katrina) and chronic (e.g., prolonged drought or sea level rise) environmental hazards influence patterns of outmigration and return migration shortly after the event and also in the long term (McLeman and Smit 2006; Fussell, Sastry, and VanLandingham 2009; Gutmann and Field 2009). Thus, a hazard perspective is focused on the hazard per se in which migration is considered an adaptation strategy. On the other hand, the relationship between migration and environmental hazards can also be examined from a migration perspective, which focuses on causes and consequences of migration to origin and destination areas. Using a migration perspective shifts the focus from migration as a part of social vulnerability to environmental hazards to one that focuses on how hazards might or might not affect changing spatial patterns of migration. These two perspectives are linked and can be thought of as two sides of the same coin, but the distinction here allows us to ask slightly different questions than those typically asked in the environmental hazards literature.

Amenity Migration and Environmental Hazards

According to Hunter's (2005) review of the literature on migration and environmental hazards, there are two principal relationships: The first views hazards as a part of the amenity package of the area, thus affecting how desirable or undesirable a place is perceived, which in turn affects to what degree economic wages or opportunities can mitigate or enhance the local amenity level. Second, migration is seen as a mechanism that either increases or decreases social vulnerabilities through the size and composition of the migration flow. Net positive migration into an area, particularly one with perceived high amenity levels, increases population density and thus the number of people exposed to possible hazards. Migration also increases or decreases levels of social vulnerability by changing the demographic and economic composition of the local population. For example, in an area with large amounts of net outmigration, generally dominated by the young, the local population grows older, increasing levels of social vulnerability (McLeman 2009).

Relationships between environmental hazards and migration tend to fit well within an amenity migration conceptual framework. Amenity migration is the idea that people move to areas to take advantage of specific natural amenities (Graves 1979; Deller et al. 2001; Rappaport 2007). Whether searching for locations offering desirable natural amenities, for a specific climate type, or an aspiration to leave behind negative environmental factors (e.g., air pollution), the pursuit of specific environmental quality is a primary factor in interregional migration (Hsieh and Liu 1983).

The amenity approach to migration is basically an extension of the general economic spatial equilibrium idea. In a spatial equilibrium, labor and capital move in opposite directions, causing a reduction in per capita income differences, which enhances per capita income convergence across regions (Partridge 2005). Locations with lower incomes are more likely to have less capital and thus labor productivity is lower. Capital, hoping to increase its returns, seeks out poorer places to benefit from potential increased productivity. As capital converges, so does productivity and hence incomes, all else being equal. The key to such spatial equilibrium is the mobility of capital and labor. Ganong and Shoag (2012) illustrated that during the strongest period of income convergence across states in the United States (1890–1980), migration accounted for approximately one third of that convergence. When convergence is achieved, individual residential preferences for different locations diminish, at least relative to wages or income.

Graves (1979) used this approach attempting to explain migration and regional adjustment in terms of regional nonequilibrium conditions. In post–World War II America, Graves recognized that even with increased income convergence across states, people were moving to areas and regions that were unexpected (the South and the West). Using the insights provided by Tiebout (1956; i.e., people vote with their feet), the amenity framework as developed by Graves and his colleagues suggests that households seek out locations that maximize their overall utility levels—including natural amenities. Thus, migration responds to both economic factors and quality of life factors (primarily based on environmental amenities). Recent research also suggests the importance of environmental amenities as the primary explanation of interregional migration in the United States. Partridge (2010) outlined three theoretical frameworks for exploring interregional movements in the United States during the twentieth century: agglomeration/new economic geography (Tabuchi and Thisse 2002; Glaeser and Resseger 2010), the people follow jobs/agglomeration approach of Storper and Scott (Storper and Scott 2009; Scott 2010; Kemeny and Storper 2012), and an amenity migration approach first put forth by Graves (1976, 1979, 1980). Partridge's conclusion is that the amenity migration framework best explains interregional migration and suggests that "Amenity migration has led to a fundamental transformation of American geography" (Partridge 2010, 533).

Within the amenity migration framework, the concept of migration selectivity provides additional insight. Migration selectivity suggests that migrants are not randomly selected out of the population but have specific characteristics that set them apart from nonmigrants, including, among others, age, skills and education, and income. The concept of migrant selectivity is crucial, as it proposes that the composition of the migrant flow is as important as the number of people moving.

Natural Amenities as a Superior Good

The context within which people evaluate levels of stress, and particularly environmentally-induced stress, has dramatically changed in the United States due to increases in overall wealth. Although not normally seen in this context, Zelinsky's (1971) article on the mobility transition provides a broad framework for understanding how and why amenities became such a driving force of migration after World War II. The basic concept is that as regions "developed," the magnitude, type, and direction of migration would change. The idea has its roots in Engle's Law, which demonstrated that as household income increases, a smaller proportion of income is spent on basic necessities. As applied to migration, the idea is that as a society increases in income and wealth, people's choices concerning where to live also expand. Graves (1979) followed this up by demonstrating that the value of natural amenities has increased as the United States has become wealthier. Amenities are now considered as a normal or superior good because of the increased ability to take advantage of geographic differences in environmental conditions. As a result, much of the late-twentieth-century migration in the United States can be explained by residential preferences for location-specific environmental amenities. Relative to environmental hazards, an increase in both overall and individual wealth allows people to act on preferences for less risky natural environments. Nevertheless, there is a threshold at which amenities become less influential in terms of residential preferences. As high-amenity areas attract more people, supply-side costs (housing, congestion, growth limits) start to capture the values of the amenities such that households must be willing to trade off lower real incomes to access a location's specific environmental amenities (Rappaport 2005; Withers and Clark 2006).

Other Important Influences

Although the United States is a relatively wealthy country, it still exhibits a high degree of spatial variability in income (Galbraith and Hale 2008). Relative to environmental hazards, household wealth and assets are important because assets represent the stock of wealth necessary to recover from unexpected events. Devastating, prolonged, or frequent environmental hazards can reduce household wealth and destroy or significantly reduce capital infrastructure, thus limiting the ability of some households to sufficiently recover. Clearly, environmental hazards act as stressors causing households to reevaluate current residential preferences; however, these preferences are tempered by such factors as household wealth, governmental transfer programs and insurance, private insurance, and local and extended social networks (McLeman 2009). Given the scope of this article, we are limiting ourselves to one aspect of this large area of research. The question we are asking is this: What is the difference between the numbers and income characteristics of in- and out-migrants from counties with differing impacts of prolonged or intense natural hazards? Instead of examining the effect of a dramatic or particularly intense hazard on short-term or return migration, we examine how migration varies among sets of counties that experience significantly different exposures to all environmental hazards.

Data and Methods

We use two data sets in our analysis. The first is the University of South Carolina's Spatial Hazard Events and Losses Database for the United States (SHELDUS), which tallies natural disasters by U.S. county (Hazards & Vulnerability Research Institute 2012). The second is the Internal Revenue Service (IRS) County to County migration files from 2000 to 2010 (U.S. Internal Revenue Service 1999–2000; 2009–2010). We first categorized counties according to their exposure to natural hazards over the fifty years of available data, and then we used those categories to analyze population and income migration over the recent decade from 2000 to 2010.

Hazards Data and Methods

The hazards database has more than 680,000 records of events classified by date, length of endurance, and impact. These occurrences include severe winter or summer weather, hail, drought, tornadoes, earthquakes, and flooding. We created an environmental hazards impact index (HII) using techniques developed by Peduzzi et al. (2009) for the United Nations Development Programme's Reducing Disaster Risk program. Instead of measuring risk, our index attempts to measure the impacts of environmental hazards at the county level over a period of years. Our HII is based on three components: the frequency of hazards, the number of people exposed, and the number of injured or loss of life. Because of changes in the South Carolina hazards database with regard to financial losses, we did not include that aspect in our index (see the SHELDUS metadata).

We standardized the three variables using z scores and then summed the results for each county for each year between 1960 and 2010. After mapping this initial index, we noticed that some states (e.g., Iowa) tended to report relatively more hazards than their neighbors, so we added a variable that increases or decreases the final hazards index according to the relative number of events in the state when compared with the national average. This weighting did not eliminate but it did minimize some of the variance in reporting across states. Due to space constraints we chose not to disaggregate by type of hazard and thus all hazard types are included in the index. Although disaggregating impacts by type of hazard is interesting and important, it falls outside the scope of this article. Persistent hazards such as extreme temperatures show up at much higher rates than do more periodic events such as hurricanes. As such, persistent hazards have more weight in the index than periodic events do. This is somewhat balanced out because injuries and deaths tend to be higher in the less frequent events. There is little guidance, either theoretically or empirically, on weighting the variables in our index. Given the purpose of this article, we decided to weight frequency at 40 percent because more people are likely to experience an event than either be injured or killed in one, thus having a greater impact on their migration decision. We weighted injuries and deaths at 30 percent, a spatial weighting variable that averages a county's score by using its ten closest neighbors (15 percent) and the state average (15 percent) with the remainder.

The resulting natural hazards impact index is shown in Figure 1. It is easy to see here the impacts of severe weather (Great Lakes/Northeast and the Southwest), and although it might seem like extremes in heat and cold are not "real" environmental hazards, they account for approximately 15 percent of the injuries and deaths from all hazards. Other areas of concentration can be found in Florida, the Gulf Coast, along major rivers and Tornado Alley, and in the some of the mountain areas in Utah and Colorado (avalanches). Because the index increases due to numbers exposed to the hazards, large urban areas are also more apparent.

After creating the HII, we then used it to assign each county to one of six clusters based on their standard deviation (see Table 1). These clusters form the basis for the rest of the analysis as we explore how patterns of income migration vary among clusters.

Income Migration Data and Methods

For the migration data we use the combined IRS/Census County-to-County migration file (U.S. Internal Revenue Service 1999–2000; 2009–2010). Starting in the early 1990s, the U.S. Census Bureau and the IRS started producing data sets containing information on migration between states and counties (based on exemptions claimed for spouses and dependents) and aggregate income (based on adjusted gross income) associated with those flows. Migrants are determined by matching Social Security numbers and addresses in consecutive years. As Plane (1999) pointed out, what the IRS data set is measuring is not actually the movement of earned income but, rather, the income earned by individuals in two different locations in successive years; thus the measure is one of human capital (and unearned income). Additional shortcomings of the IRS data include an underrepresentation of those who do not file income taxes and the inability to include other important characteristics of migrants such as age, gender, and education. Nevertheless, these data do provide a unique data set that allows the examination of population movement and its impact on aggregate and per capita income levels of U.S. states and counties. Using the IRS data and Plane's decomposition measures (see Plane 1999), we measure the impact of income migration on county clusters that were created using our HII.

Migration alters the relative income among the clusters of counties in two main ways. First, assuming similar incomes of migrants, if more people move to a specific cluster than leave, the result is a net transfer of income. This will be shown as an aggregate income increase or decrease. Second, per capita incomes of in-migrants and out-migrants are different. If in-migrants have higher or lower per capita incomes than out-migrants, there will be a net shift in per capita income.

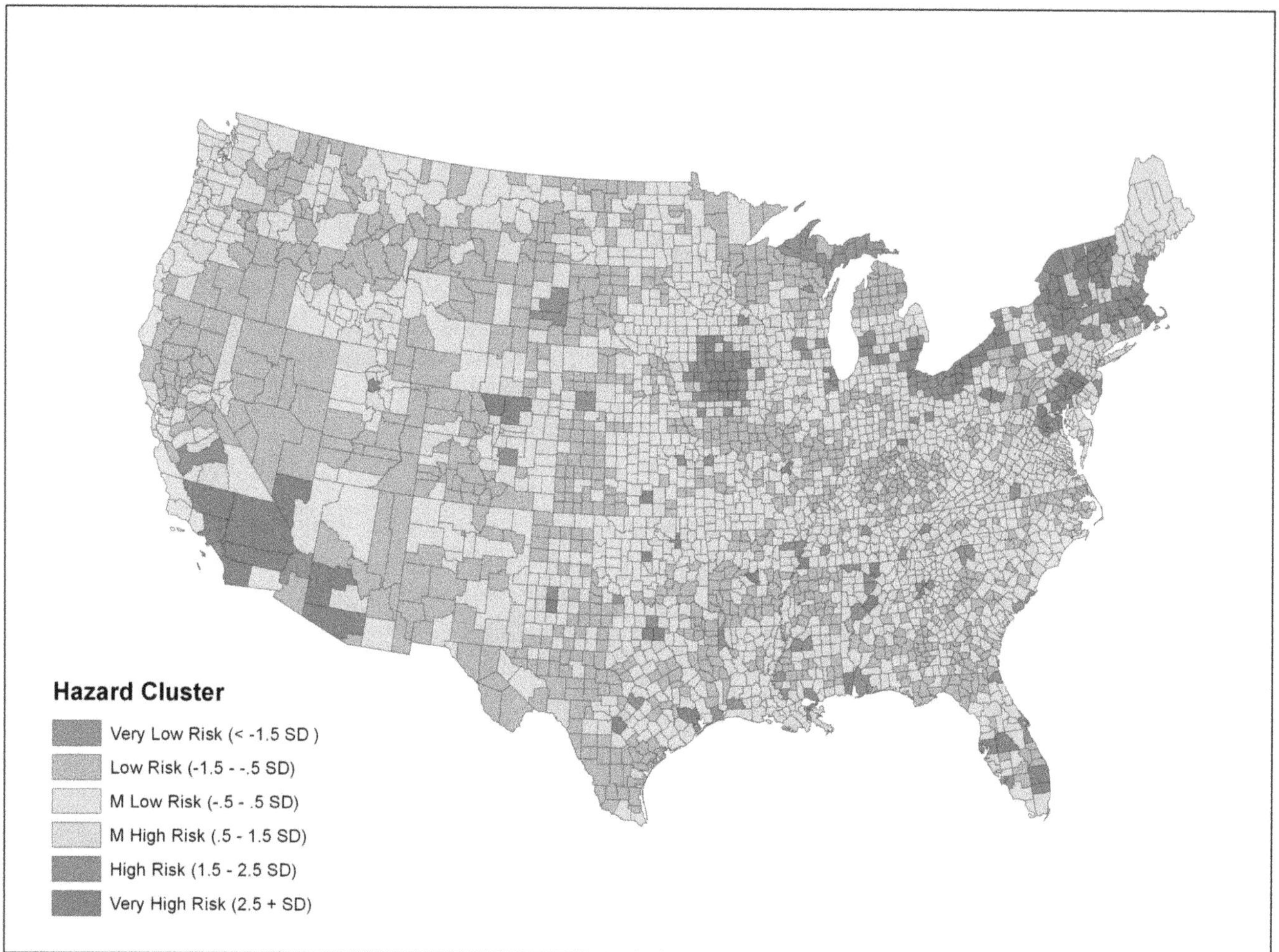

Figure 1. Hazards impact index measure of hazard impacts by county: 1960–2010. (Color figure available online.)

For each year and county in the contiguous United States, we extracted total numbers of in-migrants, out-migrants, and stayers along with the aggregate incomes for each group. Aggregate income migration is separated into two components: net migration and differential income. The net migration component is the difference in average per capita income between in- and out-migrants multiplied by the net number of migrants. This measures the change in aggregate income because of the net flow of migrants. The differential income component is the amount of aggregate income attributable to the relative differences in per capita income between in- and out-migrants.

We also derived the average per capita income change in each cluster due to migration by disaggregating income change into its three income

Table 1. County characteristics by environmental impact cluster

Cluster	*SD*	Number of counties	% Urban	Population size 2010	% Population change 2000 to 2010
Very low impact	< -1.5	21	5%	351,670	15%
Low impact	-1.5–$-.5$	1,007	9%	22,719,537	7%
Medium low impact	$-.5$–.5	1,473	26%	102,985,078	10%
Medium high impact	.5–1.5	392	55%	68,420,132	12%
High impact	1.5–2.5	141	65%	40,241,029	10%
Very high impact	2.5 +	83	80%	71,957,560	7%

Table 2. Income effectiveness and net migration by impact cluster, 2000–2010

Impact group	All counties	Metro counties	Nonmetro counties	Net migration
Very low	1.97	7.21	0.92	1,901
Low	2.55	3.17	2.19	578,876
Medium low	0.42	−0.37	2.15	1,050,264
Medium high	−0.07	−0.29	1.44	1,272,209
High	−1.79	−1.68	−3.68	132,786
Very high	−5.50	−5.54	−2.88	−1,685,285

components: in-migrants/stayers, out-migrants/stayers, and in-migrants/out-migrants.

Results

Income Migration Effectiveness

The first measure we use is income migration effectiveness (net migration/total migration), which shows the standardized size of the income flow from migration so that it can be compared across different-sized regions (Plane 1999). The calculated income migration effectiveness measure by hazard impact group for the entire time period in question is shown in Table 2.

Between 2000 and 2010, migration has consistently shifted income from counties with high and very high impacts of environmental hazards to counties with lower impacts. The very-high-impact counties lost approximately 5.5 percent and the high-impact counties lost approximately 2 percent of their income between 2000 and 2010 due to income migration; however, only the very-high-impact cluster actually had more out-migrants than in-migrants. On the other side, the very-low-impact and low-impact counties gained income due to migration by approximately 2 percent a year. There appears to be some significant movement of income away from counties subject to intense or prolonged environmental hazards and thus migration could be affecting levels of social vulnerability. The patterns are similar when disaggregated by metro–nonmetro status with a couple of exceptions. For example, nonmetro counties in the high-impact cluster lose relatively more money than high hazards metro counties, and most of the income migration flowing into the very-low-impact cluster is going to metro, not nonmetro, counties.

Aggregate Income Change

Income migration effectiveness does indicate how effectively migration exchanges income and people between places, but we also want to see the total amount of income change due to migration. To see this, first look at aggregate income flows through a decomposition technique first described by Plane (see Plane 1999 for the decomposition formula), where aggregate income is decomposed into its net migration and differential income components. The beauty of this technique is that we can see whether the principal driver of income change is due to the net numbers of people changing locations (net migration effect) or from income differences between in- and out-migrants (differential income effect). It is important to note that the effects could either reinforce each other (both are positive or both are negative) or work against each other (net effects are positive and differential income effects are negative or vice versa).

The final column in Table 3 shows aggregate income change by cluster between 2000 and 2010. Column 1 is income change due to net migration and column 2 is income change due to differences in income between in- and out-migrants. Panel A shows the results for all counties. For all counties, both the high- and very-high-impact clusters experienced a net decrease in income due to migration, although only the very high cluster lost income through both net migration and differential income—meaning that they are losing income because more people are leaving than are moving into these counties, and those who are leaving have, on average, higher incomes than those moving in. The medium high cluster had an overall net aggregate income gain, but this was all from net migration. This cluster actually lost income because out-migrants had more income than in-migrants. The three low-impact clusters all gained net income through the net migration and differential income components. Both the medium low and low clusters, however, gained more net income due to differential income migration. This is important because it suggests, at least in the aggregate, that people with higher earning capacities are concentrating in lower impact areas, all other things being equal. On the other hand, all three high-impact clusters lost income from differential income migration. The very high set of counties lost income from both the net migration and

Table 3. Income migration components by county impact cluster and metro status

County impact clusters	Net migration component	Differential income component	Net income migration
A (all counties)			
Very low	115,434,888	202,166,855	317,601,743
Low	12,364,298,720	10,018,655,680	22,382,954,400
Medium low	12,410,133,942	13,090,008,061	25,500,142,003
Medium high	21,826,073,753	−6,233,021,435	15,593,052,317
High	4,191,049,707	−4,891,449,882	700,400,175
Very high	−49,333,946,054	−9,963,910,159	−59,297,856,212
B (metro)			
Very low	195,489,377	77,741,242	273,230,618
Low	7,389,402,385	1,771,776,000	9,161,178,385
Medium low	1,188,833,312	2,796,119,091	3,984,952,403
Medium high	18,744,655,698	−8,262,995,698	10,481,660,000
High	4,819,108,366	−4,729,476,263	89,632,103
Very high	−49,006,623,517	−9,855,320,872	−58,861,944,389
C (nonmetro)			
Very low	−80,054,489	124,425,614	44,371,125
Low	4,974,896,335	8,246,879,679	13,221,776,014
Medium low	11,221,300,630	10,293,888,969	21,515,189,599
Medium high	3,081,418,054	2,029,974,263	5,111,392,317
High	−628,058,659	−161,973,619	−790,032,278
Very high	−327,322,537	−108,589,286	−435,911,823

differential income, but these losses were dominated by net out-migration of income. The counties in the very high cluster together experienced a loss of almost $60 billion due to migration between 2000 and 2010, a significant amount. Some of these patterns might be the result of age; that is, migration up and down the urban hierarchy as Plane, Henrie, and Perry (2005) have shown, but without age data in the IRS/Census county-to-county data set we cannot determine the effect of age on these patterns with these data. Nevertheless, it is clear that the migration process is shifting people with higher (or lower) income-earning capacities away from (or toward) high-impact counties.

We also disaggregated the income migration data between metro and nonmetro counties (see Table 3, panels B and C). Income migration in and out of metro and nonmetro counties follows a similar pattern to the overall county pattern except for the nonmetro very low clusters, which lost income through net migration but gained income through differential income migration. Nevertheless, the dominant overall pattern of income migration is one where income is increasingly being concentrated in lower impact counties. This suggests that those areas with the greatest impacts from environmental hazards are likely to not just lose population through net out-migration and their income but lose an increasing proportion of higher income earners as well. This in turn could affect the ability of high-impact counties to prepare for, mitigate the effects of, and recover from environmental hazards.

Changes in Per Capita Income

Aggregate income paints one picture of the effects of migration on income migration change, and per capita income provides a slightly different perspective. For example, a county could gain (or lose) aggregate income but decline (or increase) in per capita income if in-migrants have significantly lower (or higher) incomes than stayers and out-migrants. Even if an area gains in total income, its level of social vulnerability could increase if per capita income is declining due to compositional change among the local population.

Table 4 contains data on total per capita income change between 2000 and 2010. The final column is the sum of per capita income change for each year between 2000 and 2010. The first three columns contain the disaggregated components; that is, differences in per capita income between in-migrants/stayers, between out-migrants/stayers, and between in-migrants/out-migrants (columns 1, 2, and 3, respectively). Both columns 1 and 2 confirm that in- and out-migrants in every cluster have, on average, lower incomes than stayers, which is not

Table 4. Per-capita income by component, county impact group, and metro status

County impact clusters	In-migrant -stayer component	Out-migrant -stayer component	In–out-migrant component	Per capita income change
A (all counties)				
Very low	−647	1,269	55	676
Low	−992	1,454	42	504
Medium low	−1,415	1,514	13	112
Medium high	−1,827	1,655	−8	−180
High	−1,412	1,217	−10	−205
Very high	−919	734	−10	−196
B (metro counties)				
Very low	−2,124	3,118	195	1,189
Low	−1,415	1,557	24	166
Medium low	−1,527	1,480	3	−43
Medium high	−1,951	1,712	−13	−251
High	−1,421	1,216	−11	−216
Very high	−920	733	−11	−198
C (nonmetro counties)				
Very low	−298	822	26	550
Low	−729	1,388	53	713
Medium low	−1,166	1,585	36	454
Medium high	−997	1,271	25	299
High	−1,244	1,224	−5	−25
Very high	−849	800	−6	−56

unexpected. Thus, overall increases or decreases in per capita income will be determined by whether or not the in-migrant/stayer or the out-migrant/stayer difference is greater (together with the difference in income between in- and out-migrants, although this tends to be relatively small).

For the low-impact clusters, out-migrants have, on average, lower incomes than in-migrants, thus causing per capita income to increase. This is just the opposite for the high-impact county clusters, where out-migrants have higher incomes than in-migrants, resulting in a decrease in per capita income. This suggests that lower income migrants, on average, are moving into areas that have higher impacts from environmental hazards, which in turn might increase the level of community social vulnerability for those counties. The pattern is similar when disaggregated by metro status except for the loss of overall per capita income for the metro medium low cluster and an increase in per capita income for the nonmetro medium high cluster.

Hazards, Job Growth, and Migration

The amenity migration framework emphasizes the push and pull of natural amenities and disamenities, but it does not ignore economic aspects of migration. It could be that what we are seeing is job growth taking place in less hazardous counties and people might be moving to take advantage of the growth in employment opportunities with the side benefit of being in a less hazardous location. There is not any easy way to disentangle these effects using the IRS migration data or without using more sophisticated methodologies than we are using here in our initial examination of this question. Nevertheless, to get a slightly clearer picture of what is going on, we reclassified all counties into a three-by-three matrix based on high, medium, and low hazard impacts and job growth. The job growth data comes from the Bureau of Labor Statistics (BLS) and covers the same period as the migration data (BLS 2013).

Figures 2 and 3 show flows (net difference between in- and out-migrant incomes) of aggregate and per capita income, respectively. Although the figures look complex, there are a few basic patterns that stand out. First, for aggregate income the flows move primarily right to left (low job growth to high job growth county clusters) and upward (more hazardous county clusters; see Figure 2). This pattern suggests that economic opportunities as measured by growth in total employment counteract the effect of persistent or intense hazards on population and income redistribution. The pattern is a little more complex than this, though. For example, the low hazard/high job growth category gained aggregate income from every other group of counties—including

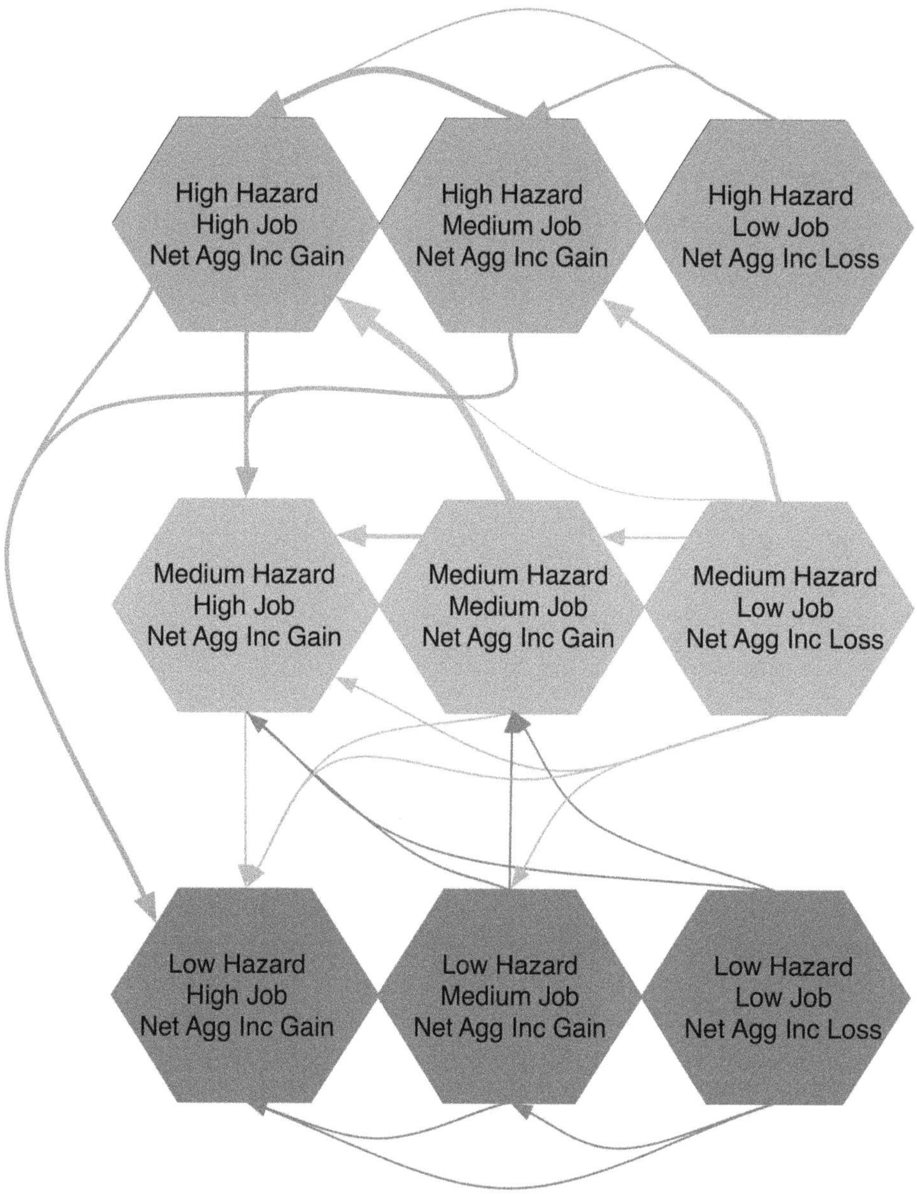

Figure 2. Changes in aggregate income migration by hazard and employment categories. (Color figure available online.)

the other two high job growth categories. The medium hazard/high job and medium hazard/medium job growth groups both received net aggregate income from their counterparts in the high hazard groups. By far the largest increases in aggregate income went to the high hazard/high and medium job growth county groups, but most of these gains came from within the same hazard group or from lower job growth groups.

The patterns shown in Figure 3 for per capita income flows are quite different than those for aggregate income. For instance, there is not a general pattern of movement from low job growth groups to high job growth groups that was found in the aggregate income migration flow. A second difference is the high hazard/high job growth category "lost" per capita income to not only the less hazardous but high job growth groups but also to all three medium job growth county groups. Per capita income change occurs because of differences in income between in-, out-, and nonmigrants, but because, on average, nonmigrants tend to have the highest incomes, the principal cause is whether or not in-migrant incomes are higher than out-migrant incomes relative to nonmigrants. In almost every case (the low job categories being the exception), per capita income "losses" were due to in-migrants having lower incomes than out-migrants (not shown), and those losses were "gains" in per capita income of less hazardous counties. This finding supports the amenity migration

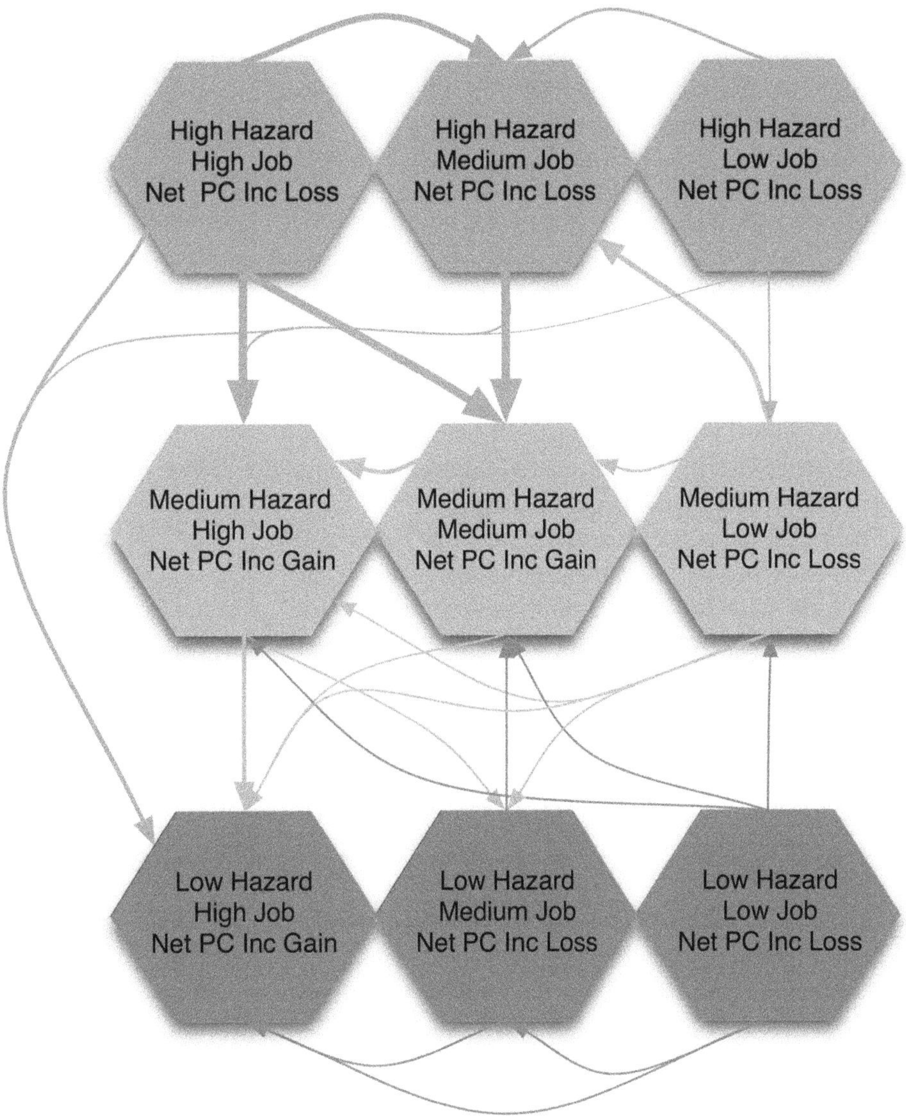

Figure 3. Changes in per capita income migration by hazard employment categories. *Note:* PC = per capita. (Color figure available online.)

notion of higher income leading to increasing importance of amenities as determinants of spatial patterns of migration. It also suggests that counties with lower hazard impacts are attracting migrants with higher incomes and exchanging these with counties that have higher impacts from hazards, thus changing the spatial balance of social vulnerability to environmental hazards.

Discussion and Conclusion

The previous analyses illustrate how migration is affecting the spatial distribution of both people and income as measured by impacts from environmental hazards. The principal finding of this article suggests that counties that experience the greatest impacts from environmental hazards are losing income as a result of migration. In counties with the highest impacts, income is lost through both net out-migration and income loss through out-migrants having higher incomes than in-migrants.

Shifts in both the total number of people and the loss of income associated with net migration have a number of important implications. First, those counties with the greatest impacts from environmental hazards are the very ones that can least afford the loss of income. According to the work of Cutter and colleagues (Emrich and Cutter 2011), household income levels are an important determinant of the overall level of social vulnerability. Although the work here is descriptive and only highlights the general pattern of income migration across counties, it does suggest that the very-high- and high-impact counties appear to be exchanging higher income households for lower income households.

If this is, in fact, what is happening, it indicates a possible divergence in the ability to develop high levels of community resilience. Migration can act as a mechanism of increasing spatial equilibrium or as one that exacerbates already spatially uneven patterns of income distribution. Previous research on income migration (see Manson and Groop 2000; Nelson 2005; Newbold 2008) indicates that divergence, at least in the United States and Canada in the past few decades, is the more likely outcome. This has the potential of putting increased pressure on local community groups, private and government insurance, and other hazard relief programs. This is particularly true for the high-impact group, as differential income effects dominate income losses occurring in these counties.

Due to the descriptive nature of the techniques used in this article, we are only able to emphasize some of the general income migration patterns as they relate to environmental hazards. As a next step, it will be necessary to undertake analyses using more sophisticated multivariate techniques. It is important to include variables associated with housing and land characteristics, labor market conditions, and other types of natural amenities to get a more complete picture of income migration and environmental hazards. This is one of the principal limitations of using the IRS income and migration database, as it only includes income and a surrogate for household size.

Finally, an additional insight is the need to include the disamenity of environmental hazards within the amenity migration framework. This analysis indicates that exposure to intense, frequent environmental hazards affects household decisions concerning migration. Thus, comprehensive analyses of the motivations of migrants and their movements' effects on communities will be improved through a thoughtful inclusion of environmental hazards issues in amenity migration research.

References

Bureau of Labor Statistics, U.S. Department of Labor. 2013. Local area unemployment statistics. http://www.bls.gov/lau/ (last accessed 7 June 2013).

Cutter, S. L., L. Barnes, M. Bery, C. Burton, E. Evans, E. Tate, and J. Webb. 2008. A place-based model for understanding community resilience to natural disasters. *Global Environmental Change* 18 (4): 598–606.

Deller, S. C., T. Tsung-Hsiu, D. W. Marcouiller, and D. B. K. English. 2001. The role of amenities and quality of life in rural economic growth. *American Journal of Agricultural Economics* 83 (2): 352–65.

Emrich, C. T., and S. L. Cutter. 2011. Social vulnerability to climate-sensitive hazards in the southern United States. *Weather, Climate, and Society* 3 (3): 193–208.

Fussell, E., N. Sastry, and M. VanLandingham. 2009. Race, socioeconomic status, and return migration to New Orleans after Hurricane Katrina. *Population & Environment* 31 (1–3): 20–42.

Galbraith, J., and J. Hale. 2008. State income inequality and presidential election turnout and outcomes. *Social Science Quarterly* 89 (4): 887–901.

Ganong, P., and D. Shoag. 2012. Why has regional convergence in the US stopped? http://ssrn.com/abstract=2081216 (last accessed 12 November 2012).

Glaeser, E. L., and M. G. Resseger. 2010. The complementarity between cities and skills. *Journal of Regional Science* 50 (1): 221–44.

Graves, P. E. 1976. A reexamination of migration, economic opportunity and the quality of life. *Journal of Regional Science* 16 (1): 107–12.

———. 1979. A life-cycle empirical analysis of migration and climate by race. *Journal of Urban Economics* 6 (2): 135–47.

———. 1980. Migration and climate. *Journal of Regional Science* 20 (2): 227–37.

Gutmann, M. P., and V. Field. 2009. Katrina in historical context: Environment and migration in the U.S. *Population & Environment* 31 (1–3): 3–19.

Hazards & Vulnerability Research Institute. 2012. The spatial hazard events and losses database for the United States, Version 10.0 [Online database]. Columbia: University of South Carolina. http://www.sheldus.org (last accessed 11 April 2012).

Hsieh, C., and B. Liu. 1983. The pursuance of better quality of life: In the long run, better quality of social life is the most important factor in migration. *American Journal of Economics and Sociology* 42 (4): 431–40.

Hunter, L. M. 2005. Migration and environmental hazards. *Population & Environment* 26 (4): 273–302.

Johnson, C. A., and K. Krishnamurthy. 2010. Dealing with displacement: Can "social protection" facilitate long-term adaptation to climate change? *Global Environmental Change: Human and Policy Dimensions* 20 (4): 648–55.

Kemeny, T., and M. Storper. 2012. The sources of urban development: Wages, housing, and amenity gaps across American cities. *Journal of Regional Science* 50 (1): 85–108.

Manson, G. A., and R. E. Groop. 2000. U.S. intercounty migration in the 1990s: People and income move down the urban hierarchy. *The Professional Geographer* 52 (3): 493–504.

McLeman, R. 2009. Impacts of population change on vulnerability and the capacity to adapt to climate change and variability: A typology based on lessons from "a hard country." *Population & Environment* 31 (5): 286–316.

McLeman, R., and L. Hunter. 2010. Migration in the context of vulnerability and adaptation to climate change: Insights from analogues. *Wiley Interdisciplinary Reviews: Climate Change* 1 (3): 450–61.

McLeman, R., and B. Smit. 2006. Migration as an adaptation to climate change. *Climatic Change* 76 (1–2): 31–53.

Nelson, P. B. 2005. Migration and the regional redistribution of nonearnings income in the United States:

Metropolitan and nonmetropolitan perspectives from 1975 to 2000. *Environment and Planning A* 37 (9): 1613–36.

Newbold, K. 2008. Interprovincial migration and retirement income transfers among Canada's older population: 1996–2001. *Environment and Planning A* 40 (6): 1501–16.

Partridge, M. D. 2005. Does income distribution affect US state economic growth? *Journal of Regional Science* 45 (2): 363–94.

———. 2010. The duelling models: NEG vs amenity migration in explaining US engines of growth. *Papers in Regional Science* 89 (3): 513–36.

Peduzzi, P., H. Dao, C. Herold, and F. Mouton. 2009. Assessing global exposure and vulnerability towards natural hazards: The Disaster Risk Index. *Natural Hazards and Earth System Sciences* 9 (4): 1149–59.

Plane, D. 1999. Geographical pattern analysis of income migration in the United States. *International Journal of Population Geography* 5 (3): 195–212.

Plane, D., C. Henrie, and M. Perry. 2005. Migration up and down the urban hierarchy and across the life course. *Proceedings of the National Academy of Sciences* 102 (43): 15313–18.

Rappaport, J. 2005. How does labor mobility affect income convergence? *Journal of Economic Dynamics and Control* 29 (3): 567–81.

———. 2007. Moving to nice weather. *Regional Science and Urban Economics* 37 (3): 375–98.

Scott, A. J. 2010. Jobs or amenities? Destination choices of migrant engineers in the USA. *Papers in Regional Science* 89 (1): 43–63.

Storper, M., and A. J. Scott. 2009. Rethinking human capital, creativity and human growth. *Journal of Economic Geography* 9 (2): 147–67.

Tabuchi, T., and J. F. Thisse. 2002. Taste heterogeneity, labor mobility and economic geography. *Journal of Development Economics* 69 (1): 155–77.

Tiebout, C. 1956. A pure theory of local expenditures. *Journal of Political Economy* 64 (5): 416–24.

U.S. Internal Revenue Service. 1999–2000; 2009–2010. *IRS area to area migration and county income data*. Washington, DC: IRS Statistics of Income Division.

Withers, S., and W. Clark. 2006. Housing costs and the geography of family migration outcomes. *Population, Space and Place* 12 (4): 273–89.

Zelinsky, W. 1971. The hypothesis of the mobility transition. *Geographical Review* 61 (2): 219–49.

Migration Amidst Climate Rigidity Traps: Resource Politics and Social–Ecological Possibilism in Honduras and Peru

David J. Wrathall,* Jeffrey Bury,† Mark Carey,‡ Bryan Mark,§ Jeff McKenzie,# Kenneth Young,¶ Michel Baraer,£ Adam French,¥ and Costanza Rampini†

*Institute for Environment and Human Security, United Nations University
†Department of Environmental Studies, University of California Santa Cruz
‡Clark Honors College, University of Oregon
§Department of Geography, The Ohio State University
#Department of Earth and Planetary Sciences, McGill University
¶Department of Geography and the Environment, University of Texas, Austin
£École de technologie supérieure, Université du Québec
¥Energy and Resources Group, University of California, Berkeley

According to dominant narratives about adaptation to climate change, those facing worst-case scenarios, without means at their disposal to adapt in situ, face an ineluctable set of adaptation strategies that ultimately includes the permanent abandonment of geographic spaces rendered uninhabitable and unproductive for human use. Yet environmental stress and adaptive capacity are distributed unevenly, and power structures play a role in fashioning them. It is argued here that when access to land and water are impacted by environmental stress, the structures that mediate their access are reinforced, even as the adaptive alternatives for smallholders are undermined. In this way, dominant resource regimes set up migration as the primary viable alternative for adaptation among a dwindling set of choices. This framework is applied to two early analogues of climate change impacts: flooded Garífuna villages of Honduras's North Coast and communities enduring glacier recession and shifting hydrologic regimes in Peru's Cordillera Blanca. In both cases, stress motivates new forms of migration that reinforce dominant power structures. In Honduras, migrants from wealthier social strata are moving on a more permanent basis, and in Peru, the once historical pattern of labor migration is becoming a practical necessity. These cases underscore the role of political economy in adaptation to climate change and adaptive migration in particular.

根据适应气候变迁的主流叙事, 面对最恶劣的情势却未能掌握适应现况工具者, 面临着一连串不可避免的调试策略, 这些策略终将包含永远遗弃那些成为不适合人类居住、且无法进行生产的地理空间。但环境压力与适应能力呈现出不均分配, 而权力结构在形塑此般不均分配中, 扮演着特定的角色。本文主张, 当获取土地与水的管道受到环境压力影响时, 调节这些管道的结构会受到进一步强化, 即便当小农的另类适应方法受到损害时。主流的资源体制, 便藉此在逐渐缩小的选择组合中, 将迁徙设定为适应的首要可行方案。此一架构被应用于气候变迁冲击的两个早期相似案例: 洪都拉斯北部海岸遭受洪泛的加利福纳 (Garífuna) 村落, 以及秘鲁布兰卡山脉 (Cordillera Blanca) 中饱受冰川后退与水文体制变迁所苦的社区。在两个案例中, 压力同时驱动了巩固主流权力结构的新迁徙形式。在洪都拉斯, 来自较为富裕的社会阶层的迁徙者, 其迁徙有着较为永久、稳固的基础。而在秘鲁, 历史中的劳动迁徙模式正逐渐成为实际的必要性。这些案例凸显出适应气候变迁, 特别是适应性的迁徙中, 政治经济的重要性。

De acuerdo con las narrativas corrientes acerca de la adaptación al cambio climático, quienes se enfrentan a los escenarios de la peor clase, sin medios a su disposición para adaptarse *in situ*, tienen que encarar un conjunto ineludible de estrategias de adaptación que en últimas incluyen el abandono permanente de espacios geográficos que se han vuelto inhabitables e improductivos para uso humano. Pero el estrés ambiental y la capacidad adaptativa se distribuyen desigualmente, y las estructuras de poder juegan un papel importante en su formación. Aquí se sostiene que cuando el acceso a la tierra y al agua es impactado por estrés ambiental, las estructuras que intervienen en su acceso son reforzadas, aun a expensas de las alternativas adaptativas para los pequeños propietarios. De esa manera, los regímenes de recursos dominantes colocan a la migración como la alternativa viable primaria de adaptación entre un conjunto de opciones cada vez más limitado. Este esquema se aplica a dos análogos tempranos impactados por el cambio climático: aldeas garífunas afectadas por inundación en la Costa Norte de Honduras y comunidades que padecen cambiantes regímenes hidrológicos en la Cordillera Blanca del Perú por la recesión de glaciares. En ambos casos el estrés motiva nuevas formas de migración que refuerzan las dominantes estructuras de poder. En Honduras, los migrantes de los estratos sociales pudientes se desplazan de una manera más permanente, y en el Perú, el que en otros tiempos fuera un patrón histórico de migración laboral se está convirtiendo en una necesidad práctica. Estos casos descubren el papel de la economía política en la adaptación al cambio climático y sobre la migración adaptativa en particular.

Global climate change scenarios warn that human landscapes could become uninhabitable, unproductive, or otherwise unfit for human use due to increasing storm surges, melting glaciers, and changing rainfall regimes (Intergovernmental Panel on Climate Change 2012; World Bank 2012). In worst-case scenarios, environments are more hazardous, negative impacts occur in livelihoods and economic opportunities, and conflicts develop over dwindling resources. Looking for early analogues of things to come, new research is focusing on migration dynamics taking place in climate "hot spots," or vulnerable regions across the planet where environmental stresses are spatially concomitant with poverty and rapid demographic change (Hugo 2011). Two key conclusions of this research are that migration is either occurring because it is a strategy for adapting to adverse conditions (McLeman and Smit 2006; Bardsley and Hugo 2010) or because local populations lack the capacity to adapt (Renaud 2007; Tacoli 2009).

Although climate change scenarios often assume that some populations must adaptively migrate as their landscapes are made unproductive by forces beyond their control, this brand of environmental determinism is contested by human geographers and political ecologists who point to the complex and uneven geographies of human vulnerability across scales of analysis as well as the ways in which adaptation is limited by institutional rigidity, governance inefficiencies, or pathway dependency (Adger et al. 2009; Liverman 2009; Peet, Robbins, and Watts 2011). The unequal distribution of these adaptation alternatives highlights the political mechanisms that constrain or expand agency and raise questions about who produces what kinds of social–environmental relationships and for whom (Heynen, Kaika, and Swyngedouw 2006; Eakin et al. 2009; Carey, Huggel et al. 2012). In this sense, environmental migration should not be considered environmentally predetermined but rather as an expression of political economy as well.

This article builds on these discussions by arguing for an adaptive systems framework that more effectively incorporates political economy. It then comparatively examines how political structures influence adaptation alternatives in two distinct case studies where climate stresses are motivating migration and where these stresses are advancing rapidly. These include disappearing glaciers in Peru and the intensification of tropical storms at sea level in Honduras. Climatic stresses are undermining the availability of coastal land (in Honduras) and water resources (in Peru), but also in each case study, political economies have developed around these key resources and mediate access to them. This article hypothesizes that a systems approach to adaptation that accounts for the structures that govern access to changing resources is appropriate regardless of the particular setting; to the extent that we can apply such an approach to these highly contrasting environments (with distinct social, political, and economic histories), this substantiates its applicability in understanding migration as adaptation to climate change.

Political Economy as the Basic Structure of Adaptive Systems

Decisions about migration are influenced not only by climatic and other environmental changes but also by a constellation of other drivers beyond the household level, such as economic scarcity, political pressure, socioeconomic aspirations, and cultural values (Black, Bennett, et al. 2011). Climate stress also feeds back into these social processes by impacting a range of socioeconomic factors that present opportunities for migration, such as education, access to services, access to health care, remittances, income decline, and income variability (Hugo 2008; Feng, Krueger, and Oppenheimer 2010; Raleigh and Jordan 2010; Lilleør and Van den Broeck 2011). Simultaneously, a clearer picture is emerging of the context-specific constraints on human migration behavior: Women possess different migration alternatives from men, young people from the elderly, and the relatively wealthy from the poor (McLeman and Hunter

2010; Gray and Mueller 2012). In addition, these constraints also lead people to remain in hazardous areas when short-term benefits outweigh the risks associated with potential hazards or when they identify risks associated with mobility (Carey 2010; Black, Adger, et al. 2011).

Adaptive migration is commonly framed as a question of livelihood resilience, and it is argued here that resource inputs for livelihoods are linked to dynamics at higher scales. Under "normal" conditions, household livelihood strategies depend on social–ecological relationships that channel the essential inputs to human life, like water and fertile soil, that enable productive livelihoods and permit relatively stable modes of human settlement (Hummel et al. 2012). These fundamental relationships confer capacity to resist change or resilience to reorganize in the aftermath of a disturbance (Folke 2006; Loarie et al. 2009), but environmental disturbance (including climate stress) can also suddenly and catastrophically overwhelm livelihood systems (Anderies, Walker, and Kinzig 2006; Walker et al. 2006; Scheffer et al. 2009). When livelihood systems can no longer absorb stress, people migrate (Wrathall 2012).

The availability of land and water are stabilizing influences in rural livelihood systems, but these are embedded into a set of power relations, practices, and institutions that mediate access (Leach, Mearns, and Scoones 1999; Rocheleau 2008). Of particular concern are the institutional structures of domination and the authorities, politics, and mechanisms that determine resource distribution in society (Giddens 1984; Pelling and Manuel-Navarrete 2011). Dominant structures, for example, might include a water allocation regime or a land tenure policy, which, over the course of time, order specific resources and modes of resource accumulation. These structures create periods of systemic stability, and despite a tendency to be excessively rigid, they are often able to withstand significant stress because they are maintained through the exercise of coercive power (Scheffer et al. 2002; Carpenter and Brock 2008). In this way environmental stress might in fact reinforce and amplify dominant structures, even while altering or dissolving the resource relationships that undergird livelihood stability at household levels. This increasingly inflexible systemic state, in which adaptation alternatives rapidly diminish, is called a *rigidity trap* (Allison and Hobbs 2004). In this way, political economy sets up migration as an increasingly viable adaptation alternative among a dwindling set of adaptive choices.

Resource Politics and Climate Stress in Honduras and Peru

To more carefully examine the relationship between dominant political structures and adaptive migration, case studies in Honduras and Peru were selected for comparative analysis. The rationale for selecting these locations is based on the similarly strong influence that institutional structures exert over access to essential resources—land in the case of Honduras and water in Peru. The resource regimes in these contexts—embodied in Honduras's 1992 Law for the Modernization and Development of the Agricultural Sector, and Peru's 2009 Water Resources Law—shifted over the last few decades, with most of Latin America in a neoliberal alignment toward a commoditization and marketization of natural resources (Jansen and Roquas 1998; Budds and Hinojosa 2012). Additionally, in both cases, climate stresses—intensification of storms and sea level rise in Honduras and melting glaciers in Peru—also alter access to these essential resources, undermining the bases of rural livelihoods and motivating migration. For these two converging reasons, a multimethod quasi-experimental comparative approach was used to examine the role of dominant political structures in ordering alternatives for adaptive migration. The sites are presented as a continuum of exposure, forming the basis for a quasi-experimental comparison of migration behaviors by analogy (McLeman and Hunter 2010).

Tropical Storms, Dominant Land Tenure Structures, and Migration in Honduras

We are currently in the midst of an active decadal cycle of Atlantic cyclones, and more energetic storms are linked to increasing average sea surface temperatures (Bender et al. 2010). Named tropical storms have made landfall in northern Honduras in twelve of the last fourteen years, and villages at sea level among estuarial zones are experiencing a pattern of catastrophic flooding and permanent, irreversible erosion of land, rendering population centers uninhabitable. One particular Afro-indigenous group, the Garífuna, arrayed in over fifty settlements, is on the frontline of exposure (Cochran, Reese, and Liu 2009; Wrathall 2012). Almost without exception, villages share common environmental features that increase local susceptibility to land loss: immediate proximity to sea level; a geological substrate of loamy sand; proximity to fresh water sources; proximity to the steep, (de)forested mountain ranges

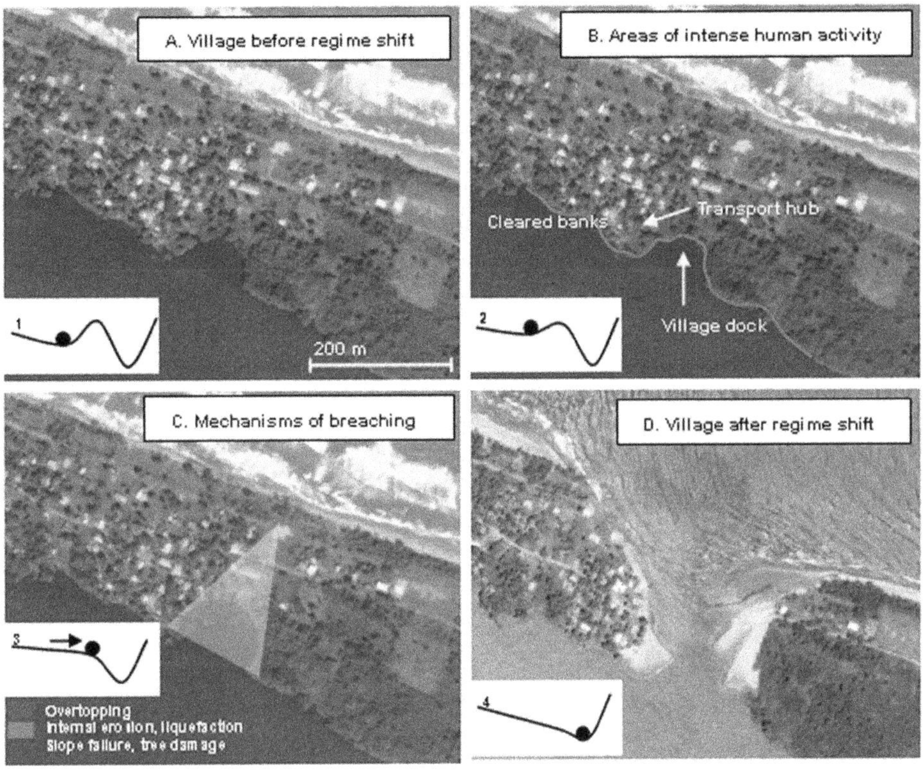

Figure 1. (A) and (B): Batalla, Honduras, in 2003 moving toward a social–ecological threshold. (C) The compounding environmental stressors associated with Topical Depression Gamma that breached thresholds in November 2005. (D) Batalla, Honduras, in 2006 in a new stable state. (Color figure available online.)

of Honduras's interior; and exposure to wind, rainfall, and storm surges associated with tropical storms. A combination of cascading biophysical mechanisms (including slope failures, overtopping, tree damage, and internal erosion and liquefaction) together produce a sudden, catastrophic shift that can occur within hours (Figures 1C and 1D). These catastrophic shifts have occurred already in eight villages and more villages have been identified as vulnerable. Mixed-method vulnerability mapping and ethnographic research were conducted in three of these communities at different stages of exposure: Santa Rosa de Aguán (acute, progressive, and repetitive flooding with > 300 households displaced between 1998 and 2009), Batalla (acute, albeit more recent flooding with > 30 households displaced between 1998 and 2009), and Iriona (general flooding with < 10 households displaced between 1998 and 2009).

The Garífuna of northern Honduras have a complex and storied historical relationship to land tenure and migration, involving pre-Columbian ancestry and traditions, the African slave trade, mercantilism in the eastern Caribbean, and forced exile to Honduras at the hands of the British in 1797 (Gonzalez 1988). At various stages, en route to Honduras, the Garífunas have been marginalized by (and have fiercely resisted) the dominant land tenure regimes of historical periods. Subsequent to the their arrival in Honduras, where this analysis focuses, the extant land tenure regimes relegated Garífuna settlements to the estuaries and small river deltas of the coastal littoral (García 2001). During the colonial *hacienda* period of the nineteenth century and the neocolonial era of the North American banana companies of the twentieth century, land use conventions constrained access to land in Honduras's interior, confining the geographical extent of Garífunas' farming practices to a narrow fringe of marginal mainland immediately adjacent to settlements (García 2004; Soluri 2005). Responding to land scarcity and insecurity, the Garífuna engaged in mobile livelihoods based around fishing and maritime trade across the Bay of Honduras through the nineteenth and twentieth centuries. By the end of the twentieth century, the Garífuna's transoceanic livelihoods situated them as longshoremen, navigators, and merchant marines throughout North America and beyond, and this transformed them into a community in diaspora—albeit one with an eye fixed homeward (England 2006; Anderson 2009). Migration historically served the primary goal of remitting incomes to the village. Ricardo explained:

> To achieve upward mobility, to save money, to build a house, it has always been thus that young people migrate. And the idea traditionally is to return. Always! You would work until you earned enough to establish a base: maybe a

> restaurant, or a ranch, or an import–export business. Then you could live very well in the village. You may not have a fat life here, but you would be very comfortable.

The dominant structures around land ownership in northern Honduras amplified during the latter half of the twentieth century and have articulated with environmental stress to influence forms of adaptive migration that are essentially distinct from historical patterns. The new dynamic interrupts the linkages between migrants (with their earnings) and their ancestral villages.

In 1992, a new neoliberal policy, the Law for the Modernization and Development of the Agricultural Sector, aligned with and amplified long-standing structures around land ownership. The policy mechanism intended to formalize informal land tenure across the country, exalting land commodification and even imposing privatization on land cooperatives (Jansen and Roquas 1998). The policy enforced a prior constitutional entitlement to communal lands for ethnic communities, like the Garífuna, but the impact of the policy varied among communities. Communally titled lands generally came under immediate threat from both powerful economic actors, as well as marginalized *campesinos*. The law provided industrial plantation owners a favorable policy environment for monopolizing farmland on the coastal plains, easing restrictions on the amount of land that a legal entity could possess. In the cases of the Garífuna communities of Limón and Punta Piedra, ancestral lands were effectively surrounded by large industrial plantations and midscale cattle ranches (García 2004).

Land competition driven by economically powerful actors has also created conditions for conflict among marginalized social strata. Although conflicts over land between Garífunas and local Miskitus have occurred historically (Mollett 2006), new land scarcity enflamed tensions between Garífunas and immigrant *campesinos* from western Honduras. Under the 1992 law, although land invasions were prohibited, nonindigenous *campesinos* who could show peaceful homesteading occupancy were eligible for formal land rights. In short, this produced a number of invasions within indigenous territories (Mollett 2011) and began a series of sustained challenges to communal control of Garífuna territory, such as the ongoing conflicts in Iriona, and near Punta Piedra, where each party is justifying its respective claims on the basis of the 1992 land laws.

The most troubling feature of this Honduran brand of neoliberalism is the extent to which land transfers were not only mediated by the private market but also through coercion, impunity, and naked violence. The recent bloody conflict over land in Bajo Aguán—adjacent to one of the study sites but not directly affecting Garífunas—is one manifestation of this tendency (Bird 2011). In the words of one respondent reporting on this common theme near Iriona, midscale cattle ranchers would "cut down the Garífunas' fence posts, and introduce cattle in our land. They wait for you to complain with a pistol in their belt." There is little recourse in the courts, as the judicial process favors enterprises with legal representation and is highly corruptible regardless. Maria reported:

> People often see themselves as obligated to cede or sell their lands, sometimes under threat of violence, for concern about their lives and their families. Municipal authorities even lend themselves to these misdeeds because land is lucrative. Threats or not, the thing is, if someone takes a liking to your property for one reason or another, they will make your life impossible until they obtain it.

In the cases of Laguna de Zambuco, Rosita, Nueva Go, and Cayo Venado, landowning estates have allegedly fenced traditional Garífuna plots or introduced cattle (García 2004).

This insecurity and land scarcity has transformed land into cultural obsession and a centerpiece of Garífuna political solidarity (Anderson 2007), but meanwhile at local levels, because land is limited, Garífuna livelihoods are increasingly concentrated in and around the villages, putting significant environmental pressure on the geological substrate and local ecology. Fishing, trading, intensive subsistence agriculture, and small-scale cattle ranching are the core of village economic systems, and these activities require docks, markets, and transport hubs, and in some areas, mangroves are cleared to provide space for pasture (Cochran, Reese, and Liu 2009). Moreover, the intensification that accompanies these activities—foot, boat, car, and livestock traffic—serves to disarticulate loose soil structures that prevent erosion. Thus, the overall environmental state of Garífuna villages is increasingly intense activity on increasingly small pieces of sensitive terrain (Figure 1B). So the dominant structures around access to land—the 1992 land policy regime and the subsequent land scarcity—exacerbate the flooding risk at local levels.

At the same time, dominant structures influence the available alternatives for adaptive migration. First and foremost, tropical storms are driving displacement, and yet the total number of migrants from affected communities well exceeds the number displaced. For

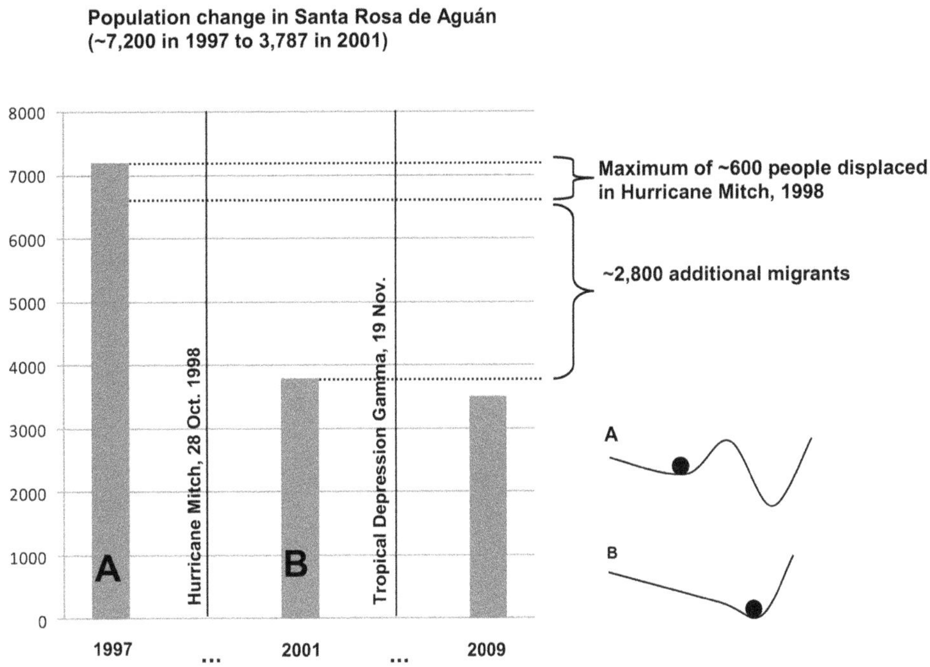

Figure 2. Catastrophic flooding and population change in Santa Rosa de Aguán, Honduras. Although a maximum of ∼600 were displaced in the initial storm, a much larger population migrated (A) and (B). This population profile remains relatively stable through time despite other major storm impacts and massive investment in resettlement projects.

example, according to cadastral records in Santa Rosa de Aguán, a village that was catastrophically flooded in 1998 with Hurricane Mitch, the predisaster population was approximately 7,200. And yet, although a maximum of 600 individuals were displaced, the census taken in 2001 found 3,787 individuals in the municipality. Nearly 3,000 additional people migrated. In 2009, eleven years following the initial disaster, the cadastre reported a population of about 3,500—half the predisaster population—despite an increase in the overall number of domiciles in the municipality, including 445 new homes in four different resettlement projects (Figure 2).

At closer inspection, the trend in migration is different from historical modes for a number of reasons. First, from the perspective of human timescales, land is permanently disappearing, and there is no substantive regeneration. Displaced households can obtain land within the village in increasingly vulnerable areas or further away from the village and their fields. So in new circumstances, a majority of displaced households sold plots because, like Paulo, "we no longer had access to it anyway. Ours was on the far side of the river, and so after relocating here, we couldn't get to it easily." Obtaining land for resettlements, another adaptation strategy, was fraught with the same problems of land conflict outlined earlier. Four resettlement organizers (in each of the communities in the study) reported being physically threatened for their efforts to obtain land for resettlement projects, and one organizer in Santa Rosa de Aguán, the son of a respondent, was murdered as he petitioned land from a rancher. Because adaptation alternatives within the villages are limited by land scarcity, people adapted to environmental stress by liquidating vulnerable assets in and around the village, investing in strategies outside of the villages: more permanent modes of migration. One respondent described the general effect:

> People used to leave here by the dozen and go to New York. At that time, people used to bring money back and it would enrich everybody, and there was work for everybody in the village. I would build three houses, and earn USD$10,000! But now, that money is gone. You don't see the investment in the village anymore, and now you can't find jobs. This is a problem that has affected the whole community.

This new migration dynamic led a period of economic decline and thus reinforced the persistent need for adaptive migration. "There were no jobs here!" respondents expressed, "So, I had to leave." In this way, land scarcity, migration, and the cascading effects have led to the emergence of a demographic profile that is relatively stable through time: Those with suitable alternatives left, whereas others with diminishing prospects stayed behind (Figure 2B). One respondent elegantly described the process, "In those first months after the storm, I would have stayed, but everybody was planning to leave, and so I had to decide to leave, too."

Glacier Recession, Dominant Water Provisioning Structures, and Migration in Peru

Just as land availability is critical for livelihood function, so is water provisioning, and climate change and glacier recession are changing water availability in the Tropical Andes. In Peru's Santa River catchment, glaciers function to regulate water flow interannually, maintaining stability in ecological systems and supplying water to a host of regional users through both wet and dry seasons. Historic and multiscalar observations, however, identify an acceleration of glacier melt across several decades, a nonlinear trend in the downstream hydrological regime, and diminishing water supplies to dependent smallholders during the dry season (Mark et al. 2010). During the twentieth century, glacier cover over the Cordillera Blanca has declined from 800 to 850 km^2 in 1930 to slightly less than 600 km^2 at the end of the twentieth century (Georges 2004). During the past few decades, water resources in the Cordillera Blanca crossed a threshold of "peak water" (Baraer et al. 2012). New hydrological assessments based on hydrological data analysis and numerical modeling of nine glacierized tributary watersheds of the Santa River indicate an evolution of glacier hydrology through four impact phases of climate change (Figure 3). The overall state of studied watersheds indicates that, on average, glacier-fed rivers are experiencing an increase in discharge variability and a decrease in dry-season discharge (Phase 3), and it is likely that the Santa River's peak discharge occurred around 1980. One of the Phase 3 watersheds, La Balsa, likely crossed the threshold from Phase 2 near 1970. These nine subcatchments form the basis for quasi-experimental comparison of migration behaviors, and hydrological modeling was paired with survey and ethnographic research in settlements along three of these subcatchments, the districts of Yungay (Phase 2), Huaraz (Phase 3), and Catac (Phase 4).

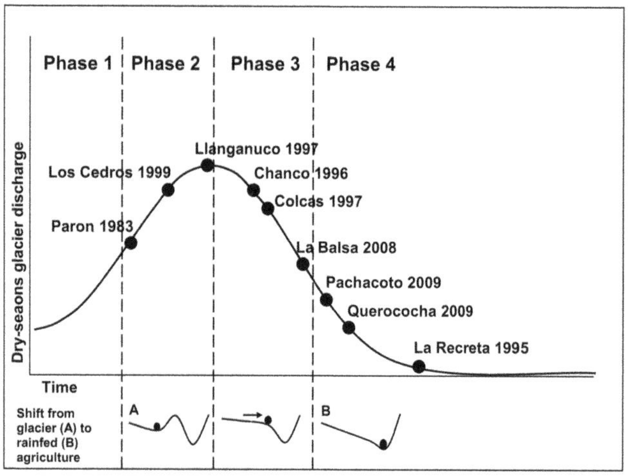

Figure 3. Four climate change impact phases of dry-season glacial discharge. Phase 1: Increasing dry-season and yearly average discharges (as glaciers begin to recede); Phase 2: Annual average discharge reaches a "peak" (due to reduced glacier masses and diminishing dry season discharges); Phase 3: A pronounced decrease in discharges (as glacier recession stabilizes); Phase 4: The end of glaciers' influence on dry-season outflows. Points indicate the allocation of nine subcatchments in the Cordillera Blanca across the four phases.

In Peru, water-provisioning systems have been a critical basis of human settlements across the Western Pacific slope for millennia, largely because only 2 percent of the country's total fresh water flows through the arid west. Highland smallholders in the Cordillera Blanca historically utilized glacial meltwater to provide year-round water for the cultivation of staples and market crops. Since the conquest of Peru in the mid-1500s, the structures of domination that govern hydraulic relations in the region have been based on highly unequal access to water resources and distribution systems. Generally, through the colonial and neocolonial periods, access to water resources (and jointly arable land resources) was mediated by *hacienda* institutional control in the highlands and by large sugarcane plantations along the coast (Galeano 1973; Thorp and Bertram 1978). Nevertheless, during this time, smallholders in the upper catchments independently relied on often decentralized and ad hoc access to irrigation systems and springs. Beginning in the late 1960s, though, new agrarian reform movements and successive revolutionary military dictatorships partially restructured land and water institutions across the highlands, fostering the growth of smallholder agriculture and livelihood systems and the creation of new water use and distribution institutions (Cameron 1994). This evolution in water governance regimes coincided with a national economic strategy beginning in the 1990s that prioritized export-led growth (Bury 2005). Heavy concentrations of infrastructural investments on the coast began a process of rapid emigration from the rural highlands and urbanization of the coast that is sustained through the present, so that today Peru is one of the most urbanized countries in South America (Álvarez-Berríos et al. 2013). This relatively recent pattern of demographic transformation is still unfolding in the Cordillera Blanca as populations move coastward (Bury, Mark, and McKenzie 2013).

During this revolutionary period and through the mid-1970s, water governance remained spatially discrete across the steep Andean coastal escarpment that links coastal and highland altitudinal zones. Water resource use along the coast intensified significantly as new large-scale national agricultural initiatives and water distribution systems such as *Chinecas* and *Chavimochic* were developed (Kus 1987). The growing water needs of coastal plantation and urban centers shifted the locus of water management upstream. Over the past decade, due to the limits to freshwater availability and increasing intensification of water use, coastal and highland water users and management institutions have become increasingly linked across geographic and political scales.

In the Santa River watershed of the Cordillera Blanca, a number of new tensions related to water resources have recently begun to emerge. In its more than 300-km course, the Santa River articulates the water demands of highland agro-pastoralists, growing urban centers, a U.S.-owned hydropower company, extractive industries, and numerous agro-exporters producing crops like asparagus, tamarind, and paprika for distant markets (Mark et al. 2010; Carey, Huggel et al. 2012; Bury, Mark, and McKenzie 2013). Facing mounting competition and conflicts between economic sectors, Peruvian lawmakers passed new water legislation in 2009 premised on the concept of integrated water management. Conceptually, the law is attractive for its emphasis on balancing different environmental, cultural, and economic values of water, as well as for its ambitious reconfiguration of the institutional landscape of hydrologic resource management in Peru. Yet, the institutional flexibility necessary to implement the law's ambitious vision demands collaboration and compromise across the factious political and economic sectors, including broad public participation (Autoridad Nacional del Agua 2009). This increasing institutional interconnectedness has begun a period of conflict over access to water resources and distribution with several high-profile disputes, including the conflict between *campesino* organizations and Duke Energy at Lake Paron (Carey, French, and O'Brien 2012). Water laws present opportunities for powerful economic actors to control water resources even further, increasing the rigidity of water provisioning systems across the region.

Climate-driven hydrological trends in the Santa River watershed are already decades old and, in conjunction with other climate-change-related stresses, are driving new patterns of migration behavior. Migration is motivated by livelihood stress, but the form that it takes is shaped by dominant institutional forces. Semistructured interviews conducted with households across all nine watersheds support the "peak water" hypothesis and its implicit risks to livelihoods. For example, respondents from communities in Phase 2 subcatchments reported risks from a superabundance of water as canal water levels were at maximum during 2012 and threatened land and crops through flooding and soil erosion. Conversely, communities in Phase 4 subcatchments reported in 2012 that canals were empty halfway through the dry season, which led to significant crop mortality. In addition, respondents reported switching completely from irrigation to rain-fed agriculture, which has increased the risks to crops from increasing weather and temperature extremes. Households across the upper Santa watersheds have widely reported increasing extremes in daily and seasonal temperature and weather variability. This includes factors such as colder nighttime temperatures, warmer and drier daytime conditions, and precipitation events that have become more extreme and are occurring more frequently during the dry season (see also Bury et al. 2011). Whereas historically smallholders irrigated fields once weekly, because of warmer daytime temperatures, crops must be irrigated at more frequent intervals. At the same time, heavy frosts have been more intense and have occurred more frequently during the growing season. Frosts pose a risk to household livelihoods in all catchments, regardless of water abundance. As Victor, a rural smallholder, described the challenge to livelihoods in Phase 2 catchments, "We still have water, and we can adapt to the pests and poor quality soil with pesticides and fertilizer, but there is no adaptation to the cold nights."

Households facing risks have begun to pursue new resource diversification strategies, including an increasing reliance on migration and remittances. Semistructured interviews conducted in 2012 with migrants across the region indicated that increasing livelihood risks are intensifying historical population dynamics and migration behavior. Relative to respondents in Phase 2 subcatchments, respondents in Phase 4 subcatchments reported increasing expectations that young members of their household will migrate to Peru's cities and remit money to contribute to household budgets back home. Population data illustrate this relationship between livelihood risk and migration: Communities in subcatchments in the more advanced stages of hydrologic change where households are switching to rain-fed agriculture (Phases 3 and 4) have the most fragmented populations (Figure 4). Environmental stress is concomitant with higher concentrations of elderly people and absent working-age populations relative to those areas with more regular water supplies. This trend was consistently

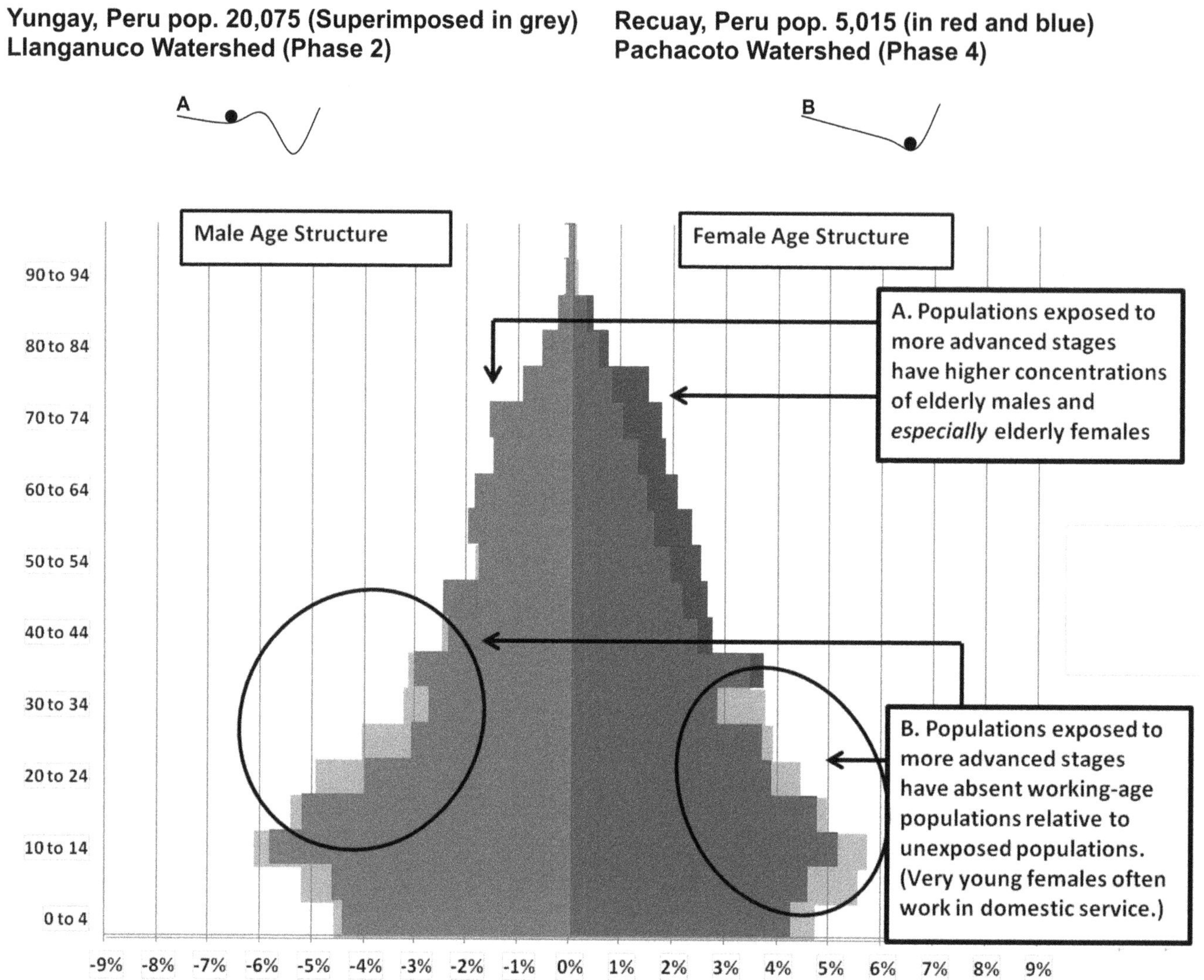

Figure 4. A demographic shift: A comparison of populations in different climate impact phases. Age of population (y axis) against total population (x axis). (Color figure available online.)

observed in our study areas, and ground-truthed in field visits.

Households also indicated widespread preoccupation with larger, and more sudden, climate change–related risks such as glacier lake outburst floods (GLOFS) and avalanches, both of which have occurred repeatedly over the past few decades with devastating consequences for the lives and livelihoods of people beneath the glaciers of the region (Carey 2010; Carey, Huggel et al. 2012). In past decades, GLOFs and glacier landslides have killed more than 20,000 residents, and there have been subsequent population relocations, both in the immediate areas destroyed by these disasters and more widespread emigration from the effected regions. After the 1941 GLOF, for example, the downtown section of the destroyed city of Huaraz was left empty for decades while people resettled and rebuilt elsewhere in the city. After the zone's catastrophic earthquake and Yungay avalanche in 1970 that killed a combined 55,000 people, thousands of middle- and upper-class urban residents emigrated to Lima. As those residents fled the region's urban areas in the disaster's aftermath, rural farmers inhabiting upland slopes migrated to the partially vacated urban areas in search of aid, food, and shelter so that one third of the Callejón de Huaylas population inhabited urban areas by the early 1970s (Walton 1974).

Like Honduras, the dominant structures for water resource provisioning limit the available adaptation alternatives and accelerate existing modes of migration as the most desirable among a dwindling set of possibilities. Cesar, a smallholder in Catac, reported that in the switch to rain-fed agriculture, water provisioning is as much a function of stream flows as it is about new restrictions on use.

> Because of the shortages, we're required to share the small amount of water that we have. There are lots of restrictions and rules about water use, more than in the past. We're taking lessons from the communities that already rely on rain-fed agriculture.

New restrictions meet headlong with aspirations of the younger generation, who "don't want to farm because there are easier, more productive alternatives" like mining, manual labor, and construction. What appears as the expression of agency in adaptation, however, is also a reflection of limited alternatives. As Rafael stated, "Young people couldn't stay here even if they wanted to because there is less water and more rules."

Discussion

The most significant difference between the sites is the nature of climate disturbances, the time frame of exposure, and forcing: Glacier melting is a secular change (nonperiodic, long term) with progressive unfolding (albeit with enhanced likelihood of secondary shocks like GLOFs), whereas tropical storms are near instantaneous from a resource change perspective and difficult to predict with specificity but will be increasingly likely given progressive intensification of the boundary forcings (higher sea surface temperatures, enhanced tropospheric heat, and water cycling). Furthermore, within each context, often subtle but significant differences existed in the way dominant structures were expressed between communities. For example, some Garífuna villages were surrounded by African palm, whereas others were surrounded by cattle ranches. Likewise, some of the study areas in Peru were home to mining operations, which introduced specific dynamics in water competition. Despite these differences, in both cases, structures of domination generally shaped the new forms that migration took in the presence of climate stress (Figure 2 and Figure 4). In Honduras, migrants from wealthier social strata are moving on a more permanent basis, and in Peru, the once historical pattern of labor migration is becoming a practical necessity.

The analysis so far has focused on adaptation within community scales, but structures of dominance have significant implications for broader scales. First, it is tempting to praise emigration and increased reliance on remittances as a novel form of adaptation, but this dynamic reinforces inequitable structures. In Honduras, for example, displaced Garífuna households reported selling lands to mid- and large-scale *mestizo* cattle ranchers on the peripheries of the villages to cope with economic losses, repay debts, and manage future risks. Environmental stress and migration become a mechanism by which indigenous controlled lands are inserted into the private market and ceded to nonindigenous tenants. This has implications for the social and political cohesiveness of the entire indigenous group. In Peru, migration was reported to serve as a pressure valve, loosening households' dependence on water, such that competition over remaining resources favors vested stakeholders in the Santa River. In the context of decreasing water availability and ever more inflexible and interconnected water provisioning, smallholder households must compete with powerful water consumers to secure alternatives for in situ adaptation. Several high-profile cases of water conflict between communities and water-intensive industries, such as hydropower and mining operations, are evidence of rigidity in the system (Carey, French, and O'Brien 2012; Lynch 2012). When young people migrate, it reduces demand and resolves the conflict.

Conclusion

Environmental stress and adaptation alternatives, like migration, are historically derived and politically embedded. This conceptualization goes beyond viewing migration as adaptation or lack thereof but rather as an aspect of a shift produced by fundamental structures of power and domination that circumscribe options for local adaptations and instead promote alternatives, like migration, that displace adaptation to other geographical spaces. Reinserting an analysis of power relations at the heart of adaptive systems approaches coincides with the agenda of political ecology by encouraging a formulation of social–ecological structures that delimit access to resources (Heynen, Kaika, and Swyngedouw 2006; Pelling 2010). Moreover, by shedding light on the politics that order inequalities in the opportunity to adapt, scholarship on environmental change and human and political ecology can provide insight into the interaction of human agency and social (ecological) structure (Tompkins and Adger 2004; Eriksen et al. 2011).

In particular, more research must be devoted to the capacities of vulnerable people to adapt in new social–ecological states—the rigidity traps—that emerge as environmental stress and migration feedback against each other and how those traps can be mitigated. This article comes at a timely policy moment in the United Nations Framework Convention on Climate Change negotiations, with international agreement to clarify modalities of human mobility vis-à-vis climate change in future conferences of parties. It is imperative to move discourses about climate change adaptation—and particularly climate-induced displacement and migration—away from environmental determinism and toward social–ecological possibilism. Political economies constrain adaptive alternatives and force the undesirable scenarios in which migration occurs. In this respect, geography, for its multidisciplinary emphasis on both environmental change and resource politics, is particularly well situated for advancing this area of policy and research at a critical juncture in human history.

References

Adger, W. N., S. Dessai, M. Goulden, M. Hulme, I. Lorenzoni, D. Nelson, L. Naess, et al. 2009. Are there limits to social adaptation to climate change. *Climatic Change* 93 (3–4): 335–54.

Allison, H. E., and R. J. Hobbs. 2004. Resilience, adaptive capacity, and the lock-in trap of the Western Australian agricultural region. *Ecology and Society* 9 (1): 3.

Álvarez-Berríos, N. L., I. K. Parés-Ramos, and T. M. Aide. 2013. Contrasting patterns of urban expansion in Colombia, Ecuador, Peru, and Bolivia between 1992 and 2009. *Ambio* 42 (1): 29–40.

Anderies, J. M., B. H. Walker, and A. P. Kinzig. 2006. Fifteen weddings and a funeral: Case studies and resilience-based management. *Ecology and Society* 11 (1): 21.

Anderson, M. 2007. When Afro becomes (like) indigenous: Garífuna and Afro-indigenous politics in Honduras. *The Journal of Latin American and Caribbean Anthropology* 12 (2): 384–413.

———. 2009. *Black and indigenous: Garífuna activism and consumer culture in Honduras*. St. Paul: University of Minnesota Press.

Autoridad Nacional del Agua. 2009. Ley de Recursos Hidricos #29338; M.T. Oré, L. Castillo, S. Van Orsel and J. Vos. *El agua, ante nuevos desafios: Actores e iniciativias en Ecuador, Perú, y Bolivia, Lima*.

Baraer, M., G. Bryan, J. McKenzie, T. Condom, and S. Rathay. 2012. Glacier recession and water resources in Peru's Cordillera Blanca. *Journal of Glaciology* 58 (207): 134–50.

Bardsley, D. K., and G. J. Hugo. 2010. Migration and climate change: Examining thresholds of change to guide effective adaptation decision-making. *Population & Environment* 32 (2–3): 238–62.

Bender, M. A., T. R. Knutson, R. E. Tuleya, J. J. Sirutis, G. A. Vecchi, S. T. Garner, and I. M. Held. 2010. Modeled impact of anthropogenic warming on the frequency of intense Atlantic hurricanes. *Science* 327 (5964): 454–58.

Bird, A. 2011. Drugs and business: Central America faces another round of violence. *NACLA Report on the Americas* 45 (1): 35–36.

Black, R., W. N. Adger, N. W. Arnell, S. Dercon, A. Geddes, and D. Thomas. 2011. The effect of environmental change on human migration. *Global Environmental Change* 21:S3–S11.

Black, R., S. R. G. Bennett, S. M. Thomas, and J. R. Beddington. 2011. Climate change: Migration as adaptation. *Nature* 478 (7370): 447–49.

Budds, J., and L. Hinojosa. 2012. Restructuring and rescaling water governance in mining contexts: The co-production of waterscapes in Peru. *Water Alternatives* 5 (1): 119–37.

Bury, J. 2005. Mining mountains: Neoliberalism, land tenure, livelihoods, and the new Peruvian mining industry in Cajamarca. *Environment and Planning A* 37 (2): 221–39.

Bury, J., B. G. Mark, and J. M. McKenzie. 2013. New geographies of water and climate change in Peru: Coupled natural and social transformations in the Santa River watershed. *Annals of the Association of American Geographers* 103 (2): 363–74.

Bury, J. T., B. G. Mark, J. M. McKenzie, A. French, M. Baraer, K. I. Huh, M. A. Zapata Luyo, and R. J. Gómez López. 2011. Glacier recession and human vulnerability in the Yanamarey watershed of the Cordillera Blanca, Peru. *Climatic Change* 105 (1): 179–206.

Cameron, M. A. 1994. *Democracy and authoritarianism in Peru: Political coalitions and social change*. London: Palgrave Macmillan.

Carey, M. P. 2010. *In the shadow of melting glaciers: Climate change and Andean Society*. New York: Oxford University Press.

Carey, M., A. French, and E. O'Brien. 2012. Unintended effects of technology on climate change adaptation: An historical analysis of water conflicts below Andean glaciers. *Journal of Historical Geography* 38 (2): 181–91.

Carey, M., C. Huggel, J. Bury, C. Portocarrero, and W. Haeberli. 2012. An integrated socio-environmental framework for glacier hazard management and climate change adaptation: lessons from Lake 513, Cordillera Blanca, Peru. *Climatic Change* 112 (3–4): 733–67.

Carpenter, S. R., and W. A. Brock. 2008. Adaptive capacity and traps. *Ecology and Society* 13 (2): 40.

Cochran, D., C. Reese, and K. Liu. 2009. Tropical Storm Gamma and the Mosquitia of eastern Honduras: A little-known story from the 2005 hurricane season. *Area* 41 (4): 425–34.

Eakin, H., E. L. Tompkins, D. R. Nelson, and J. M. Anderies. 2009. Hidden costs and disparate uncertainties: Trade-offs involved in approaches to climate policy. In *Adapting to climate change*, ed. W. N. Adger, I. Lorenzoni, and K. L. O'Brien, 212–26. London: Cambridge University Press.

England, S. 2006. *Afro-Central Americans in New York City: Garífuna tales of transnational movements in radicalized space*. Gainesville: University Press of Florida.

Eriksen, S., P. Aldunce, C. S. Bahinipati, R. D. A. Martins, J. I. Molefe, C. Nhemachena, K. O'Brien, et al. 2011. When not every response to climate change is a good one: Identifying principles for sustainable adaptation. *Climate and Development* 3 (1): 7–20.

Feng, S., A. B. Krueger, and M. Oppenheimer. 2010. Linkages among climate change, crop yields and Mexico–US cross-border migration. *Proceedings of the National Academy of Sciences* 107 (32): 14257.

Folke, C. 2006. Resilience: The emergence of a perspective for social-ecological systems analyses. *Global Environmental Change* 16 (3): 253–67.

Galeano, E. H. 1973. *Open veins of Latin America*. New York: Monthly Review Press.

García, S. C. 2001. *Historia del pueblo negro caribe y su llegada a las Hibueras el 12 de abril de 1797* [History of the Black Caribs and their arrival to Hibueras, Honduras the 12th of April 1797]. Universidad Nacional Autónoma de Honduras, Editorial Universitaria.

———. 2004. *Adhesión étnica* [Ethnic adhesion]. Universidad Nacional Autónoma de Honduras, Editorial Universitaria.

Georges, C. 2004. 20th-century glacier fluctuations in the Tropical Cordillera Blanca, Peru. *Arctic, Antarctic, and Alpine Research* 36 (1): 100–107.

Giddens, A. 1984. *The constitution of society: Introduction of the theory of structuration*. Cambridge, UK: Polity Press.

Gonzalez, N. L. 1988. *The sojourners of the Caribbean: Ethnogenesis and ethnohistory of the Garífuna*. Urbana: University of Illinois Press.

Gray, C. L., and V. Mueller. 2012. Natural disasters and population mobility in Bangladesh. *Proceedings of the National Academy of Sciences* 109 (16): 6000–05.

Heynen, N., M. Kaika, and E. Swyngedouw. 2006. Urban political ecology. In *The nature of cities: Urban political ecology and the politics of urban metabolism*, 1–20. London and New York: Taylor & Francis.

Hugo, G. 2008. *Migration, development and environment*. Geneva, Switzerland: International Organization for Migration.

———. 2011. Future demographic change and its interactions with migration and climate change. *Global Environmental Change* 21:S21–S33.

Hummel, D., S. Adamo, A.deSherbinin, L. Murphy, R. Aggarwal, L. Zulu, J. Liu, et al. 2012. Inter-and transdisciplinary approaches to population–environment research for sustainability aims: A review and appraisal. *Population & Environment* 34 (4): 481–509.

Intergovernmental Panel on Climate Change. 2012. *Managing the risks of extreme events and disasters to advance climate change adaptation. A special report of working groups I and II of the Intergovernmental Panel on Climate Change*, ed. C. B. Field, V. Barros, T. F. Stocker, D. Qin, D. J. Dokken, K. L. Ebi, M. D. Mastrandrea, K. J. Mach, G.-K. Plattner, S. K. Allen, M. Tignor, and P. M. Midgley. Cambridge, UK: Cambridge University Press.

Jansen, K., and E. Roquas. 1998. Modernizing insecurity: The land titling project in Honduras. *Development and Change* 29 (1): 81–106.

Kus, J. S. 1987. *Chavimochic: A Peruvian irrigation project*. Syracuse, NY: Syracuse University Press.

Leach, M., R. Mearns, and I. Scoones. 1999. Environmental entitlements: Dynamics and institutions in community-based natural resource management. *World Development* 27 (2): 225–47.

Lilleør, H. B., and K. Van den Broeck. 2011. Economic drivers of migration and climate change in LDCs. *Global Environmental Change* 21:S70–S81.

Liverman, D. 2009. Conventions of climate change: Constructions of danger and the dispossession of the atmosphere. *Journal of Historical Geography* 35 (2): 279–96.

Loarie, S. R., P. B. Duffy, H. Hamilton, G. P. Asner, C. B. Field, and D. D. Ackerly. 2009. The velocity of climate change. *Nature* 462 (7276): 1052–55.

Lynch, B. D. 2012. Vulnerabilities, competition and rights in a context of climate change toward equitable water governance in Peru's Rio Santa Valley. *Global Environmental Change* 22 (2): 364–73.

Mark, B. G., J. Bury, J. M. McKenzie, A. French, and M. Baraer. 2010. Climate change and tropical Andean glacier recession: Evaluating hydrologic changes and livelihood vulnerability in the Cordillera Blanca, Peru. *Annals of the Association of American Geographers* 100 (4): 794–805.

McLeman, R. A., and L. M. Hunter. 2010. Migration in the context of vulnerability and adaptation to climate change: Insights from analogues. *Wiley Interdisciplinary Reviews: Climate Change* 1 (3): 450–61.

McLeman, R., and B. Smit. 2006. Migration as an adaptation to climate change. *Climatic Change* 76 (1): 31–53.

Mollett, S. 2006. Race and natural resource conflicts in Honduras: The Miskito and Garifuna struggle for Lasa Pulan. *Latin American Research Review* 41 (1): 76–101.

———. 2011. Racial narratives: Miskito and colono land struggles in the Honduran Mosquitia. *Cultural Geographies* 18 (1): 43.

Peet, R., P. Robbins, and M. Watts. 2011. *Global political ecology*. London and New York: Routledge.

Pelling, M. 2010. *Adaptation to climate change: From resilience to transformation*. London and New York: Taylor & Francis.

Pelling, M., and D. Manuel-Navarrete. 2011. From resilience to transformation: The adaptive cycle in two Mexican urban centers. *Ecology and Society* 16 (2): 11.

Raleigh, C., and L. Jordan. 2010. Climate change and migration: Emerging patterns in the developing world. In *Social dimensions of climate change: Equity and vulnerability in a warming world*, ed. R. Mearns and A. Norton, 103–31. Washington, DC: The World Bank.

Renaud, F. 2007. *Control, adapt or flee: How to face environmental migration?* Bonn, Germany: United Nations University—Environment and Human Security.

Rocheleau, D. E. 2008. Political ecology in the key of policy: From chains of explanation to webs of relation. *Geoforum* 39 (2): 716–27.

Scheffer, M., J. Bascompte, W. A. Brock, V. Brovkin, S. R. Carpenter, V. Dakos, H. Held, et al. 2009. Early-warning signals for critical transitions. *Nature* 461 (7260): 53–59.

Scheffer, M., F. Westley, W. A. Brock, and M. Holmgren. 2002. Dynamic interaction of societies and ecosystems—Linking theories from ecology, economy and sociology. In *Panarchy: Understanding transformations in human and natural systems*, ed. L. H. Gunderson and C. S. Holling, 195–240. London: Island.

Soluri, J. 2005. *Banana cultures: Agriculture, consumption, and environmental change in Honduras and the United States*. Austin: University of Texas Press.

Tacoli, C. 2009. Crisis or adaptation? Migration and climate change in a context of high mobility. *Environment and Urbanization* 21 (2): 513.

Thorp, R., and G. Bertram. 1978. *Peru, 1890–1977: Growth and policy in an open economy*. New York: Columbia University Press.

Tompkins, E. L., and W. N. Adger. 2004. Does adaptive management of natural resources enhance resilience to climate change? *Ecology and Society* 9 (2): 10.
Walker, B. H., L. H. Gunderson, A. Kinzig, C. Folke, S. R. Carpenter, and L. Schultz. 2006. A handful of heuristics and some propositions for understanding resilience in social-ecological systems. *Ecology and Society* 11 (1): 13.
Walton, N. K. 1974. Human spatial organization in an Andean valley: The Callejón de Huaylas, Peru. PhD dissertaion, University of Georgia, Athens, GA.
World Bank. 2012. *Turn down the heat: Why a 4°C warmer world must be avoided.* Washington, DC: World Bank.
Wrathall, D. J. 2012. Migration amidst social-ecological regime shift: The search for stability in Garífuna villages of northern Honduras. *Human Ecology* 40 (4): 583–96.
Young, O. R., F. Berkhout, G. C. Gallopin, M. A. Janssen, E. Ostrom, and S. van der Leeuw. 2006. The globalization of socio-ecological systems: An agenda for scientific research. *Global Environmental Change* 16 (3): 304–16.

The Amenity Principle, Internal Migration, and Rural Development in Australia

Neil Argent,* Matthew Tonts,† Roy Jones,‡ and John Holmes§

*Division of Geography and Planning, University of New England
†School of Earth and Environment, University of Western Australia
‡School of Built Environment, Curtin University
§School of Geography, Planning and Environmental Management, University of Queensland

Arguably, rural land markets in Australia are currently in a high state of flux, with a panoply of competing interests seeking to capitalize on both the traditional and a range of newly emergent values anchored in land. Amenity-led migration is, we argue, an important strand of this renewed interest in rural lands, albeit one that is highly spatially selective. Employing a predictive and synoptic model of migration attractiveness across southeastern and southwestern Australia, we test its associations with migration currents into and out of rural communities for the 1990s and 2000s. This article finds that communities with high relative accessibility—to metropolitan and urban centers and the coast—and an established or emerging tourism industry have been most likely to experience net migration gains. Yet amenity migration also intersects with more traditional rural land uses and, in particular, irrigated agriculture. Farming, and the biophysical environment and cultural landscape it both draws on and produces, is an important attractor of amenity migration.

澳大利亚当前的农村土地市场，可以说正处于不断变动的状态，各种相互竞争的利害关係，正企图同时从根植于土地的传统价值与各种新兴的价值中获利。我们主张，追求舒适环境的移民，是对农村土地新兴兴趣的一个重要构成，但此类移民却也具有高度的空间选择性。我们将移民吸引力的预测及概要模型运用至澳大利亚的东南及西南部，测试在 1990 与 2000 年代中，其与移出和移入农村社区趋势的关联性。本文发现，距离大都会与城市中心及海岸具有高度相对可及性的社区，以及旅游业发展完备或正在发展的社区，最有可能经历人口淨增长。但环境移民亦同时与更为传统的农村土地使用，特别是灌溉农业相互交错。耕种，及其同时仰赖并生产的生物物理环境与文化景观，是环境移民的重要吸引因素。

Es evidente que los mercados de la tierra en el ámbito rural de Australia se encuentran actualmente en un estado de alto flujo, con un abanico de intereses en competencia que buscan capitalizar tanto en los valores tradicionales apegados a la tierra, como en una gama de valores emergentes de reciente data. Nuestro argumento sostiene que la migración impulsada por la disponibilidad de equipamientos es un importante elemento explicativo de este renovado interés por las tierras rurales, si bien es algo altamente selectivo desde el punto de vista espacial. Con el uso de un modelo predictivo y sinóptico sobre atracción migratoria a lo ancho del sudeste y sudoeste de Australia, analizamos sus asociaciones con las corrientes migratorias actuales hacia y afuera de las comunidades rurales durante los decenios de los 1990 y 2000. Este artículo revela que las comunidades dotadas de alta accesibilidad relativa—a centros metropolitanos y urbanos y a la costa —y de una industria turística ya establecida o emergente, muy probablemente experimentan netas ganancias migratorias. Sin embargo, la migración en busca de comodidades también intersecta con usos del suelo más tradicionales y, en particular, con la agricultura basada en riego. La agricultura—y el medio ambiente biofísico y el paisaje cultural, de los cuales depende y a la vez transforma y construye—es un importante factor atrayente de la migración impulsada por disponibilidad de equipamientos.

Much of the rapidly expanding literature on rural amenity and amenity-led migration has reached a broad consensus on two important points. First, in high-amenity rural areas, large-scale commercial agriculture's long-term future appears doomed, caught in a powerful pincer movement between, on one side, cashed-up ex-urban migrants bidding up the price of scenically attractive farmland and, on the other, local governments happy to accommodate this demand to generate lucrative income streams from local taxes, growing populations, and a range of development and regulatory levies, thereby guaranteeing their own long-term financial viability (Smith and Krannich 2000). The second point of broad agreement centers on which population subgroups make up these in-flows. Overwhelmingly, these are seen to be ex-urban wealthy and upper middle-class households, seeking to pursue comfortable lifestyles in idyllic landscapes, effectively gentrifying the countryside (Curry, Koczberski, and Selwood 2001; Larsen et al. 2007; Nelson and Nelson 2010; Gosnell and Abrams 2011). A third point gaining purchase in the international literature is, however, the recognition of the independent influence that a suite of biophysical, economic, and locational attributes of rural place is playing in drawing migrants into select locales (McGranahan 1999; Deller et al. 2001; Hunter, Boardman, and Saint Onge 2004; Argent, Smailes, and Griffin 2007).

In this article, we explore the complex temporal and spatial dimensions of post-1990s migration into and out of nonmetropolitan areas of southeastern and southwestern Australia through the lens of rural amenity. The remainder of the article is structured as follows. The next section briefly but critically reviews the rural amenity migration literature as it pertains to postindustrial nations. The third major section tests the relationships between amenity and recent internal migration trends across the study area. The results are discussed in the fourth section, followed by the conclusions.

Amenity-Led Migration into Rural Areas: A Review

Rural amenity is a quality relating to two aspects of a locale: (1) the relative attractiveness of the general environment in which it is set and (2) more specifically, the qualities or facilities of the locale itself. The first of these is a locational attribute, the latter a site attribute, and both are likely to influence where people choose to spend time (Argent, Smailes, and Griffin 2007). As a component of internal migration (i.e., permanent population movement within national boundaries), amenity-led migration to favored rural areas is frequently considered as a key subset of counterurbanization. In the Australian academic press and popular media, this conflation is achieved through the evocative labels of "sea change" (for coastal counterurbanization flows) and "tree change" (for more inland counterurbanization destinations; Burnley and Murphy 2004; Costello 2007). Together, these labels conjure up an implicit set of environmental, locational, economic, and sociocultural variables that act as vital pull factors for this migration current, regardless of its participants' origins (see Burnley and Murphy 2004; Hugo 2005; Walmsley et al. 2006). This we term the *amenity principle*.

Despite some recent valuable contributions to the literature concerning the motivations of counterurbanization migrants (Walmsley, Epps, and Duncan 1998; Smailes 2002; Burnley and Murphy 2004), there has been little concerted effort to specify the ensemble of environmental attributes that comprise rural amenity and the ways in which these attributes interact with ex-urban migrants' locational decision-making processes. Nevertheless, a number of papers concerned with rural population growth and regional economic development in North America have shed light on this issue (McGranahan 1999; Deller et al. 2001; Hunter, Boardman, and Saint Onge 2004).

In a pioneering study, McGranahan (1999) measured rural amenity using an index that incorporated three seasonal climatic variables (warm winter, winter sun, and summer humidity), a single topographic variable, and a surface water indicator, together with a proxy accessibility or remoteness variable. Detailed analysis of this index revealed that the individual climate attributes achieved the highest correlation coefficients, suggesting that the index was better able to predict winter migration than other types of population movement. Deller et al. (2001) adopted a very similar approach, using a composite range of climatic variables, a surface water resource indicator, a land resource indicator (percentage of land in wilderness, forest, farms, and state parks), a composite winter recreation variable, and a recreational facilities indicator. This allowed them to identify high- and low-amenity rural regions and the relationships among amenity, population growth, and income. Hunter, Boardman, and Saint Onge (2004) similarly developed a composite amenity scale, incorporating measures of climate, topography, and water area, in their quest to establish whether or not high-amenity zones tended to experience higher levels of in-migration, gentrification, or both. More recently, rural amenity-focused migration research in the United States has extended into critiques of Florida's city-centric, creative class thesis (McGranahan and Wojan 2007; Wojan, Lambert, and McGranahan 2007; McGranahan, Wojan, and Lambert 2011).

In Australia, Barr (2002, 2005, 2009) has assessed the increasingly critical role of amenity values in shaping rural landscapes and thereby rural societies and economies. Of particular value is his depiction of the "rural amenity landscape," which is largely directed toward an examination of the pivotal role of what he terms an amenity premium in entrenching the presence and retention of undersized farms, facilitating and even enforcing pluriactivity (i.e., the creation of additional on-farm—but often nonagricultural—or off-farm income streams) and part-time farming. In his Australia-wide study, Barr (2005) noted that farm incomes are generally lower in amenity landscapes than in high-production zones but that farm family incomes are higher in the former areas because, in such areas, they are tied to a higher component of (earned or unearned) off-farm (or nonagricultural on-farm) income. Furthermore, the rate of decline of farmer numbers in these high-amenity areas is lower than that in the agricultural heartland. In contrasting these two landscapes, Barr (2005, 68) commented that "a consolidation of the division of rural Australia into high amenity and low amenity locations" seems to be occurring. He further suggested that the demand for landscape amenity is having a significant impact on the geography of structural change in Australian agriculture and therefore on rural Australia in more general terms. The influence of amenity is "manifest in the high price of land in the more amenable and accessible parts of the rural landscape. These higher land prices restrict the capacity of agriculture to adjust to and maintain competitiveness and inexorably drive the path of adjustment to a non-commercial agricultural future" (Barr 2002, 107).

For Holmes (2006), too, consumption-based values meet amenity landscapes head-on in Australia's multifunctional countryside. His detailed conceptualization, and regionalization, of the "multifunctional rural transition" revolves around seven "modes of occupance," including an amenity mode (Holmes 2006, 143). This is confined mainly to the metropolitan and regional city fringes, together with prime tourism locations. Echoing Barr, this mode's land use trajectories are governed by the capacity of consumption-related interests to easily outbid commercial farmers in the competition for land. Smailes, Griffin, and Argent (2005) also found rural amenity to exert a powerful influence over the growing heterogeneity of rural communities, especially those in the mixed agricultural and coastal zones. In particular, high rural amenity was strongly and positively correlated with recent in-migration, total population growth, industrial diversity, and aging populations and strongly negatively correlated with younger age structures and agriculture's proportional share of the local workforce.

This literature review suggests a number of topics that warrant further research. First, in the Australian context, what complex of place-based attributes best encapsulates the notion of rural amenity? Second, what is the nature of the association, if any, between rural amenity and in-, out-, and net migration trends and processes over the past decade or so? Finally, to what extent is the development of an "amenity mode of occupance" (Holmes 2006, 143) threatening commercial farming in Australia's agricultural heartlands? The remainder of this article seeks to address these questions.

Amenity and Internal Migration in Australia

Extending this research, Argent, Smailes, and Griffin (2007) developed an amenity index for local areas of nonmetropolitan southeastern Australia. This index included a number of environmental, locational, and socioeconomic variables, including annual rainfall, terrain and altitude, remoteness, duration of settlement, irrigation water resources, distance from a beach, and employment in recreation and related services (Table 1). For a detailed discussion and rationale for the use of these indicators, see Argent, Smailes, and Griffin (2007). This set of physical and social environmental attributes was argued to capture the essence of rural amenity and overlaps strongly with the complex of attributes that also make areas attractive for natural-resource-dependent industries, including farming. Consequently, as new, consumption-oriented interests interpret and value these attributes in different ways from established production-oriented land users, resource conflict can ensue, and land prices are bid up, leading to pressure for displacement of the latter by the former. The research demonstrated that, for southeastern Australia, there was a relatively strong statistical association between amenity and in-migration, the model achieving an r^2 of 0.53 for the 1996 to 2001 intercensal period (Argent, Smailes, and Griffin 2007, 229). For the analysis in this article, the index has been expanded to incorporate southwestern Australia, with Figure 1 showing the distribution of amenity across 479 social catchments in New South Wales, Victoria, South Australia, and Western Australia. Smailes et al. (2002) detailed the rationale for, and method of, designing bespoke spatial units. (Basically, each one centers on a significant country town with a surrounding area corresponding as closely as possible to its immediate social catchment, with overlaps split along median lines to give a full but mutually exclusive coverage of the study area.) Census collection district units were then allocated to these catchments for the 1981, 1996, 2001, and 2006 censuses, using a

Table 1. Amenity index indicators

Indicator	Source
Median altitude range (m above sea level)	The terrain variable used in this analysis was compiled using a digital elevation model (DEM) applied to a 1-km graticule of the Australian continent.
Median slope (%)[a]	As for median altitude range.
Median annual rainfall (mm)	Average of records available from local recording stations. The vast majority of community centers possess records reaching back over 100 years. Data available from the Australian Bureau of Meteorology Web site (http://www.bom.gov.au).
Settlement duration (years)	Calculated as the period since the date of proclamation (or, in some cases, survey) of the community's principal town.
Accessibility	Derived from the Accessibility/Remoteness Index of Australia (ARIA-plus 1996 version; Department of Health and Aged Care and the National Key Centre for Social Applications of Geographical Information Systems 1999). As a high ARIA-plus score indicates a high degree of remoteness, all scores were multiplied by −1 prior to testing.
Irrigation water resources (% farm income from irrigation)	The proportion of a community's total value of agricultural produce attributable to irrigation, calculated from the Australian Bureau of Statistics Agricultural Census for 2001.
Beach proximity (km)	Calculated road distances between a rural community's main town and the nearest beach or coastal center. All scores multiplied by −1 prior to testing.
Employment in tourism and related services employment (% of total workforce)	Proportion of the 2001 workforce employed in the accommodation, cultural, and personal services industrial category for each community.

[a]Median slope is a percentage-of-slope value derived by the Zevenbergen and Thorne Method, using a 3 × 3 window (Burrough and McDonnell 1998, 190).

concordance. The dependent variables are drawn from cross-classified tables of usual place of residence at Census night and five years prior for the 1996, 2001, and 2006 Censuses (Australian Bureau of Statistics 2008).

These same spatial units were also used to map net migration rates across the southern parts of rural Australia (Figure 2). This reveals a distinctive pattern, with many of the nation's inland, broad-acre agricultural regions experiencing net migration losses. This reflects long-standing processes of adjustment in Australian agriculture, with farm amalgamation and expansion contributing to the out-migration of farm families, the contraction of local economies, and the loss of jobs and people from rural service industries in what Sorensen (1990) has described as a vicious cycle of decline. In contrast, many of the nation's coastal and periurban areas have experienced net in-migration. Yet, the pattern is complex. Some coastal and periurban areas have seen losses, whereas in a number of cases more remote inland areas have experienced net migration gains. So, whereas a broad-scale analysis permits starkly dichotomous regional categorizations of migration patterns into, for example, a declining dry inland versus a booming, well-watered coastal and periurban zone, a more scalar-sensitive approach better captures the geographical complexity of contemporary demographic change in rural Australia (Smailes, Griffin, and Argent 2005).

Table 2 shows the relationships between the amenity index and in-, out-, and net migration rates. It shows that, for the intercensal periods from 1991 to 1996, 1996 to 2001, and 2001 to 2006, the coefficients for in-migration are less stable and overwhelmingly lower than for the other migration measures. To a substantial degree this can be attributed to the nature of the dependent variable itself. In-migration rates might be regarded as a reasonably reliable indicator of the attractiveness and general utility of a place, but many Australian places characterized by relatively high in-migration are also places of high population turnover, especially more remote centers that host mineral and resources extraction. Given the hypothesized positive relationship between the amenity index and in-migration, the consistent moderately negative correlations between the amenity index and out-migration

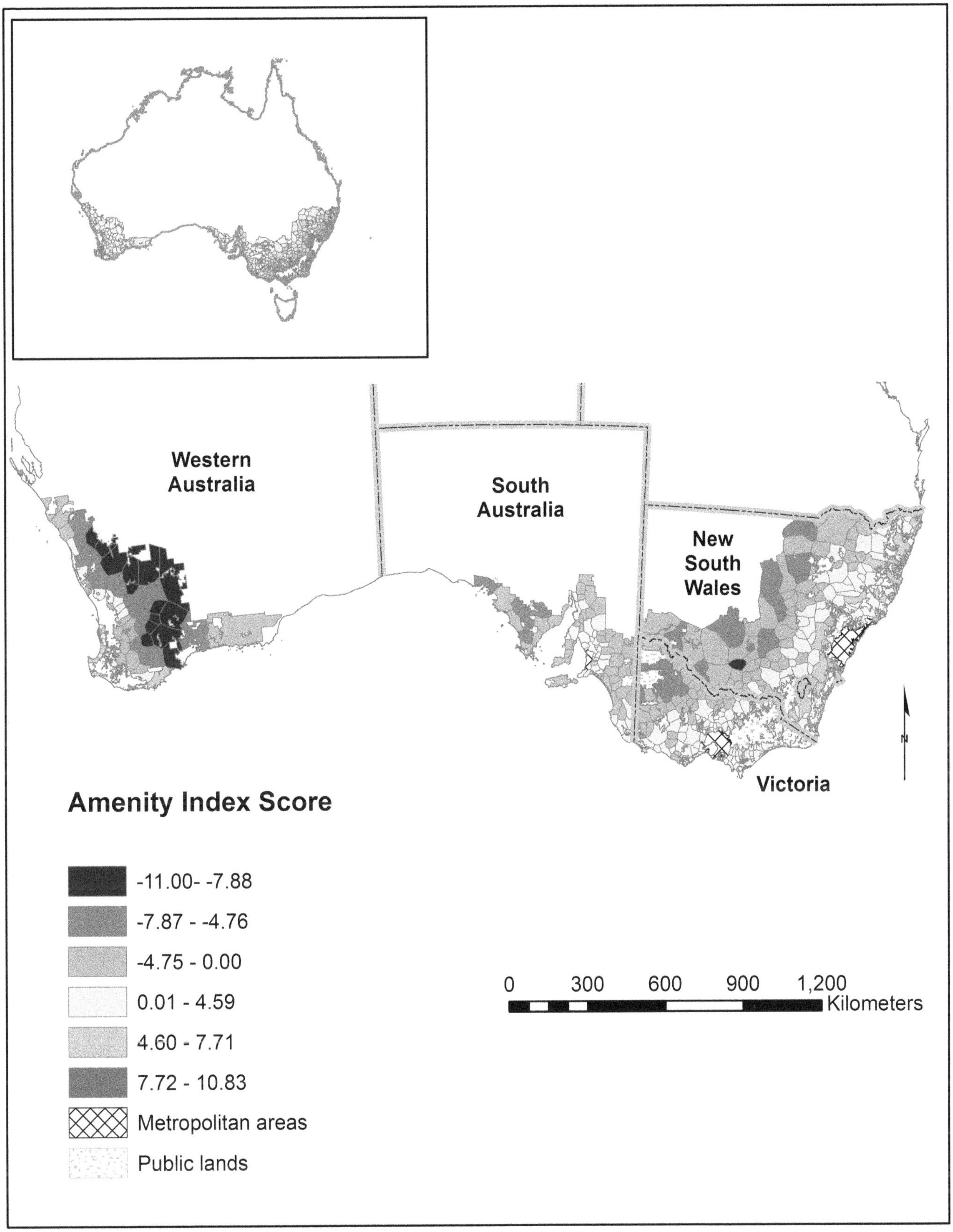

Figure 1. Rural amenity, southeastern and southwestern Australia. (Color figure available online.)

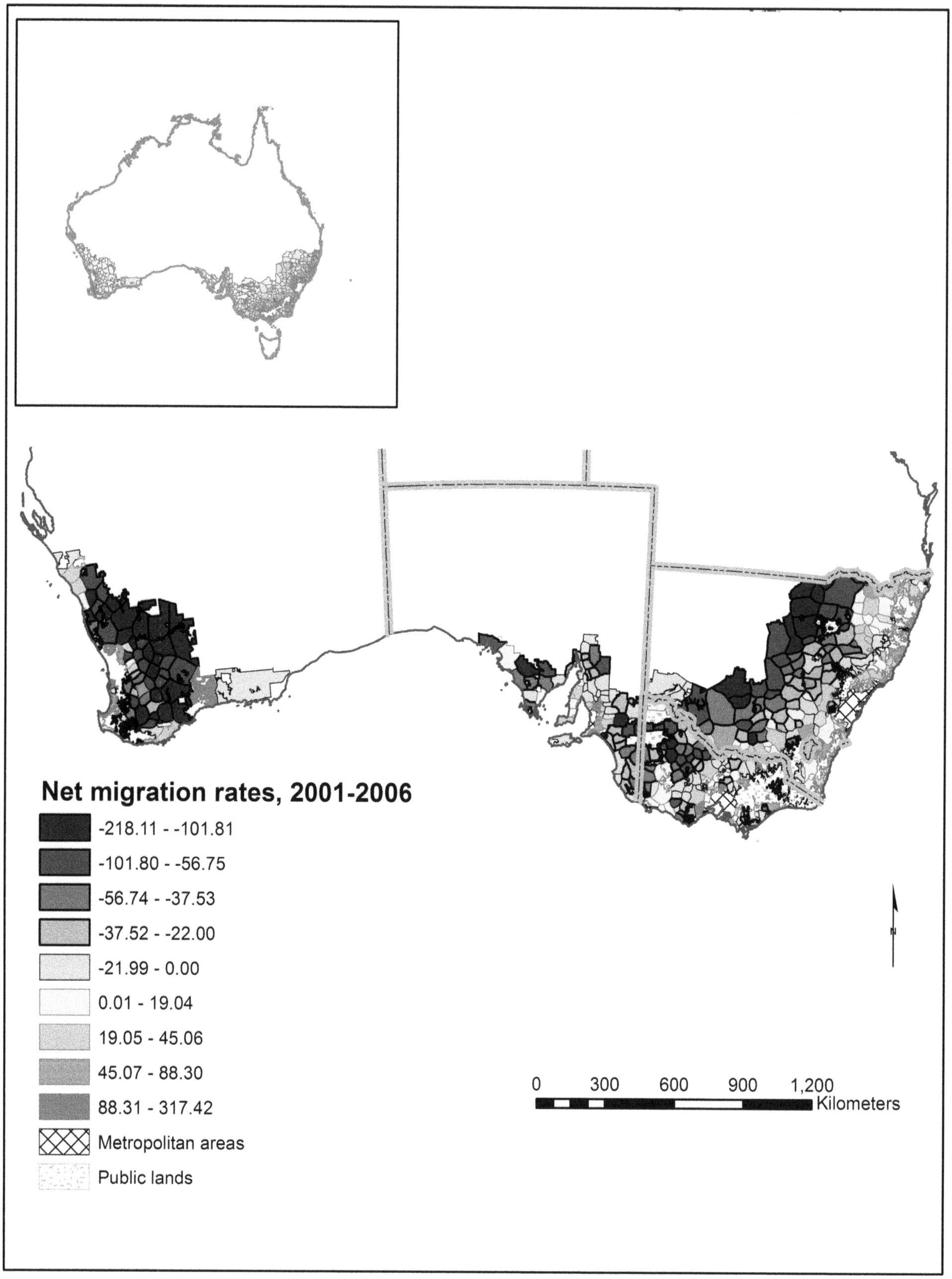

Figure 2. Net migration rates, southeastern and southwestern Australia, 2001–2006. (Color figure available online.)

Table 2. Correlation coefficients (Pearson's r) between the amenity index and net in- and out-migration rates

	Net migration rate			In-migration rate			Out-migration rate		
	1991–1996	1996–2001	2001–2006	1991–1996	1996–2001	2001–2006	1991–1996	1996–2001	2001–2006
Amenity index	0.57*	0.58*	0.64*	0.29*	–0.14*	0.32*	–0.37*	–0.32*	–0.43*

*Correlation is significant at the 0.001 level.

provide some confirmation of the applicability of the amenity model. People, however, are attracted to and driven from new rural destinations for a plethora of reasons and, therefore, we argue that a better proxy for the authentic attractiveness of a rural locale is net migration. The amenity index is a good predictor of net migration, with coefficients ranging between 0.57 and 0.64 over the period. Moreover, the evidence suggests that amenity has become an increasingly important component of rural migration patterns in recent decades. Net migration gain is at least to some extent an expression of the new settlers' commitment to and satisfaction with their new locales, whereas the obverse is true for areas experiencing net migration loss.

Table 3 provides a finer grained analysis of the relationship between rural amenity and migration. Those communities with the highest net migration gains almost all have relatively high amenity scores. These are generally located in perimetropolitan or coastal locations and in many cases have a long history of tourism and related activities. The sole exception here is Ravensthorpe in Western Australia, which gained migrants largely due to a major new nickel mine development during the late 2000s (subsequently closed in 2009). Those communities that experienced the greatest net migration losses all had low amenity scores and were typically located in areas dominated by extensive agriculture.

This synoptic mapping of rural amenity, and the exploration of its relationships with migration processes, provides insights into the spatial coincidence of the causal environmental, locational, economic, and cultural factors on the one hand and rural migration hot and cold spots on the other. In addition, it provides some much needed empirical validation of the inevitably spatially uneven development of the multifunctional countryside in Australia, especially in relation to Holmes's (2006) amenity mode of occupance. Employing the index in this aggregated fashion obscures, however, an understanding of the ways in which more specific elements of amenity are shaping migration patterns. A more robust test of the index and its explanatory capabilities therefore lies in multiple regression analysis.

Tables 4 and 5 explore the relationship between amenity and in- and out-migration flows into nonmetropolitan southeastern and Western Australia during the 1990s and early 2000s. Tables 4, 5, and 6 give results for both the full model with all variables included and a final backward elimination model. Table 4 shows that there is, overall, a relatively weak and unstable relationship between amenity and in-migration. Four of the eight variables that make up the amenity index were eliminated during model testing with the duration of European settlement and median annual rainfall discarded for every intercensal period. In a country

Table 3. Highest ten net migration gains and losses, 2001–2006

Community	Net migration rates, 2001–2006	Amenity index
Capel (WA)	317.42	2.24
Lorne (Vic.)	160.94	9.99
Kapunda (SA)	135.48	1.94
Goolwa (SA)	135.21	3.16
Broadford (Vic.)	127.46	2.09
Kilmore (Vic.)	127.46	2.99
Wallan (Vic.)	127.46	1.99
Wandong-Heathcote (Vic.)	127.46	2.36
Ravensthorpe (WA)	127.11	–7.45
Anglesea (Vic.)	124.35	3.90
Perenjori (WA)	–144.80	–9.22
Dowerin (WA)	–155.13	–6.53
Carnamah (WA)	–158.82	–7.16
Dumbleyung (WA)	–175.29	–7.95
Koorda (WA)	–178.82	–9.34
Mukinbudin (WA)	–186.32	–10.27
Southern Cross (WA)	–187.31	–9.21
Gnowerangerup (WA)	–188.69	–6.50
Dalwallinu (WA)	–190.53	–9.01
Mingenew (WA)	–218.11	–6.18

Note: WA = Western Australia; Vic. = Victoria; SA = Southern Australia.
Source: Australian Bureau of Statistics (2008).

Table 4. Results of full and backwards linear regression analysis, amenity index, and in-migration, 1991–1996, 1996–2001, and 2001–2006

	1991–1996 (β)		1996–2001 (β)		2001–2006 (β)	
	Full model	Final model	Full model	Final model	Full model	Final model
Irrigation water resources	11.29	**11.28**	7.28	**7.86**	4.62	4.71
Settlement duration	–1.94	(x)	.94	(x)	2.49	(x)
Beach proximity	20.03	**19.57**	16.30	**17.00**	16.15	**15.9**
Tourism workforce	9.68	**9.32**	6.94	**7.42**	3.84	(x)
Median slope	11.67	**11.36**	12.90	**13.45**	9.56	**8.81**
Median annual rainfall	–2.46	(x)	2.71	(x)	–3.46	(x)
Median altitude range	–14.82	**–15.38**	–11.19	**–10.78**	–8.12	**–7.29**
Accessibility	2.36	(x)	5.98	**6.83**	8.38	**8.85**
Model summary						
R	0.44	0.44	0.46	0.46	0.41	0.40
R^2	0.19	0.19	0.21	0.21	0.17	0.16
F	14.35*	22.89*	15.93*	21.19*	12.08*	18.51*

Note: (x) = variable removed in backward elimination linear regression testing. The variables yielding the largest, statistically significant coefficients, following the elimination of non-influential variables, have been highlighted in bold.
*$p < 0.001$.

world-renowned for its aridity, the last-mentioned result is perhaps surprising until it is recognized that the dry inland zones through which some of Australia's major rivers flow are key internal migration destination zones, including for counterurbanization migrants (Burnley and Murphy 2004). This also helps to explain the reasonably strong contribution that the irrigation water resources variable plays in the model for all three intercensal periods.

Given the equally legendary pull of the beach in contemporary Australian society and culture (Drewe 1983; Salt 2001), the beach proximity variable plays a predictably strong role in the model. More surprising, is the substantial although dwindling explanatory power of the two terrain variables—median slope and altitude range. Combined, these indicators suggest that (1) flat, featureless landscapes of the kind typically found in the inland broad-acre cropping and farming belt

Table 5. Results of full and backward linear regression analysis, amenity index, and out-migration, 1991–1996, 1996–2001, and 2001–2006

	1991–1996 (β)		1996–2001 (β)		2001–2006 (β)	
	Full model	Final model	Full model	Final model	Full model	Final model
Irrigation water resources	–3.84	(x)	–6.0	–6.13	–6.1	–6.38
Settlement duration	–2.96	(x)	–1.41	(x)	–2.49	(x)
Beach proximity	6.64	7.37	–1.52	(x)	4.08	3.92
Tourism workforce	–8.53	**–9.10**	–11.58	**–12.41**	–9.77	**–9.89**
Median slope	11.67	**12.66**	12.65	**11.77**	13.54	**14.05**
Median annual rainfall	–16.01	**–16.47**	–2.45	(x)	–4.79	–4.39
Median altitude range	–3.97	–4.80	–4.46	–4.97	–10.83	**–11.73**
Accessibility	–10.46	**–13.43**	–10.18	**–11.57**	–13.12	**–14.06**
Model summary						
R	0.49	0.49	0.44	0.44	0.58	0.58
R^2	0.24	0.24	0.19	0.19	0.33	0.33
F	19.18*	24.87*	13.47*	22.69*	30.09*	34.14*

Note: (x) = variable removed in backward elimination linear regression testing. The variables yielding the largest, statistically significant coefficients, following the elimination of non-influential variables, have been highlighted in bold.
*$p < 0.001$.

Table 6. Results of full and backward linear regression analysis, amenity index, and net migration change, 1991–1996, 1996–2001, and 2001–2006

	1991–1996 (β)		1996–2001 (β)		2001–2006 (β)	
	Full model	Final model	Full model	Final model	Full model	Final model
Irrigation water resources	14.48	**14.90**	18.98	**20.14**	11.08	**11.32**
Settlement duration	3.33	(x)	4.34	(x)	5.91	**6.42**
Beach proximity	13.09	**13.26**	21.36	**21.40**	13.21	**12.90**
Tourism workforce	17.85	**17.90**	17.12	**17.80**	13.10	**13.10**
Median slope	.10	(x)	–.80	(x)	–3.91	(x)
Median annual rainfall	13.82	**13.13**	4.25	(x)	.10	(x)
Median altitude	–10.94	**–10.03**	–7.01	**–5.70**	1.89	(x)
Accessibility	11.48	**12.79**	16.53	**18.88**	20.78	**20.95**
Model summary						
R	0.64	0.64	0.64	0.64	0.69	0.69
R^2	0.42	0.41	0.41	0.41	0.48	0.48
F	42.54*	56.57*	41.78*	66.29*	54.91*	87.62*

Note: (x) = variable removed in backward elimination linear regression testing. The variables yielding the largest, statistically significant coefficients, following the elimination of non-influential variables, have been highlighted in bold.
*$p < 0.001$.

discourages in-migration, whereas more varied local relief has a positive effect on it; and (2) conversely, very rugged landscapes, such as mountain ranges, also militate against in-migration, whether for farming or the pursuit of amenity.

Another interesting feature of Table 4 is the growing influence of accessibility over in-migration rates. Eliminated from the model for the 1991 to 1996 intercensal period, it was the second most influential indicator in 2001 to 2006. Also of interest is the dramatically declining explanatory power of the tourism and related services employment indicator. In many respects, this variable sits at the core of the amenity index, serving to encapsulate a place's draw on both temporary visitors and permanent residents in search of recreational, artistic, and cultural pursuits.

Table 5 displays the results of the testing of the amenity model against out-migration rates for the same area and the same periods as for Table 4. This shows generally stronger relationships between amenity and out-migration for the entire model and for many of the individual variables. For the final intercensal period—2001 to 2006—the amenity model explained one third of the variance in out-migration rates across what is, by any definition, a vast and highly internally heterogeneous area. There are other notable points of contrast between Tables 4 and 5. Key among these is the very minor influence that beach proximity has over out-migration flows. Also, the tourism and related services employment indicator exerts a consistently strong influence (and in the direction hypothesized) over out-migration rates. As with Table 4, though, the terrain variables and accessibility are consistently powerful predictors and are increasing in strength over time. The results for the terrain variables, however, are a little difficult to interpret. Regressed against out-migration rates, the results suggest that communities with undulating local relief experience high out-migration, whereas those areas containing the most rugged land forms (e.g., mountain ranges excised by deep river valleys) experience the lowest out-migration rates. Obviously, these results conflict with the findings presented in Table 4. The results for accessibility, in contrast, accord with a commonsense understanding of the Australian space economy. As in Table 4, settlement duration and median annual rainfall are relatively weak indicators, although rainfall was a very powerful predictor of out-migration rates for the period from 1991 to 1996.

Table 6 provides the results of a multiple regression analysis that disaggregates each of the indicators of the amenity index as independent variables and uses net migration rates for the respective intercensal periods as the dependent variables. As noted earlier, net migration is a more statistically stable and conceptually reliable measure of place attractiveness than in-migration rates and is therefore preferred for this round of testing. The results emphasize the fitness for purpose of the amenity index, which explains more than 40 percent of the variance in the first two intercensal periods, rising to just under 50 percent for the period from 2001 to 2006.

The modeling shows that only a few of the independent amenity variables are consistently important in explaining net migration. Those that remained in the final model for all three census periods were the value of irrigated agriculture, beach proximity, tourism workforce, and remoteness. Altitude was important in the 1991 to 1996 and the 1996 to 2001 periods but was absent for 2001 to 2006, and rainfall was present in 1991 to 1996 but absent thereafter. The assumedly critical environmental variables median slope and annual rainfall were absent in all final models, and settlement duration was present only in 2001 to 2006.

In broad terms, access to the beach and local tourism potential had very good explanatory value, reflecting Australians' strong sense of attachment and attraction to the coast. Perhaps it also suggests that tourism centers are often likely to make the transition to places of more permanent settlement by the process of amenity-led migration. The consistent presence of irrigation water resources in the models reflects a landscape's attractiveness for both productivist agricultural uses and more inland-oriented amenity pursuits. Strangely, perhaps, median rainfall had only a weak and diminishing influence across the entire study period. This seemingly anomalous relationship could be attributed to the fact that migration decision making generally involves perceptions that encompass longer term trends in key environmental features, rather than year-to-year fluctuations in a single variable. Another important result relates to accessibility. When other variables are controlled for, accessibility is one of the major influences over the pattern of net migration. Those places relatively close to large metropolitan areas or urban centers typically experience the strongest net migration gains. Accessibility also appears to be becoming increasingly significant over time, largely as result of the extension of urban fields.

The relative robustness of the final regression models, in particular for 2001 to 2006, is evident in Figure 3, which maps the net migration residuals. This shows that, for large parts of rural southern Australia, the residuals are relatively small, indicating that the model is able to offer a good explanation for the geography of net migration rates. There are, however, a number of communities where the model tends to either over- or underpredict net migration. This is particularly evident in Western Australia, where levels of net migration are higher than expected in a number of rural communities. Most evident here are Ravensthorpe and Esperance in the far eastern agricultural regions and parts of the coast north of the state capital, Perth. In all of these areas, mining-led development has contributed substantially to the pattern of net migration. The periurban area around Perth also tends to have higher levels of net migration gain than would be expected, largely due to the outward expansion of one of Australia's fastest growing cities (Kennewell and Shaw 2008) and to the growing role of fly-in/fly-out work in the Western Australian labor force.

In eastern Australia, the residuals map is more complex. The model tends to underpredict net migration in some of the New South Wales irrigation areas, and particularly those along the Murray River bordering Victoria. Similarly, in the Yorke Peninsula of South Australia and parts of the Eyre Peninsula, rates of net migration gain are higher than expected. In the former case, in spite of their relatively high degree of remoteness, flat terrain, and low rainfall, the riverine landscapes of these areas have, for some time, lured in-migrants keen to enjoy the water views and recreation opportunities that the Murray River (and its many tributaries) offers but at a fraction of the cost of most beach real estate. In the latter case, the rugged coastal scenery, good beaches, and small country town atmosphere of the Yorke Peninsula has equally made it a favored location for holidays for generations but, with the expanding urban field of Adelaide, the South Australian capital, parts of the Peninsula have come to fall within the outer limits of the city's commuting belt, making it an increasingly attractive choice for occasional physical and telecommuters and retirees (Smailes 2002). The southern portion of the Eyre Peninsula offers similarly attractive natural environments within which to either consume the landscape or attempt to make a living from it, underscoring the double-edged character of the amenity index. Intensive aquaculture and agriculture ventures recently established on the Eyre Peninsula have drawn in interregional migrants. Lower than expected net migration gains were recorded in parts of northern New South Wales and around the Victorian and South Australian borders.

As previously discussed, rural amenity can be seen to overlap significantly with the physical characteristics necessary for primary production and natural resource extraction, including farming. Indeed, the farming landscapes, and the aesthetic they create, are critical in the production of rural amenity. Yet there is often a somewhat paradoxical process at work here, with this very attribute that contributes to in-migration being radically transformed as a result of this process. As new interests interpret and value these attributes in different, albeit, at times, highly conservative ways (Jones

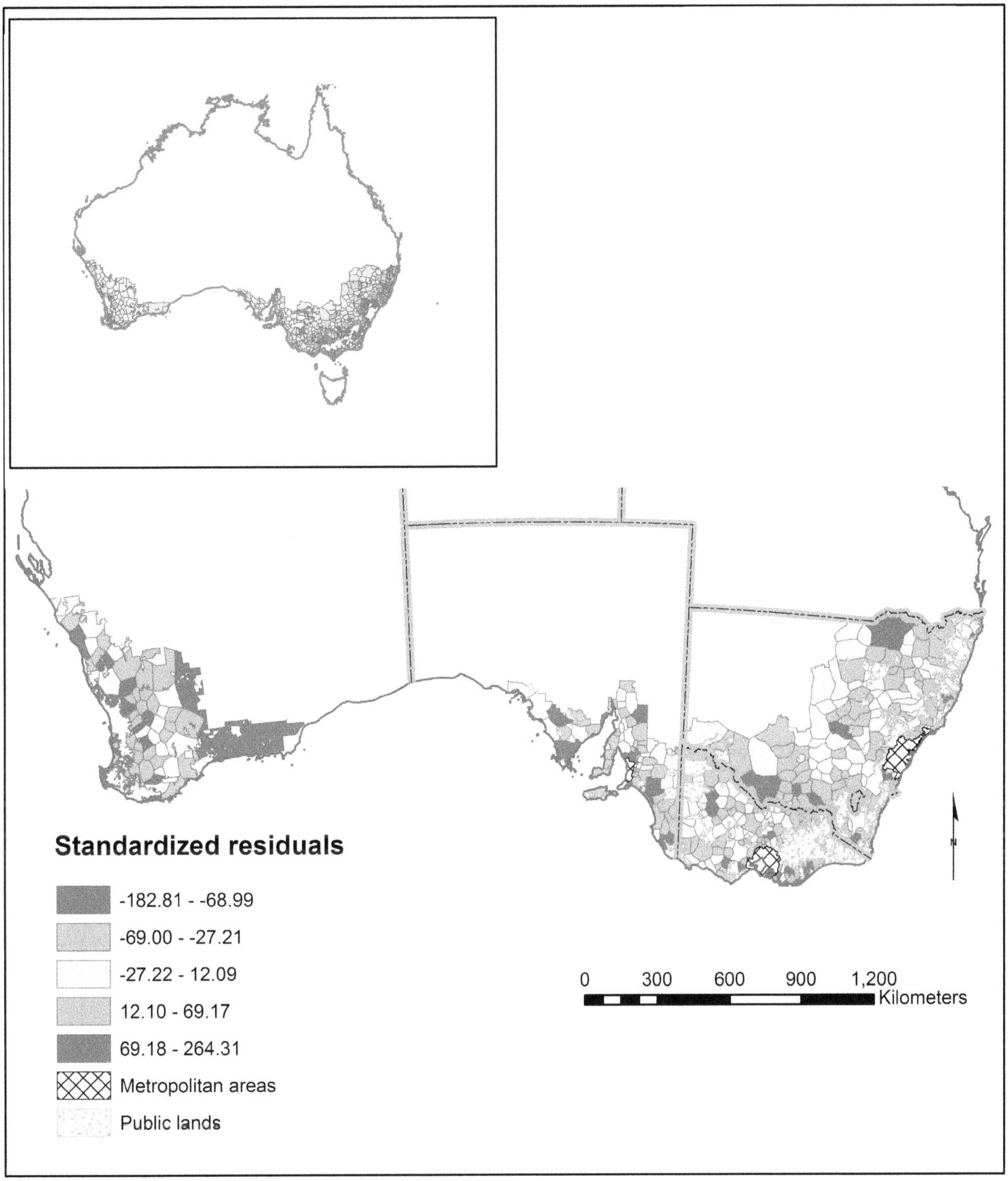

Figure 3. Residuals from regression analysis of net migration rates, southeastern and southwestern Australia, 2001–2006. (Color figure available online.)

and Dolin 2012), resource conflict can ensue as land prices are bid up, leading to pressure for displacement, characteristically of production by consumption. These forces are encapsulated by Barr's notion of the "amenity premium," which is expressed here as the ratio of the market value of land to agricultural production value per unit area. Figure 4 displays the amenity premium for the study area using 2006 land sales and agricultural

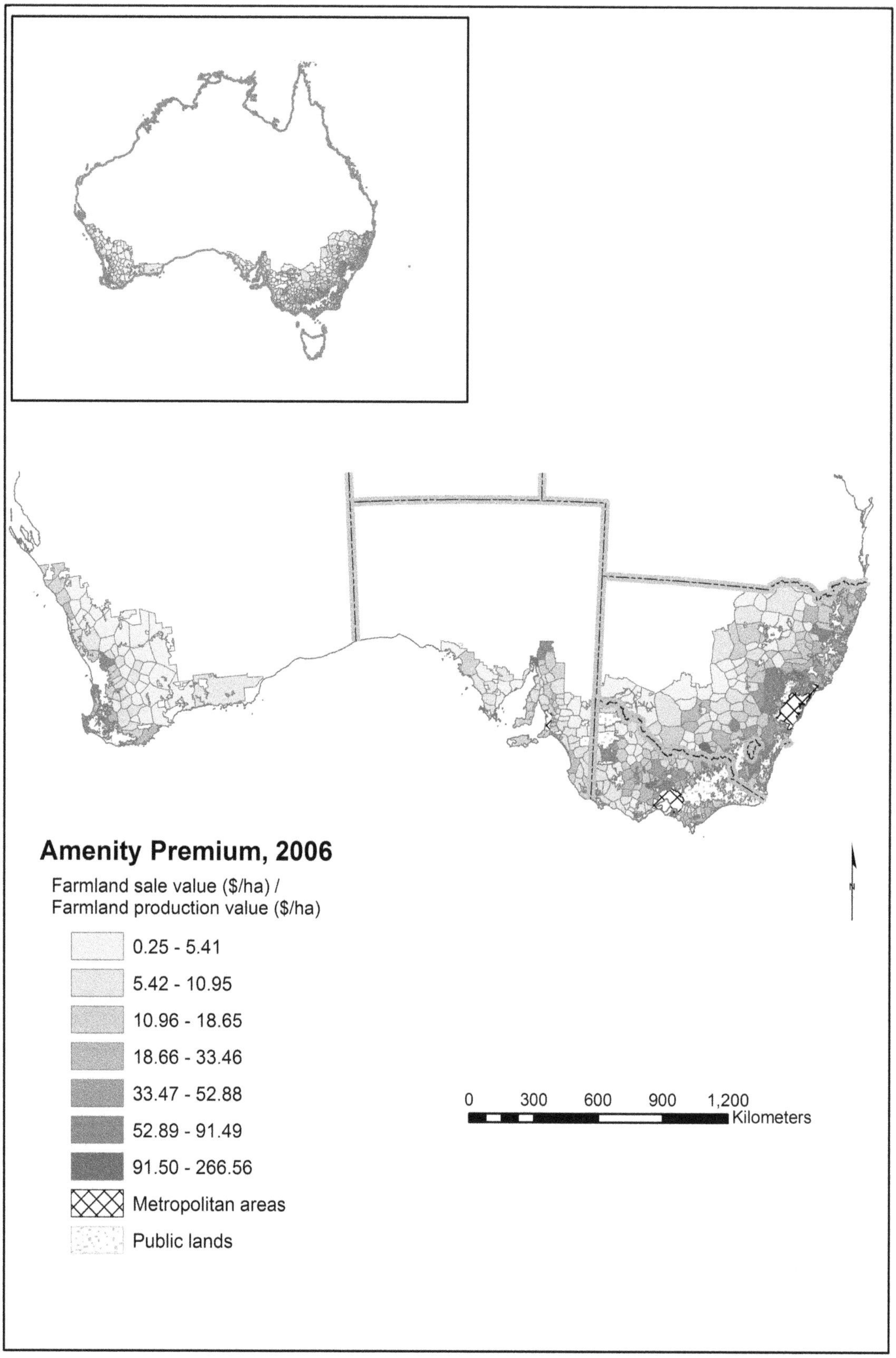

Figure 4. The amenity premium, southeastern and southwestern Australia, 2006. (Color figure available online.)

commodity sales data. Generally speaking, there is a visually strong relationship between the amenity premium and accessibility, with high-amenity premium values occurring in the urban fringes of the capital cities and along the popular coastal belts on the west and east coasts of the continent. The correlation between the amenity index and the amenity premium is moderately positive (Pearson's $r = 0.56$, $p < 0.001$), broadly confirming the association, although some surprising outliers can be found in the far north of South Australia and in the Wimmera region of Victoria. Consistent with the central tenets of the multifunctional countryside, the growing intrusion of amenity values into rural spaces does not appear to spell doom for agricultural production, with the amenity premium having no association with the value of agricultural production ($r = 0.07$). We would suggest a number of possible reasons for this. The first is that, rather than agriculture being fully displaced, it is being restructured in ways that favor higher value production, though often on smaller parcels of land. Indeed, semicommercial (and even fully commercial) hobby or, perhaps more accurately, microfarms in high-value niche sectors such as viticulture, horticulture, and specialist livestock production have emerged in some of Australia's amenity environments. The second is the role of various policy instruments, including land use planning regulations and protective covenants designed to protect agricultural land, particularly in the urban fringe. Thus, agricultural activity often occurs alongside more intensive land uses. More broadly, however, we would suggest that further empirical work is required to better understand the relationships among amenity, land use, and agricultural production.

Conclusion

As in many other places, amenity migration is reshaping the geography of settlement in much of rural Australia. Those areas that are relatively accessible to metropolitan and urban centers are close to the coast and, with an existing or emerging tourism industry, have been the most likely to experience net migration gains. Yet it is also apparent that amenity migration is intersecting with more traditional rural land uses and, in particular, irrigated agriculture. The evidence suggests that outcome is not simply one of rural gentrification but a more complex process consistent with a multifunctional rural transition and Holmes's (2006, 143) notion of an "amenity mode of occupance." This conceptualizes rural space as dominated by consumption values but does not preclude agricultural production. Indeed, in many respects, agricultural production, and the landscape it has produced and is producing, is an important part of what is being consumed by amenity migrants. Moreover, as Holmes (2006) pointed out, the interplay of consumption and production needs to also be understood in the context of protection. This is particularly evident in discourses related to the protection of agricultural lands, not just on the basis of their productive value but also in terms of the preservation of cultural landscapes. Collectively, of course, these changes are not simply reshaping rural land use but also the social, cultural, and political dynamics of rural areas. Indeed, there is evidence to suggest that the multifunctional transition that is occurring across much of rural Australia is contributing to growing levels of sociocultural, economic, and political heterogeneity. Although this article has pointed to some of the key drivers of this, further progress requires a new phase of empirically grounded research as a means of both improving our understanding of the processes underpinning amenity migration and rural change and our understanding of how these are playing out in terms of rural community and economic life.

References

Argent, N., P. Smailes, and T. Griffin. 2007. The amenity complex: Towards a framework for analysing and predicting the emergence of a multifunctional countryside in Australia. *Geographical Research* 45:217–32.

Australian Bureau of Statistics. 2008. *Cross-classified migration tables from the 1996, 2001 and 2006 Censuses of Population and Housing*. Canberra, Australia: ABS.

Barr, N. 2002. Victoria's small farms. CLPR Research Report No. 10, Department of Natural Resources and Environment, Epsom, Australia.

———. 2005. *The changing social landscape of rural Victoria*. Melbourne: Department of Primary Industries.

———. 2009. *The house on the hill*. Melbourne: Land and Water Australia.

Burnley, I., and P. Murphy. 2004. *Sea change: Movement from metropolitan to Arcadian Australia*. Sydney: UNSW Press.

Burrough, P., and R. McDonnell, R. 1998. *Principles of geographical information systems*. Oxford, UK: Oxford University Press.

Costello, L. 2007. Going bush: The implications of urban–rural migration. *Geographical Research* 45:85–94.

Curry, G., G. Koczberski, and J. Selwood. 2001. Cashing in, cashing out: Rural change on the south coast of Western Australia. *Australian Geographer* 32:109–24.

Deller, S., S. Tsai, D. Marcouller, and D. English. 2001. The role of amenities and quality of life in rural economic growth. *American Journal of Agricultural Economics* 83:352–65.

Department of Health and Aged Care and the National Key Centre for Social Applications of Geographical

Information Systems. 1999. Measuring remoteness: Accessibility/Remoteness Index of Australia (ARIA). Occasional Papers, New Series No. 6, Department of Health and Aged Care, Canberra, Australia.

Drewe, R. 1983. *The bodysurfers*. Darlinghurst, Australia: James Fraser.

Gosnell, H., and J. Abrams. 2011. Amenity migration: Diverse conceptualizations of drivers, socioeconomic dimensions and emerging challenges. *Geojournal* 76:303–22.

Holmes, J. 2006. Impulses towards a multifunctional transition in rural Australia: Gaps in the research agenda. *Journal of Rural Studies* 22:142–60.

Hugo, G. 2005. The state of rural populations. In *Sustainability and change in rural Australia*, ed. C. Cocklin and J. Dibden, 56–79. Sydney: UNSW Press.

Hunter, L., J. Boardman, and J. Saint Onge. 2004. The association between natural amenities, rural population growth and long-term residents' economic well-being. Working Paper EB 2004-0005, Institute of Behavioural Science, University of Colorado, Boulder, CO.

Jones, R., and T. Dolin. 2012. Literary heritage or national heritage? Landscape preservation and change in Dorset. In *Heritage 2012: Proceedings of the 3rd International Conference on Heritage and Sustainable Development*, ed. R. S. Amoeda, S. Lira, and C. Pinheiro, 1119–25. Porto, Portugal: Green Lines Institute for Sustainable Development.

Kennewell, C., and B. Shaw. 2008. Perth, Western Australia. *Cities* 25:243–55.

Larsen, S., C. Sorensen, D. McDermott, J. Long, and C. Post. 2007. Place perception and social interaction on an exurban landscape. *Professional Geographer* 59: 421–33.

McGranahan, D. 1999. *Natural amenities drive rural population change*. Agricultural Economic Report No. 781. Washington, DC: Food and Rural Economics Division, Economic Research Service, U.S. Department of Agriculture.

McGranahan, D., and T. Wojan. 2007. Recasting the creative class to examine growth processes in rural and urban counties. *Regional Studies* 41:197–216.

McGranahan, D., T. Wojan, and D. Lambert. 2011. The rural growth trifecta: Outdoor amenities, creative class and entrepreneurial context. *Journal of Economic Geography* 11:529–57.

Nelson, L., and P. Nelson. 2010. The global rural: Gentrification and linked migration in the rural USA. *Progress in Human Geography* 35:441–59.

Salt, B. 2001. *The big shift: Welcome to the third Australian culture*. South Yarra, Australia: Hardie Grant Books.

Smailes, P. 2002. From rural dilution to multifunctional countryside: Some pointers to the future from South Australia. *Australian Geographer* 33:79–96.

Smailes, P., N. Argent, T. Griffin, and G. Mason. 2002. *Rural community social area identification*. Adelaide, Australia: National Centre for Social Applications of Geographical Information Systems.

Smailes, P., T. Griffin, and N. Argent. 2005. The changing social framework. In *Sustainability and change in rural Australia*, ed. C. Cocklin and J. Dibden, 80–102. Sydney: UNSW Press.

Smith, M., and R. Krannich. 2000. Culture clash: Newcomer and longer term residents' attitudes toward land use, development and environment in rural communities in the Rocky Mountain West. *Rural Sociology* 65:396–421.

Sorensen, A. 1990. Virtuous cycles of growth and vicious cycles of decline: Regional economic change in Northern NSW. In *Change and adjustment in Northern NSW*, ed. D. Walmsley, 41–59. Armidale, Australia: Department of Geography and Planning, University of New England.

Walmsley, D., N. Argent, F. Rolley, and M. Tonts. 2006. Inland Australia. Paper presented at Australian on the move: Internal migration in Australia. Academy of the Social Sciences in Australia Symposium, Canberra, 21 November.

Walmsley, D., R. Epps, and C. Duncan. 1998. Migration to the New South Wales North Coast 1986–1991: Lifestyle motivated counterurbanisation. *Geoforum* 29:105–18.

Wojan, T., D. Lambert, and D. McGranahan. 2007. Emoting with their feet: Bohemian attraction to creative milieu. *Journal of Economic Geography* 7:711–36.

"Under the Radar": Undocumented Immigrants, Christian Faith Communities, and the Precarious Spaces of Welcome in the U.S. South

Patricia Ehrkamp* and Caroline Nagel†

*Department of Geography, University of Kentucky
†Department of Geography, University of South Carolina

This article examines the limits of welcome that Christian communities of faith in the U.S. South extend to recent immigrants. We argue that churches are political spaces in which pastors and lay members weigh faith-based conceptions of hospitality against law-and-order discourses and in which notions of universal membership confront racialized immigration politics. Drawing on sixty interviews with pastors and lay ministers in thirty-five churches in Greenville–Spartanburg, South Carolina; Atlanta, Georgia; and Charlotte, North Carolina, we show how hospitality within individual churches often operates quietly, "under the radar," producing a politics of invisibility. This invisibility, although providing some shelter from harsh law enforcement practices, does little to fundamentally alter the precarious situation of immigrants. We show that as Christian ethics of hospitality come up against worldly social boundaries of race and legal status, the actual practice of hospitality in these churches falls short of biblical ideals. Our analysis furthers understandings of political and faith-based membership and the dynamic articulations between them.

本文检视美国南方的基督教信仰社群在欢迎晚近移民时的限制。我们主张，教堂是政治的空间，其中牧师与教会成员秉持着以信仰为基础的悦纳异己之概念，而非根据法律与规范的论述，其中普世的成员资格遭遇了种族化的移民政治。我们在南卡罗来纳州格林威尔—斯巴坦堡、乔治亚州亚特兰大，以及北卡罗来纳州夏洛特的三十五座教会中，对牧师与教会神职人员进行了六十个访谈，以此展示个别教会如何经常暗中、"秘密"进行悦纳异己，生产了不可见的政治。此一不可见性，儘管对严峻的执法工作提供了部分庇护，但却几乎无法根本性地改变移民的不安定处境。我们指出，当基督教悦纳异己的伦理遭遇世俗的种族与法律身份的社会边界时，这些教会悦纳异己的实际作为却难以达到圣经的理想。我们的分析得以进一步理解以政治和信仰为基础的会员资格，以及两者之间的动态接合。

Este artículo examina las limitaciones en la acogida que las comunidades de fe cristiana del Sur de los EE.UU. brindan a inmigrantes recientes. Nuestro argumento es que las iglesias son espacios políticos donde pastores y feligreses laicos sopesan las concepciones de hospitalidad basadas en la fe, frente a los discursos de ley y orden, y donde las nociones de membrecía universal tienen que confrontar políticas de inmigración racializadas. Con base en sesenta entrevistas con pastores y ministros laicos de treinta y cinco iglesias de Greenville-Spartanburg, Carolina del Sur; Atlanta, Georgia; y Charlotte, Carolina del Norte, hacemos ver que la hospitalidad a nivel individual de iglesias opera silenciosamente, "por debajo del radar," lo cual produce una política de invisibilidad. Si bien provee alguna protección contra las duras prácticas de aplicación de la ley, esta invisibilidad poco hace para alterar significativamente la precaria situación de los inmigrantes. Mostramos que en cuanto la ética cristiana de la hospitalidad tiene que vérselas con los mundanos límites sociales impuestos por raza y estatus legal, la práctica real de la hospitalidad en aquellas iglesias queda bien alejada de los ideales bíblicos. Nuestro análisis avanza en la comprensión de una membrecía basada en política y fe y de las articulaciones dinámicas que ocurren entre éstas.

The devolution of immigration policy in the United States has expanded border control and immigration policing from the external border to include interior spaces of the United States and has rendered the lives of immigrants—especially undocumented immigrants—increasingly precarious (Ellis 2006; Winders 2007; Varsanyi 2008; Coleman 2012). The recent proliferation of anti-immigration legislation at the state level, often modeled on Arizona's 2010 Senate Bill 1070, has compounded this situation. These laws variably make it a crime to be in the state without proper documentation, give the police broad authority to check individuals' immigration status during unrelated traffic stops, and criminalize the "harboring" and transportation of undocumented immigrants. Although such legislation exists in several states of the United States, some of the most far-reaching and restrictive legislation has been enacted in the U.S. South—notably Alabama, South Carolina, and Georgia.

Recent research has documented a link between the pace of demographic change and exclusionary policies geared at immigrants (Singer, Wilson, and DeRenzis 2009; Walker and Leitner 2011). And indeed, anti-immigrant fervor in the U.S. South follows on the heels of rapid social and economic transformation. The region has experienced dramatic changes since the 1950s, when Jim Crow segregation gave way fitfully to an aggressive progrowth agenda pushed by business elites (Kruse 2005). Through the 1960s, 1970s, and 1980s, federal defense spending flowed into Southern states, as did private-sector investment eager to tap into the region's nonunionized labor force (Marrow 2011). During this time, Southern metropolitan areas experienced spectacular growth as the country's population shifted southward. These economic and demographic changes, combined with federal immigration reform, contributed to the rapid increase of immigrant populations and turned cities in the region into booming immigrant gateways (Singer, Hardwick, and Brettell 2008). Between 1990 and 2010, states like Georgia, North Carolina, and South Carolina saw immigrant populations grow by 300 to 500 percent, although the foreign-born continue to constitute a relatively small proportion of state populations.[1] Assertions by local business elites and chambers of commerce that "diversity" and "internationalization" benefit local economies (Winders 2006) compete with the region's conservative political culture, which often casts minorities, the poor, and immigrants as society's takers (Lassiter 2006). Anti-immigrant legislation, accordingly, goes hand-in-hand with the reconfiguration of race in the U.S. South and the racialization of recent immigrants as "other" (Winders 2007; Chavez 2008).

In the midst of such anti-immigrant legislation, places of worship serve as important spaces of integration and welcome for recent immigrants (Ehrkamp and Nagel 2012). Because this immigration is so recent and the U.S. South lacks immigrant mutual assistance organizations, local churches and other nongovernmental institutions have become important springboards for immigrants' incorporation into wider society (Cabell 2007; Winders 2012). Individual churches offer health clinics, legal services, and English as a second language (ESL) and citizenship classes to immigrants regardless of legal status, but wider Christian organizations, such as the Catholic Archdiocese of Atlanta, have also advocated on behalf of undocumented immigrants by calling for comprehensive immigration reform (Catholic Bishops of Georgia 2011). Beyond the provision of services and advocacy work, many churches have also sought to incorporate immigrants into congregational life. Their efforts have particular significance given the South's history of racial segregation and the role of white churches in sustaining it (Wells 2004).

In welcoming immigrants, churches act on biblical directives to extend hospitality to strangers (e.g., in the Old Testament and in Matthew 25:35 "I was a stranger and you invited me in"). Methodist theologian Arias (1982) describes this ethic of hospitality as a distinctive mark of Christianity from the earliest days of the faith. Opening one's home was not just a matter of politeness or civility but rather a solemn affirmation of universality and inclusivity, intimately related to Christian life and mission. Baptist scholar Pohl (2007) associates hospitality with "breaking down social boundaries, forming community, expecting transformation, and welcoming Jesus and angels" (33). Similarly, Baptist theologian Newman (2007) described Christian hospitality as "a public way of life together" (12) that renders Christ's body visible to the world and that attaches us to "the places and people we are given" (16). Yet others insist that such welcoming of strangers is an act of neighborliness across difference (Daniel 2010) and a selfless act that emphasizes Christians' debt to others (Ahn 2010). These Christian ethics of hospitality approximate what academic scholarship terms "unconditional hospitality," which is limitless (Honig 2009, 115).

But, as Dikeç (2002, 228) reminded us, "the notion of hospitality, simply because it is almost always taken for granted as implying a desirable quality, invites critical reflection." Rather than a limitless, unconditional ideal, hospitality in practice entails conditionality and assertions of sovereignty over space, be it a home or a democratic state (Derrida 2000). Examining hospitality highlights the tensions that emerge as different constituents—guest and host—lay claim to the same space(s). Hospitality is an apt lens through which to examine the incorporation of immigrants in the U.S. South because it considers the relationships between the ethics, the politics, and the laws of belonging without presupposing unconditional welcome or outright exclusion (Derrida 2005). We argue that a Christian ethics of welcome as practiced in the U.S. South becomes enmeshed in a politics of hospitality that straddles sacred and secular spaces and that blurs the line between divine and secular law. In other words, in its focus on the public good and on changing society beyond the immediate community of faith, Christian hospitality is inherently political (Newman 2007). As such, Christian ideals of welcome come up against, and must contend with, worldly social boundaries of race, class, and legal status.

Following recent calls by geographers to examine the production of religiosities through critical engagement with politics (Olson, Hopkins, and Kong 2012), our research delves into the ways that religious identities, practices, and ethics become deeply entangled with secular[2] immigration politics as churches minister to recent immigrants in the U.S. South. Our arguments build on scholarship in geography and the wider social sciences on the role of religious institutions in mediating immigrants' integration into American society (Young 1928; Gordon 1964; Ley 2008; see Ehrkamp and Nagel [2012] for a detailed discussion). Much of this research has brought into focus the role of faith communities as sites of resistance against societal oppression (Hondangneu-Sotelo 2007; Cloke and Beaumont 2013) or as schools of citizenship in which immigrants learn civic virtues and values (Kniss and Numrich 2007). Places of worship, scholars have argued, are key spaces in which immigrants and nonimmigrants are able to develop the relationships of reciprocity and trust that are central to democratic societies (Ecklund 2005; Foley and Hoge 2007; Ley 2008). Others, however, have also cautioned that communities of faith are not always able to overcome sociospatial exclusion and prejudice (cf. Valentine and Waite 2012).

In what follows, we analyze the moral and political dilemmas that emerge as various Christian congregations in the U.S. South seek to accommodate their own beliefs, immigrants' needs, and secular politico-legal frameworks. In doing so, we contribute to geographic research on spaces of encounter (e.g., Leitner 2012) and studies of social relations between immigrants and nonimmigrants that are central to understanding processes of immigrant incorporation (Ehrkamp 2006; Ehrkamp and Leitner 2006). Contrary to publicly declared mandates of a Christian ethics of hospitality and public advocacy, we found that politics within the individual churches in our research often did not manifest themselves as overtly political, public, or even collective. Instead, we found that pastors and church congregants—although deeply politically and religiously motivated—frequently operated quietly, avoiding publicity and engaging instead in a politics of invisibility. This invisibility, on the one hand, shields immigrants from aggressive law enforcement activities but, on the other hand, does very little to change the fundamental precariousness of their situation.

Welcoming the Immigrant Stranger: Christian Hospitality in an Era of Anti-Immigrant Legislation

The U.S. South boasts above-average religiosity as compared to the rest of the United States and the highest percentage of residents that are considered "very religious" (Newport 2012). More than residents of other regions, people in the U.S. South are likely to characterize themselves as "evangelical" (Ownby 2005), and there is a widespread assumption that the public square is Christian, as seen with ongoing efforts to institute prayer in public schools and to place the Ten Commandments in public buildings in Southern states (Wells 2004). Religious worship in the U.S. South has largely taken place within racially homogenous faith communities—a pattern established after the Civil War, when emancipated slaves abandoned white-led churches and formed their own denominations (Harvey 2005). But patterns of segregated worship are changing, partly due to the growth of Roman Catholicism through Latino and Asian immigration and partly through the purposeful outreach to immigrants by predominantly white Protestant churches.

Beyond biblical mandates to welcome the stranger, churches in the U.S. South have myriad reasons for seeking out immigrants. The Southern Baptist Convention, for instance, has attempted to redress their complicity in the oppression of African Americans in the pre–Civil Rights era by publicly renouncing racism and encouraging member churches to engage with immigrants and nonimmigrant minorities (Southern Baptist Convention 1995). Others, at a more pragmatic level, are reaching out beyond their traditional constituencies to boost membership rolls. This is particularly the case for mainline Protestant churches that have been losing members for decades. A number of Baptist churches have adopted a strategy of growing by caring (Greer 1993); that is, expanding membership by ministering to those who are ethnically and culturally different. Yet other churches are adapting to demographic change in their immediate vicinity, such as the influx of immigrants into older suburbs and periurban areas in these new immigrant destinations (Ehrkamp and Nagel 2012). These strategies play out differently in terms of how immigrants are incorporated into congregational life (as we explain later). Regardless of denomination and faith tradition, however, incorporating immigrants has created new dilemmas for those Christian communities of faith that try to do so.

The arguments that follow draw on intensive, semistructured interviews with sixty pastors, priests, and lay members in thirty-five churches—Baptist, Presbyterian, Catholic, Coptic, Episcopal, nondenominational, and others—in Atlanta, Georgia; Charlotte, North Carolina; and Greenville–Spartanburg, South Carolina, conducted between December 2011 and August 2012, a period when new immigration legislation was debated and newly enacted in all three states (with parts of Georgia's immigration law blocked by court order in spring 2012). Churches in our sample varied in terms of membership, from fifty-member inner-city churches to affluent suburban megachurches with more than 15,000 members. With the exception of three nondenominational multiracial churches, these are mostly churches dominated by white members. The immigrant groups these churches serve vary: Some churches have large Latino ministries; others have refugee populations from Bhutan, Haiti, or Iran; and yet others are immigrant-only Vietnamese or Chinese churches (Baptist, Roman Catholic, Assembly of God, etc.). Reflecting the patriarchal structure of many churches and restrictions on female ordination in many denominations, most of the pastors we interviewed were men and most of them were native-born whites. A third of our interviewees were immigrants from a variety of countries (e.g., Peru, Colombia, China, Vietnam, Egypt, Cameroon, etc.) Interviews with lay members such as missions coordinators, administrators, and ESL teachers complement our data and allowed us to include more female and immigrant voices.

We asked our interviewees about their reasons for reaching out to immigrants and their aims in doing so, the challenges and opportunities they have encountered in incorporating immigrants and achieving a multiethnic or multicultural church, and the impact of recent anti-immigrant laws on their ministries. Our interviewees' answers to these questions form the basis for the following discussion of the complex ethics and politics of hospitality toward immigrant newcomers in these churches.

Differentiated Welcome

In seeking to create a welcoming, inclusive environment for immigrants, contemporary Christian faith communities often hire a full-time minister to serve immigrant communities or they offer foreign-language worship services. Although they spoke of common worship as an ideal, the pastors we interviewed argued that forcing any one group to adapt to the other was counterproductive and, indeed, would run contrary to the ideal of heaven as a multilingual, multicultural realm as described in the Book of Revelation. Hence, a more common approach among the churches in our research was to create separate immigrant congregations with different worship times and, in some cases, different worship spaces. Head pastors (usually native-born and white) described such arrangements not in terms of keeping people separate but in terms of respecting distinctive cultural practices and music preferences and being sensitive to socioeconomic disparities. As one white Baptist pastor in Spartanburg explained,

> [Latinos] don't want worship with us! And it's not because they don't like us, it's because culturally it's just different. . . . They want to sing songs in Spanish. They want to hear the sermon in Spanish. They want to do the different things that are culturally different than what we do. That's okay! (White Baptist pastor, Spartanburg, May 2012)

The immigrant pastors we interviewed generally echoed these sentiments, insisting on the desire of most immigrants to worship in their own language and style. Yet the Chinese, Vietnamese, and Latino pastors who led immigrant congregations within larger churches also expressed frustrations in our interviews. For example, while praising the head pastor's openness toward welcoming poorer Latinos, the Latino pastor in an affluent nondenominational church stated that he and members of his congregation also "felt a little bit rejected" by their separation from white congregants and the head pastor's assertion that immigrants needed to learn English. Despite their gratitude for the support offered by the wealthy white church, this Latino pastor and his staff felt that by worshipping in a smaller space, generally fully removed from the main sanctuary, and occasionally being shepherded in for joint services, they were second-class members of the church. Such frustrations were compounded by the prejudices Latino pastors and staff had encountered from white congregants. For instance, a Latino pastor reported hearing white congregants saying that Latina immigrants are seeking to "get married just for documentation" in the United States (Latino pastor, nondenominational, Charlotte, December 2011)—a testimony on the one hand to the difficulty of overcoming the racial segregation of worship and, on the other, to a pervasive public discourse of immigrant illegality that is applied liberally to Latinos regardless of their actual immigration status (Chavez 2008; Ramakrishnan, Esterling, and Neblo 2010).

These tensions indicate that in extending hospitality to immigrants, churches may well reproduce existing societal inequalities and perpetuate racial divides, a problem also observed in white evangelical churches seeking to break black–white racial barriers (Emerson and Smith 2000). Contrary to what Cloke and Beaumont (2013) suggested, shared faith, in this case, does little to successfully counter prejudice against those perceived as other, a point that Valentine and Waite (2012) made with respect to homosexuality in Christian communities of faith. The churches in our research do provide some welcome, but as the preceding example illustrates, such welcome is deeply enmeshed in secular anti-immigrant politics that stereotype Latino immigrants and taint ideals of inclusivity. This reality is thrown into sharper relief when considering how church leaders deal with the undocumented status of new congregants.

Operating "Under the Radar"

Although most national and regional organizations of the Roman Catholic Church and of the mainline Protestant denominations have advocated publicly on behalf of undocumented immigrants and have called for comprehensive immigration reform[3] (Nteta and Wallsten 2012), most of the pastors we interviewed, regardless of denomination, were not using their role for public advocacy. Instead, they emphasized serving the spiritual, social, and sometimes material needs of their congregants. But they also made clear that they were not in the business of policing immigration status, drawing a clear line between state policy and ministry. The Latino minister of a Baptist megachurch[4] in Charlotte put it this way:

> Well, . . . in our church, we don't care who has documents and who has not. We are not a law or government agency, so here we don't care about that. We care about character. We care about discipleship. We care about spiritual needs. . . . The best tool that we have is that we are under the radar.

He went on to caution us,

> Please with this information we have to be careful. . . . Our advantage is that we are part of the Anglo church and because we're a part of the Anglo church, they [immigration enforcement] don't come here and bother. (Latino pastor, Baptist, Charlotte, December 2011)

As this quote indicates, not asking about immigration status and operating "under the radar" is an important strategy for ministering to immigrants; in many ways, it does create safe, invisible spaces for undocumented congregants where they know they will not be harassed or arrested by the police (cf. Herbert 2009). But this pastor's request to tread carefully with information about undocumented immigrants in his congregation shows that this politics of invisibility also means that the welcome extended to undocumented immigrants remains precarious because immigrants remain prone to discovery.

Although the churches in our research had not (yet) become targets of immigration enforcement, almost all church leaders indicated that aggressive immigration enforcement had impinged significantly on immigrants' lives. Pastors had witnessed dropping attendance due to detentions or deportations and the unwillingness of undocumented immigrants to drive to church on Sundays for fear of being pulled over by the police and deported, the choice method of immigration enforcement as it encroaches on the spaces of immigrants' social reproduction (Coleman 2012). One pastor in a Baptist church on the outskirts of Charlotte said that his Latino congregants sensed a stronger police presence outside their weekly basketball game and in their neighborhoods. Still another pastor at a Presbyterian church in Greenville noted that law enforcement activities had discouraged undocumented immigrants from attending a church-sponsored health fair—one of the few opportunities for them to receive medical treatment.

The pastors we interviewed spoke with great compassion about the plight of undocumented immigrants, and occasionally they confronted police and asked them to refrain from their enforcement activities. Church leaders felt deeply troubled by the effects of the current immigration enforcement regime on undocumented immigrants. The pastor of the Latino congregation in a Charlotte megachurch, for instance, stated:

> [I]t bothered me in here [pointing to his heart] that the people sometimes are so blind that the compassion is going and what we have is just the legality of the law. How can you explain for me you picking up a mother who has a few kids, babies and then take this mother and deport her

> for her country? . . . And the kids stay here in foster home with somebody that they don't know. . . . And they call this law. (Latino pastor, nondenominational, Charlotte, December 2011)

Despite expressing compassion, our interviewees suggested that undocumented immigrants, having entered the United States without authorization, have no real choice other than to live within the parameters of their precarious status and to hope that at some point the law changes. The same Latino pastor of the nondenominational megachurch in Charlotte again invoked the phrase "under the radar" in describing the advice he gives to undocumented immigrants in his congregation:

> Now, in our congregation, I did not ask for anybody to stay illegal in the country. I don't encourage them. And one thing that I'd say publicly, and they are my witness, I say all the time, if you are here illegal, don't break the law one more time. Don't buy a [false] documentation to present a documentation that's not real. Second, don't drive with no license because if [the police catches up] with you, you are going to have troubles. In other words, if you are here illegal, do the best to accommodate yourself under the radar and do the best you can to provide something in the community that one day, if the law changed, you are able to say, "That's what I did. That's how I worked." (Latino pastor, nondenominational, Charlotte, December 2011)

Staying under the radar, here takes on another meaning as immigrants are meant to operate as if they were citizens by contributing to society through working and not breaking further laws. Such a strategy positions undocumented immigrants to become "good illegals" who stand a better chance of future legalization and incorporation as citizens (Chauvin and Garcés-Mascareñas 2012, 249).

Staying under the radar also serves as a political strategy through which pastors contend with hostility toward undocumented immigration within their churches. Several pastors in our research chose to operate under the radar because they were, in part, concerned with alienating their own (usually white and often conservative) congregants. One Latino Catholic deacon in a church with a large Latino congregation noted that some donors had begun withholding financial support because they thought the church had been "supporting illegals." Donors, he said, had suggested that "because we have bilingual employees, myself included, 'that's because you're helping illegals, so we're not giving any more money to you'" (Latino Catholic deacon, Greenville, July 2012). Again, regardless of their immigration status, Latinos are racialized as "illegal" and reduced to their ascribed undocumented status rather than welcomed as fellow Christians—yet another indication of just how intricately churches are connected to rather than removed from the racialized discourses and a regional political culture that prioritizes the rule of law on immigration matters (cf. Winders 2007). Clearly, even within individual churches, Christian hospitality often fails to unsettle the power disparity that inheres in the relationship between hosts and guests as institutional and faith-based practice diverge from theological ideals (cf. Olson 2008).

Given such potential backlash and difficulties, pastors were reluctant to address immigration politics in their sermons[5] despite the fact that these politics were impeding their ability to fully abide by Christian imperatives of hospitality. Several immigrant pastors reported that they quietly helped immigrants in their congregations; for example, by writing so-called character reference letters for visa overstayers from China, Cameroon, or India to help send them on a path toward legalization (or stay a deportation). One Presbyterian[6] pastor in Greenville suggested that guiding his church toward more openness toward immigrants was a longer process that he did not wish to jeopardize by preaching immigration politics from the pulpit. Instead, this pastor quietly connected individual immigrants in need of legal advice to lawyers within the congregation, or he pointed doctors and dentists interested in doing pro bono work to undocumented immigrants in need. This pastor justified his approach by referring to the Gospel of Matthew 6:3:

> And you know, there is a sense where that's biblical: "don't let your left hand know what your right hand is doing," don't get credit for it, just kind of go about it quietly. And if you want the credit for it, it's probably a bad idea. (White Presbyterian pastor, Greenville, July 2012)

While staying under the radar renders the hospitality his church extends to immigrants precarious as they remain at risk of discovery and eviction, this pastor's decision to keep his actions quiet to protect undocumented immigrants from immigration enforcement and conservative backlash by keeping them invisible (see Herbert 2009) is deeply political. This "quiet encroachment of the ordinary" (Bayat 2010, 56) often explicitly avoids publicity to fulfill needs. It is "not a politics of protest, but of redress, struggle for an immediate outcome through individual direct action" (59). Hence, our analysis reveals churches as complex political and highly politicized spaces of hospitality—even if their politics is one of invisibility that explicitly seeks to stay hidden. Such politics may simultaneously challenge and reproduce existing societal exclusions. These politics of hospitality become more evident when considering how church leaders variously interpret the relationship between secular and divine authority.

God's Word and "the Law of the Land"

In explaining their negotiations of secular and Christian authority, our interviewees presented competing interpretations of the Bible and of secular law. Referring to the Book of Revelation or the Gospel of Matthew, some pastors suggested that "the Kingdom of Christ in Heaven" knows no boundaries and that immigrants deserved welcome in their churches, with or without documents:

> At the end of the day, if we believe in a place like Heaven, and we believe that the only requirement is our faith, and we know that people who look very differently from us have faith regardless of their status, or legal documentation, or whatever, then it kind of helps people begin to think about, you know, we're kind of sort of supposed to start doing that *here*—not just wait for that. (White United Methodist pastor, Atlanta, August 2012)

This Methodist pastor's views echo Presbyterian theologian Daniel (2010, 41), whose interpretation of Romans 13 suggests that the "hierarchy of fidelity necessarily causes us to question laws even when set down by an emperor whose reign was established by God." Similarly referring to Romans 13, one Baptist pastor in Spartanburg stated,

> Peter . . . more or less says, "You tell me, who am I supposed to obey, God or the country's laws?" It hasn't come to that yet, for us, but it will. And when it does, we're going to obey God. We're not going to *not* help immigrants just because they're illegal. (White Baptist pastor, Spartanburg, March 2012)

Our research shows, however, that there are competing interpretations of a universal Christian ethic of hospitality even among pastors of the same denomination. One conservative Baptist pastor in Greenville interpreted the same biblical passages to emphasize the need to submit Christian compassion to the law of the land. His church has embarked on aggressive evangelization efforts among immigrants in surrounding neighborhoods. Well aware of the presence of undocumented immigrants in his Latino congregation, this pastor nonetheless insisted that

> Our philosophy is that the Bible clearly teaches that you are to be submitted to the authorities, being the governmental authorities, as long as they do not command you to violate a direct commandment of God. Okay, so the ultimate authority is God, but God says in Romans 13 that we are to submit to the powers that be, because the powers that be are ordained of God. . . . Peter says, "You are to submit to every ordinance of man for the Lord's sake." . . . God does not ask for anarchy from His people. (White Baptist pastor, Greenville, June 2012)

Similarly, we spoke to the missions coordinator of a Baptist church in Atlanta that prides itself as "loving the nations" (a reference to Revelations 7:9) and that has established congregations for immigrants from countries as diverse as China, India, and Haiti, in addition to a large Latino congregation. She described how the cultural diversity brought by immigrants had enriched the church and her personal life. Yet, when asked about Georgia's new immigration law, she said that church members "cannot be illegal." Asked why, she pointed at the Bible on her desk and said:

> It's a tough call, but I don't think we would ever condone that. I mean, because we can't. The Bible teaches that we are to respect our authorities, and so we couldn't condone anybody doing anything illegal! Whether it be their residency status or any other thing. (White Baptist staff, Atlanta, August 2012)

Interestingly, violating (secular) immigration law here turns into a (Christian) sin that, in turn, is used to police the boundaries of membership in the church. Neither one of these interviewees would report undocumented immigrants to law enforcement, but they made clear that undocumented immigrants' legal status limited their participation within the community of faith. As the pastor in the same Atlanta Baptist church put it, "We would not tell [undocumented immigrants] they could not come, but they would not be put into a leadership position." Clearly, the politics of hospitality is conditioned in this case and others on legal status rather than biblical imperatives of compassion to produce a "conditional hospitality," a "finite set of resources and calculable claims" (Honig 2009, 115). Rather than the limitless ideal of a Christian ethic of hospitality, immigrants are only welcome to an extent when the "politics of Jesus"—that is, compassion and unconditional welcome—becomes interwoven in complex ways with "the politics that sustains the empire or the nation-state" (Newman 2007, 17), producing spaces of encounter that are ambiguous at best.

Conclusion

Following Dikeç's (2002) invitation to interrogate critically spaces of hospitality, we have shown that Christian communities of faith in the U.S. South are spaces of precarious welcome for immigrants, especially for those with undocumented legal status. In doing so, we speak to a long tradition of scholarship in geography and beyond on the role of religious institutions in mediating immigrant integration into American society (cf. Young 1928; Gordon 1964; Ley 2008), on new spaces of immigrant belonging (e.g., Leitner 2012), as well as to critical geographies of religion interested in the entanglement of the spiritual and the secular in the spaces of everyday life (cf. Olson 2008; Wilford 2010; Olson, Hopkins, and Kong 2012; Valentine and Waite 2012).

Our research shows that across a variety of denominations, churches in the U.S. South are political spaces in which pastors and lay members weigh faith-based conceptions of welcome against law-and-order discourses and in which notions of universal membership confront racialized immigration politics (cf. Lassiter 2006; Winders 2007). Although individual churches and denominations might not exercise formal authority over the secular state, churches and their members do have a role in defining and redefining the boundaries, not only of the faith community but also of broader society, and in articulating the moral conduct of public life beyond church walls. Churches' extension of hospitality toward immigrants, in this sense, is an eminently political project that underscores the deep entanglement of secular and spiritual ideals, norms, and practices, as they circulate through multiple spaces and spheres of everyday life.

While geographers and other scholars have commented on the ambiguities of hospitality (Dikeç 2002; Derrida 2005; Honig 2009), our research has begun to tease out *how* a Christian ethics of hospitality collides with the politics of hospitality when secular and sacred understandings of compassion, law and order, and justice intersect in spaces of immigrant integration. As a result, the churches in our research emerge as sites of multilayered politics involving immigrants, congregants, and the wider public. Within these churches, public advocacy on behalf of immigrants is often abandoned in favor of a "quiet encroachment of the ordinary" (Bayat 2010, 56) that serves immigrants' immediate needs while pastors work to chip away at such views within congregations that deem infractions against immigration laws a sin. Accordingly, there is not a single way in which these Christian communities of faith engage in politics toward immigrant inclusion, nor is there a singular outcome of such engagements. A politics of invisibility allows the churches in our research to create somewhat safe spaces for immigrants in a context of racism, growing hostility, and legal action toward them that geographers have documented (Winders 2007; Herbert 2009; Coleman 2012). Yet by maintaining social and legal boundaries between congregants, these churches also reproduce existing exclusions and racialized discourses that circulate in the public realm. It is for these reasons that we argue that it is necessary to interrogate the limits and contestations of hospitality in communities of faith in the U.S. South. Such an analysis, we suggest, furthers understandings of how spaces of immigrant belonging and exclusion emerge in new immigrant destinations.

Acknowledgments

This research was supported by the National Science Foundation (NSF; BCS-1021907 and BCS-1021666). Any opinions, findings, and conclusions or recommendations expressed in this material are those of the authors and do not necessarily reflect the views of the NSF. We thank our research participants for sharing their insights, and Catherine Cottrell, Rebecca Lane, and Derek Ruez for their research assistance.

Notes

1. The foreign-born population constitutes 9.5 percent in Georgia, 8 percent in North Carolina, and 4.5 percent in South Carolina. The undocumented population is estimated at 4.39 percent in Georgia, 3.41 percent in North Carolina, and 1.19 percent in South Carolina.
2. *Secular*, as we use it here, refers to different realms of authority—separating church and state, rather than determining whether or not individual subjects are spiritual or religious (cf. Wilford 2010).
3. They were joined, in the spring of 2013, by numerous conservative evangelical Protestants previously refusing to advocate on behalf of undocumented immigrants.
4. Megachurches are defined as having a minimum of 2,000 members (Warf and Winsberg 2010).
5. Although no pastor mentioned this as a reason, churches stand to lose their tax-exempt status if they engage in explicitly political activities.
6. A member church of the conservative Presbyterian Church in America (PCA), which does not ordain women.

References

Ahn, I. 2010. Economy of "invisible debt" and ethics of "radical hospitality": Toward a paradigm change of

hospitality from "gift" to "forgiveness." *Journal of Religious Ethics* 38:243–67.
Arias, M. 1982. Centripetal mission, or evangelization by hospitality. *Missiology* 10:69–81.
Bayat, A. 2010. *Life as politics: How ordinary people change the Middle East*. Stanford, CA: Stanford University Press.
Cabell, M. G. 2007. *Mexican immigrant integration in the U.S. Southeast: Institutional approaches to immigrant integration in Owensboro, Kentucky*. Working Paper 153, Center for Iberian and Latin American Studies and Center for Comparative Immigration Studies, University of California–San Diego, San Diego, CA.
Catholic Bishops of Georgia. 2011. *Pastoral statement on immigration*. Atlanta, GA: Archdiocese of Atlanta.
Chauvin, S., and B. Garcés-Mascareñas. 2012. Beyond informal citizenship: The new moral economy of migrant illegality. *International Political Sociology* 6:241–59.
Chavez, L. R. 2008. *The Latino threat: Constructing immigrants, citizens, and the nation*. Stanford, CA: Stanford University Press.
Cloke, P., and J. Beaumont. 2013. Geographies of postsecular rapprochement in the city. *Progress in Human Geography* 37:27–51.
Coleman, M. 2012. The "local" migration state: The site-specific devolution of immigration enforcement in the U.S. South. *Law & Policy* 34:159–90.
Daniel, B. 2010. *Neighbor: Christan encounters with "illegal" immigration*. Louisville, KY: Westminster John Knox Press.
Derrida, J. 2000. *Of hospitality (De l'hospitalité): Anne Dufourmantelle invites Jacques Derrida to respond*. Stanford, CA: Stanford University Press.
———. 2005. The principle of hospitality. *Parallax* 11:6–9.
Dikeç, M. 2002. Pera peras poros: Longings for spaces of hospitality. *Theory, Culture & Society* 19:227–47.
Ecklund, E. H. 2005. "Us" and "them": The role of religion in mediating and challenging the "model minority" and other civic boundaries. *Ethnic and Racial Studies* 28:132–50.
Ehrkamp, P. 2006. "We Turks are no Germans": Assimilation discourses and the dialectical construction of identities in Germany. *Environment and Planning A* 38:1673–92.
Ehrkamp, P., and H. Leitner. 2006. Guest editorial: Rethinking immigration and citizenship: New spaces of migrant transnationalism and belonging. *Environment and Planning A* 38:1591–97.
Ehrkamp, P., and C. R. Nagel. 2012. Immigration, places of worship and the politics of citizenship in the US South. *Transactions of the Institute of British Geographers* 37:624–38.
Ellis, M. 2006. Unsettling immigrant geographies: US immigration and the politics of scale. *Tijdschrift voor Economische en Sociale Geografie* 97:49–58.
Emerson, M. O., and C. Smith. 2000. *Divided by faith: Evangelical religion and the problem of Race in America*. Oxford, UK: Oxford University Press.
Foley, M. W., and D. R. Hoge. 2007. *Religion and the new immigrants: How faith communities form our newest citizens*. Oxford, UK: Oxford University Press.
Gordon, M. M. 1964. *Assimilation in American life: The role of race, religion, and national origins*. New York: Oxford University Press.
Greer, B. A. 1993. Strategies for evangelism and growth in three denominations (1965–1990). In *Church and denominational growth*, ed. D. A. Roozen and C. K. Hadaway, 87–111. Nashville, TN: Abingdon Press.
Harvey, P. 2005. At ease in Zion, uneasy in Babylon: White evangelicals. In *Religion and public life in the South: In the evangelical mode*, ed. C. R. W. A. M. Silk, 63–78. Walnut Creek, CA: AltaMira Press.
Herbert, S. 2009. Contemporary geographies of exclusion II: Lessons from Iowa. *Progress in Human Geography* 33:825–832.
Hondangneu-Sotelo, P. 2007. *Religion and social justice for immigrants*. New Brunswick, NJ: Rutgers University Press.
Honig, B. 2009. *Emergency politics: Paradox, law, democracy*. Princeton, NJ: Princeton University Press.
Kniss, F., and P. D. Numrich. 2007. *Sacred assemblies and civic engagement: How religion matters for America's new immigrants*. New Brunswick, NJ: Rutgers University Press.
Kruse, K. 2005. *White flight: Atlanta and the making of modern conservativism*. Princeton, NJ: Princeton University Press.
Lassiter, M. 2006. *The silent majority: Suburban politics in the Sunbelt South*. Princeton, NJ: Princeton University Press.
Leitner, H. 2012. Spaces of encounters: Immigration, race, class, and the politics of belonging in small town America. *Annals of the Association of American Geographers* 102:828–46.
Ley, D. 2008. The immigrant church as an urban service hub. *Urban Studies* 45:2057–74.
Marrow, H. B. 2011. *New destination dreaming: Immigration, race, and legal status in the rural American South*. Stanford, CA: Stanford University Press.
Newman, E. 2007. *Untamed hospitality*. Waco, TX: Center for Christian Ethics at Baylor University. http://www.baylor.edu/content/services/document.php/53382.pdf (last accessed 30 April 2013).
Newport, F. 2012. Mississippi is most religious U.S. state. *Gallup Poll*. http://www.gallup.com/poll/153479/Mississippi-Religious-State.aspx?ref=more (last accessed 1 June 2012).
Nteta, T. M., and K. J. Wallsten. 2012. Preaching to the choir? Religious leaders and American opinion on immigration reform. *Social Science Quarterly* 93:891–910.
Olson, E. 2008. Common belief, contested meanings: Development and faith-based organisational culture. *Tijdschrift voor Economische en Sociale Geografie* 99:393–405.
Olson, E., P. Hopkins, and L. Kong. 2012. Introduction: Religion and place: Landscape, politics and piety. In *Religion and place: Landscape, politics and piety*, ed. P. Hopkins, L. Kong, and E. Olson, 8–37. Amsterdam: Springer.
Ownby, T. 2005. Evangelical but differentiated: Religion by the numbers. In *Religion and public life in the South: In the evangelical mode*, ed. C. R. Wilson and M. Silk, 31–62. Walnut Creek, CA: AltaMira Press.
Pohl, C. 2007. *Making a place for hospitality*. Waco, TX: Center for Christian Ethics at Baylor University. http://www.baylor.edu/content/services/document.php/53383.pdf (last accessed 10 May 2013).
Ramakrishnan, K., K. Esterling, and M. Neblo. 2010. *Illegality, national origin cues, and public opinion on immigration*. Paper presented at the annual meeting of the Midwest Political Science Association, Chicago, IL. http://polisci.osu.edu/sites/polisci.osu.edu/files/_illegality,%20national

%20Origin%20cues,%20and%20public%20Opinion%20On%20immigration_.pdf (last accessed 12 July 2013).
Singer, A., S. W. Hardwick, and C. Brettell. 2008. *Twenty-first century suburban gateways: Immigrant incorporation in suburban America.* Washington, DC: Brookings Institution Press.
Singer, A., J. H. Wilson, and B. DeRenzis. 2009. *Immigrants, politics, and local response in suburban Washington.* Washington, DC: Brookings Institution.
Southern Baptist Convention. 1995. Resolution on racial reconciliation on the 150th anniversary of the Southern Baptist Convention. http://www.sbc.net/resolutions/amresolution.asp?id=899 (last accessed 10 May 2013).
Valentine, G., and L. Waite. 2012. Negotiating difference through everyday encounters: The case of sexual orientation and religion and belief. *Antipode* 44:474–92.
Varsanyi, M. W. 2008. Rescaling the "alien," rescaling personhood: Neoliberalism, immigration, and the state. *Annals of the Association of American Geographers* 98:877–96.
Walker, K. E., and H. Leitner. 2011. The variegated landscape of local immigration policies in the US. *Urban Geography* 32:156–78.
Warf, B., and M. Winsberg. 2010. Geographies of megachurches in the United States. *Journal of Cultural Geography* 27:33–51.
Wells, T. 2004. *Growing religious diversity in South Carolina: Implications for the Palmetto State.* Cambridge, MA: Pluralism Project at Harvard University. http://www.pluralism.org/affiliates/student/wells/SC_Religious_Diversity.pdf (last accessed 10 June 2012).
Wilford, J. 2010. Sacred archipelagos: Geographies of secularization. *Progress in Human Geography* 34:328–48.
Winders, J. 2006. "New Americans" in a "New-South" city? Immigrant and refugee politics in the Music City. *Social & Cultural Geography* 7:421–35.
———. 2007. Bringing back the (b)order: Post 9/11 politics of immigration, borders and belonging in the contemporary US South. *Antipode* 39:920–42.
———. 2012. Seeing immigrants: Institutional visibility and immigrant incorporation in new immigrant destinations. *Annals of the American Academy of Political and Social Science* 641:58–78.
Young, P. V. 1928. The reorganization of Jewish family life in America: A natural history of the social forces governing the assimilation of the Jewish immigrant. *Social Forces* 7:238–44.

Enclaves of Rights: Workplace Enforcement, Union Contracts, and the Uneven Regulatory Geography of Immigration Policy

Virginia Parks

School of Social Service Administration, University of Chicago

Recent geographic research on U.S. immigration policy highlights the devolution of policy formulation and implementation to local state actors. This study extends this research by analyzing how labor unions shape the implementation of state immigration policy and innovate institutional practices that affect regulatory spaces for immigrants at the local level. Using a case study of the hotel union in Chicago and Los Angeles, this article examines the origin, content, and implementation of immigration provisions recently negotiated in the union's contracts. These contract provisions mediate the implementation of state immigration policies by specifying rules that govern employer actions in response to immigration enforcement activities by the Department of Homeland Security and other federal agencies, including admittance to the workplace, inspection of I-9 forms, and Social Security no-match letters. The contracts also establish nondiscrimination protections for immigrant workers, such as guaranteed leave to attend immigration proceedings. The legal codification of these protections and the robust practices of union enforcement yield "enclaves of rights" at the local level, further contributing to the highly uneven space of security for immigrant residents in the United States The article concludes by examining the possibility that unions—given their influence in local labor markets, their federated (or national) structures, and their role in the broader moral economy—could extend these rights beyond the confines of the enclave.

晚近美国的移民政策研究，强调政策制定与执行的权力，下放至地方政府行动者。本研究透过分析工会如何形塑州政府的移民政策施为，并革新影响地方层级移民管制空间的制度实践，以此拓展移民政策之研究。本文运用芝加哥与洛杉矶的一个酒店工会案例研究，检视晚近在工会合约中协商的移民供给之起源、内容及执行。这些合约供给透过详细载明因应国土安全部及其他联邦机构的移民执行活动之聘雇管理规范，包含进入工作场所的许可、I-9表格的查阅，以及社会安全资料不合的通知，以此中介州政府移民政策的实施。这些合约同时为移工确立了非歧视性的保护，诸如保障移工休假以进行移民诉讼。这些保障的法律编撰，以及强而有力的工会实践，在地方层级创造出"权益的飞地"，并进一步导致美国境内移居者高度不均质的安全空间。结论中，本文检视拥有对在地劳动市场的影响力、具有联邦 (或国家) 层级的结构，并在更广泛的道德经济中有所角色的工会，能够将这些权益延伸至飞地范围之外的可能性。

La investigación geográfica reciente sobre las políticas de inmigración de los EE.UU. destaca el traslado de la formulación e implementación de políticas a los actores estatales locales. Este estudio amplía esta investigación analizando la manera como los sindicatos de trabajadores configuran la implementación de políticas de inmigración estatal e innovan las prácticas institucionales que afectan los espacios regulatorios para inmigrantes a nivel local. Utilizando un estudio de caso del sindicato hotelero en Chicago y Los Ángeles, este artículo examina el origen, contenido e implementación de las medidas sobre inmigración recientemente negociadas en los contratos del sindicato. Estas provisiones contractuales median en la implementación de las políticas estatales de inmigración, especificando reglas que gobiernan las acciones del empleador, en respuesta a las actividades de control de la inmigración por el Departamento de Seguridad Nacional y otras agencias federales, incluyendo la admisión a los lugares de trabajo, la inspección de las formas I-9 y las cartas de la Seguridad Social para notificación de datos discordantes. Los contratos establecen también protecciones contra la discriminación para los trabajadores inmigrantes, tales como permiso garantizado de ausencia para cumplir con procedimientos inmigratorios. La codificación legal de estas protecciones y las sólidas prácticas de aplicación por los sindicatos producen "enclaves de derechos" a nivel local, contribuyendo así al espacio de seguridad altamente desigual para inmigrantes residentes de los Estados Unidos. El artículo concluye con un examen de la posibilidad de que los sindicatos—dadas su influencia en los mercados locales, sus estructuras federadas (o nacionales) y su roles en la más amplia economía moral—puedan llegar a extender estos derechos más allá de los confines del enclave.

The border and the workplace in the United States are locales subject to geographically targeted federal immigration enforcement activity, including surveillance, monitoring, and apprehension activities. Geographers have demonstrated the border's territorial production of illegality and the policymaking capacities of border control operations (Mountz 2010; Nevins 2010). Less attention has been paid to the politics, practice, and policy implications of worksite immigration enforcement. The ramifications of worksite immigration enforcement are multiple and complex. The border and the workplace are both embedded within a migration system shaped by global capitalism, but the worksite uniquely couples immigration enforcement with economic production. As wage laborers, U.S. immigrants occupy their workplaces as both subjects of immigration policy and protected workers. The complex interplay between immigrants' legal status and their rights under U.S. labor and employment laws constitutes the workplace as a territorial site at which these rights and restrictions intersect and often collide. Worksite immigration enforcement lays these contradictions bare.

In assessing contemporary immigration policy, Bosniak (2000) emphasized the rise in policies that make legal status more socially significant and the shift of border control functions to the territorial interior of the nation state. Recent geographic research makes similar claims, particularly studies highlighting the devolution of immigration policy formulation and implementation to local state actors (Varsanyi 2008, 2011; Walker and Leitner 2011; Coleman 2012). This study similarly attends to the local as a site of immigration policy innovation and contestation. I identify the workplace as a highly localized sphere of engagement that both generates its own practices and realizes the local manifestation of multiple scalar phenomena including global capital and migration flows, national state policy regimes, regional enforcement practices, corporate management structures, and federated labor organizations. As both a key node in global capitalism and a primary site of immigration enforcement in the United States, the worksite provides a compelling strategic research site through which to examine the complex and contested process by which immigration policy regulates the labor market and, reciprocally, the ways in which labor market practices mediate the implementation of immigration policy.

I examine the worksite as a geographic arena of engagement where the actions of state and nonstate actors intersect and affect immigration policy through an analysis of how unions, as nonstate labor market institutions, shape the implementation of federal immigration policy at the street level and innovate institutional practices that establish alternative regulatory spaces for immigrants at the local level. Drawing on a case study of a large service-sector union in Chicago and its affiliate in Los Angeles, this study analyzes how union contracts legally codify rights and protections for immigrant workers at the workplace, mediating federal immigration enforcement and the local effects of federal immigration policy.

This article focuses on the immigration provisions negotiated through collective bargaining and the shop-floor implementation and enforcement of these provisions. These provisions establish nondiscrimination protections and privileges specifically for immigrant workers and shape employer actions with regard to immigration enforcement procedures and activities by the Department of Homeland Security (DHS) and other federal agencies. I argue that the legal codification of protections and their vigilant enforcement by the union yield "enclaves of rights" at the local level, further contributing to the highly uneven space of security for immigrant residents in the United States. I then explore the possibility that unions could extend these rights beyond the confines of such enclaves.

The Workplace as Immigration Enforcement Site

The legal foundation for workplace enforcement rests with the 1986 Immigration Reform and Control Act (IRCA), which makes it unlawful for an employer to knowingly employ an undocumented immigrant. Through IRCA, workplaces became sites where illegal activity—the employment of undocumented immigrants—is produced and carried out. Enforcement previously contained at the territorial boundaries of the nation-state was brought to its interior (Bosniak 2000), extending the border's work of exclusion, differentiation, and criminalization (Mountz 2010; Nevins 2010).

Unlike the border, however, interior enforcement sites are superimposed onto preexisting social realms, filled with rights, rules, and norms separate and distinct from immigration law. Workers, regardless of documentation status, enjoy the rights and protections afforded under federal labor and employment law, such as minimum wage, overtime, occupational health and safety, and the right to organize (Gleeson 2012). Workplace immigration enforcement demonstrates the clash of illegality and legal rights for the same individuals—undocumented workers. The lines between illegality-as-outsiders and rights-as-insiders are thus blurred as the worksite simultaneously confers inclusion and engenders exclusion.

These conflicts and contradictions have become increasingly acute over the past twelve years, following differences between the DHS and the Department of Labor (DOL) in immigration and labor enforcement activities. The 2002 Supreme Court decision in *Hoffman Plastic Compounds, Inc. v. National Labor Relations Board*, in which an undocumented worker was denied back pay and reinstatement following unlawful retaliation by an employer, set the contradictory tone for this period.[1] Immediately following the *Hoffman Plastic* decision, federal labor and employment agencies, such as the DOL and the Equal Employment Opportunity Commission, reaffirmed their commitment to enforcing labor and employment law regardless of documentation status. Simultaneously, the Bush administration began ramping up worksite enforcement through the DHS.

Under the Bush administration, enforcement meant large-scale raids and arrests. In 2006, workplace arrests of undocumented immigrants by federal agents jumped by 44 percent over the prior year, from 1,116 to 3,667—the single greatest spike in the ten-year period between 2002 and 2011. The number of administrative arrests increased again in 2007 to 4,077 and peaked in 2008 to a record high of 5,185. The U.S. Immigration and Customs Enforcement (ICE) agency made 12,928 arrests in workplace raids between 2006 and 2008, nearly five times more than the 2,731 arrests made between 2002 and 2005 (U.S. ICE 2007, 2008). Although these raids accounted for only a small percentage of all deportations, they were high-profile operations that reverberated locally and nationally within immigrant communities. Workplace raids communicate to immigrants that they can be found easily and that every workday poses a deportation risk.

Under Bush, ICE targeted well-known national employers, often at multiple sites. The highest profile raids of this kind occurred at meat-processing plants across the country. On 12 December 2006, ICE agents stormed meat-processing facilities owned by Swift & Co. in six states. During these raids, 1,297 undocumented immigrants were arrested (U.S. ICE 2007). In January 2007, ICE agents raided a huge Smithfield pork-processing plant in Tar Heel, North Carolina, that employed 5,200 workers. The plant was in the midst of a unionization drive, prompting labor officials to complain bitterly that the company had collaborated with federal authorities in an effort to discourage union organizing (Preston 2007). The single largest workplace raid under the Bush administration occurred on 12 May 2008, when federal authorities arrested 389 undocumented workers at a kosher meat-processing plant in Postsville, Iowa (Hsu 2008).

Critics of the workplace raids pointed to the disproportionate prosecution of workers over employers (Greenhouse 2006); the traumatic impact of raids on individuals, families, and nearby communities (Capps et al. 2007); and their unnecessarily high cost—upward of $10 million for larger scale raids (Bennett 2011). After a series of raids in Los Angeles County in 2008, including a highly publicized raid of a Van Nuys manufacturing factory where ICE made 130 arrests, Los Angeles Mayor Antonio Villaraigosa sent a letter to Michael Chertoff, secretary of the DHS, warning that workplace raids of "non-exploitative" businesses could have "severe and lasting effects" on the Los Angeles economy (Gorman 2008).

The Bush administration agreed to stop the use of immigration enforcement under the guise of health and safety inspections in early 2006 (Greenhouse 2006). Despite this concession, criticism from unions, worker advocates, and even the DOL intensified. Labor leaders pointed to raids carried out in the midst of unionization campaigns, such as at the North Carolina Smithfield plant, or on the heels of litigation brought against companies for wage and hour violations. A union-sponsored advocacy organization, American Rights at Work, asserts, "The single-minded focus on immigration enforcement without regard to violations of workplace laws has enabled employers with rampant labor and employment violations to profit by employing workers who are terrified to complain about substandard wages, unsafe conditions, and lack of benefits, or to demand their right to bargain collectively" (Smith and Ortega 2009, 5).

The Obama administration has used more subtle "silent raids," inspecting companies' I-9 records—government forms used by employers to verify an employee's identity and eligibility for employment in the United States (Preston 2010). Although the DOL and the DHS signed a memorandum of understanding in 2011 on worksite enforcement and the roles of each agency—putting labor enforcement on equal footing with immigration enforcement—the often contradictory and opposing goals of immigration enforcement and labor protection continue (U.S. DHS and U.S. DOL 2011; These changes came long after the initial inclusion of immigration provisions in the union contracts examined in this study.) As interviews with union staff indicate, protection of immigrant workers on the shop floor and the tangible realization of workers' rights endures as a daily process of negotiation and contestation.

Unions as Local Institutional Actors

When the implementation and enforcement of immigration law moves from the border to the interior it includes a more expansive and diverse group of actors (Ridgley 2008; Mountz 2010; Coleman 2012). In workplace immigration enforcement, unions occupy a unique position as a social and political mediator of labor market processes. Because workplace immigration enforcement unfolds within the realm of work and intersects with daily labor market practices over which unions already have influence, unions have frontline access to the implementation process. Further, unions' role as a legally recognized representative agent of workers at the workplace provides them with a compelling claim to the "local bargaining process" of policy implementation (Elmore 1979, 611).

Unions actively contest, shape, and remake practices within, and expectations about, the workplace, rendering visible the social and spatial embeddedness of labor markets (Hanson and Pratt 1992; Peck 1996). Western and Rosenfeld (2011, 517) described three ways in which unions, "pillars of the moral economy," influence labor market norms: (1) culturally, through communicative power; (2) politically, by influencing social and economic policy; and (3) institutionally, by establishing rules that govern labor market processes. Most research on labor's engagement with immigration issues focuses on labor's cultural, political, and organizing activities (Milkman 2000; Varsanyi 2005).

The union examined in this study, Unite Here, engages the issue of immigrant rights in each of these ways. Culturally, it partners with immigrant rights organizations and immigrant communities in speaking out against immigrant abuses, not just on the job. Politically, it lobbies in support of federal immigration reform, engages in electoral work, and participates in local campaigns in coalition with the immigrant community. Institutionally, it governs labor market functions that directly affect immigrants through its union contracts, enforces these contracts, and empowers its members to defend their broader federal and narrower contract rights. But unlike the institutional rules most often ascribed to unions—that more explicitly address relations between employers and workers, such as hiring, promotion, and pay—the Unite Here contracts address the implementation of federal immigration policy. In so doing, the union extends its influence within the domain of immigrant rights institutionally as well as politically.

Unite Here's Immigration Reform Agenda

In 2006, Unite Here Local 1 in Chicago and Local 11 in Los Angeles won unprecedented protections for their immigrant workforces in their hotel contracts, providing immigrant workers with a transparent process of response, adjudication, and resolution in matters connected to their documentation status involving both the federal government and the employer. The contracts added to the broad and expansive nondiscrimination language of previous contracts, as well as to the equity provisions these locals had secured for their diverse workforces, such as hiring and retention commitments aimed at African-American workers (Parks and Warren 2012).

The 2006 contract provisions emerged from initiatives driven by the international body of Unite Here under the leadership of John Wilhelm. Unite Here primarily represents workers in lower wage service sector industries, such as hotels, food service, and industrial laundries. As these industries became key points of labor market entry for an expanding immigrant population during the 1990s, Unite Here was confronted with a new set of issues and concerns that arose for its growing immigrant membership. After election as president of the union's national organization (then known as the Hotel Employees and Restaurant Employees [HERE]) in 1998, Wilhelm began formulating and implementing an organizational and political program around immigration reform that included internal educational campaigns, community outreach, and political advocacy and mobilization. HERE was among the progressive labor unions that pushed the national American Federation of Labor and Congress of Industrial Organizations (AFL-CIO) to reverse its anti-immigrant position in 1995, with the election of John Sweeney as president.

Austin Lynch, an organizer at Local 11, identified Unite Here's broader immigrant rights agenda as central to understanding the impetus behind the immigration provisions in the union's contracts and the union's commitment to realizing broader immigration reform:

> Part of the picture is the union's program outside the workplace, the Immigrant Freedom Ride and the May Day rallies. We have a history of organizing around immigration. The Immigrant Freedom Ride was a moment when a number of our members took huge risks. We also had years and years of political campaigns where we did talk about [the fact] that people who are not citizens could impact politics because you can mobilize voters. And I think that has made it very clear to our members that we stand for immigrant rights but also has made it known to the employers that we are very serious about it. (Interview, 26 November 2012).

The push from above, from the international, and from below, from members, spurred Unite Here locals toward fighting for, and winning, contract language that made a concrete difference for immigrant workers on the shop floor while pushing for immigration reform in the political arena.

Bargaining Immigration Enforcement

Union contracts generally reflect two political processes: negotiation between the union and an employer and negotiation and deliberation between the union and its members. Codifying demands through a legally binding collective bargaining agreement makes the employer accountable to the union and the union accountable to its members. The Unite Here immigration provisions add a third party: the state. The contract provisions influence and mediate the federal government's immigration enforcement activities. These provisions do not preempt federal law but, rather, exploit the interstitial gaps in the implementation of federal immigration law by governing the employer's actions in three immigration-related areas: workplace immigration enforcement, reverification of status, and Social Security no-match letters (written notices issued by the Social Security Administration to an employer indicating that the name or Social Security number reported for an employee does not "match" a name or number in the Administration's records). Furthermore, they provide immigrant workers with protections beyond the guarantees of federal law: stronger nondiscrimination commitments, seniority safeguards, right to converse in a language other than English, unpaid leave for immigration-related proceedings, and a paid holiday to attend one's own citizenship ceremony.

Regarding workplace immigration enforcement, Unite Here contracts stipulate what an employer will do in the event of a federal enforcement operation. First, the contracts require that the employer do no more than is legally required. For example, the contract between the union and major hotels in Los Angeles states that the employer must "refuse admittance of any agents of the DHS who do not possess a search and/or arrest warrant, administrative warrant, subpoena or other legal process signed by a federal law or magistrate" (Section 15.2.b).[2] Second, the contract instructs the employer to "permit inspection of I 9 forms by DHS or DOL only after a minimum of three days written notice" and only when a DHS warrant or subpoena "specifically names employees or requires the production of I 9 forms" (Section 15.2.c). The employer is further restricted from revealing any additional information to the DHS, such as "the names, addresses or immigration status of any employees in the absence of a valid DHS administrative subpoena, or a search warrant or subpoenas signed by a federal judge)" (Section 15.2.c).

Notably, the employer must contact the union when it receives notification from a government agency regarding an immigration-related case:

> Unless objected to by the affected employee, notify a representative of the union as soon as practical if the Employer receives a no-match letter from the Social Security Administration, or is contacted by the Department of Homeland Security (DHS) (formerly the INS) related to the immigration status of an employee covered by this Agreement or if a search and/or arrest warrant, administrative warrant, subpoena, or other request for documents is presented in order that the Union can take steps to protect the rights of its members. The Union agrees that it shall keep confidential any information it obtains pursuant to this provision and that it will use any such information solely to represent and/or assist the affected employee(s) in regards to the DHS matter. (Section 15.2.a)

This provision guarantees a worker the option of union representation when a federal immigration inquiry is initiated at the workplace. The contract leverages the employer's structural position in the process of immigration enforcement by requiring that the employer share information to which it alone is privy as the purported "target" of enforcement. Requiring that the employer contact the union as soon as possible and wait at least three days before providing access to I-9 forms, the contract provides immigrant workers with both information and time. This counters an asymmetrical power relation that characterizes most enforcement activities of the federal government.

The Unite Here contracts guard against employers requiring reverification of status, a common intimidation tactic. Contract provisions address reverification in three ways: (1) no employee employed continuously on or before 6 November 1986 (the enactment of IRCA) will be required to reverify, (2) the employer "shall not require or demand proof of immigration status" beyond the minimum required by law, and (3) in the event of the sale of the business, requiring the transfer of I-9 forms to the new employer, who "shall maintain said forms" (Section 15.3.d).

Finally, most of the major hotel contracts specify that the employer will not use E-Verify, the federal government's online I-9 verification system that compares data from the DHS to Social Security Administration records. Employer participation in E-Verify is voluntary, except in six states (Feere 2012). The federally mandated I-9 process requires employers to obtain and record (using the I-9 form) information from employees establishing their identity and employment eligibility, primarily a Social Security or alien registration number. The I-9 form instructs employers to determine whether an identification document "reasonably appears to be genuine." Employers are not required to verify further the document's validity.

Austin Lynch described the driving logic behind these and other contract provisions:

> Most of it is, we always tell employers our position as the union is that you're not an enforcement agent. You should do what the law requires you to do and no more. For instance, we don't want you to use E-Verify unless you have to and currently you don't have to. We want you to transfer I-9s to the new employer, which they are very uncomfortable with. They say, "Well, we'll get sued and this and that for accepting other people's paperwork." Which is not true. We've had numerous fights. Ultimately, even without that language, we've prevailed in most circumstances in making employers pass all their I-9 paperwork to a new employer. (Interview, 26 November 2012)

The contract keeps employers from taking an activist or even vigilante—"you're not an enforcement agent"—position with regard to immigration law.

In the cases of E-Verify and Social Security no-match letters, the federal government tries to engage employers in more stringent enforcement of identity and documentation verification. From the union's perspective, this opened up new avenues of worker intimidation to the employer. Austin Lynch describes the union's concern with no-match letters:

> We decided to deal with [no-match letters because of cases] where the company would get a no-match letter, and they're not required to do anything, but they would. They would call people [workers] down, intimidate them, get them to admit they didn't have documentation. So we started putting in language, "No, you have to tell us first. You have to call us in. Promise not to take any action." (Interview, 26 November 2012)

These contract provisions establish legal safeguards that prohibit an employer from using federal immigration inquiries as cause for firing, intimidation, or retaliation. The contracts' no-match language specifies that "the Employer agrees that it will not require employees listed on the notice to bring in a copy of their Social Security card for the Employer's review, complete a new I 9 form, or provide new or additional proof of work authorization or immigration status, solely as a result of the receipt of a no-match letter" (Section 15.4.b). Lastly, the language prohibits an employer from contacting the Social Security Administration or any other governmental agency solely as a result of receiving a no-match letter. These provisions mitigate against employer retaliation and attempt to neutralize the employer as an enforcement partner of the state.

Given worker concerns about no-match letters, the union recognized the need to protect workers who made adjustments to their paperwork. Austin Lynch and Susan Minato, executive vice president of Local 11, described the difficulties that immigrant workers face when trying to correct paperwork:

> AL: If somebody fixes their papers, they shouldn't be fired for coming to their employer and saying, "You know, my name isn't Bill Smith, it's Joe Smith. I got a new Social Security number." In a lot of hotels today, you can be fired for that. The employer says, "Oh, you lied on your application." So we put in provisions saying you cannot be terminated for that.
>
> SM: It's huge.
>
> AL: It is huge because otherwise you have to stay in the shadows even though you've fixed your papers. (Interview, November 2012)

To address this situation, the Unite Here contracts guarantee that immigrant workers who are fired because they fail to provide adequate proof of work authorization will be reinstated immediately with seniority after providing adequate proof. Workers are given twelve months to exercise this guarantee. If additional time to prove his or her work authorization is needed, the contract stipulates "the Employer will rehire the employee into the next available opening in the employee's former classification, as a new hire without seniority, upon the employee providing proper work authorization within a maximum of twelve additional months" (Section 15.5.b).

Finally, the Unite Here contracts guarantee accommodations for the special conditions experienced by immigrant workers. For instance, the workers can speak a language other than English "when speaking amongst themselves" and "in a manner that is respectful of guests and other employees" (Section 15.6.a) and will be provided an interpreter during disciplinary interviews if necessary. Workers are granted five unpaid days of leave to attend immigration proceedings related to their own immigration matters and receive one paid citizenship holiday to attend their own citizenship swearing-in ceremony.

Unite Here engages its workers in "know-your-rights" trainings that help them become shop-floor enforcers of their contract rights, as well as their federal labor and employment rights. Karen Kent, executive vice president of Chicago's Local 1, explained:

> Workers have a tendency to huddle. They share information. They can put a place on lock-down even before the union knows anything. They can tell their co-workers to keep their mouths shut [to protect them] until the union gets involved. They know their rights and what the contract says. (Interview, 17 May 2011)

Kent described many occasions when workers were the first to invoke the union contract in immigration-related worksite incidents in Chicago.

Unite Here's contract provisions are a tactical innovation that exploits the gaps in the implementation process of federal immigration law, neutralizing the enforcement role of the employer to the greatest extent allowed under the law. Although Unite Here contracts also shape employer actions that are not explicitly addressed in immigration law, such as requiring employers to transfer their I-9 forms to a new employer, union officials described the provisions as "truly helpful to employers." Susan Minato of Local 11 said, "We gave them [employers] a play-by-play book that was helpful because many employers are scared" (Interview, 26 November 2012). While reducing the employer's uncertainty, the union engaged the employer as a local coalition partner in the bargaining, literally and figuratively, of federal policy implementation.

There are, however, limits to the union's ability to direct implementation, especially in relation to the actions of the federal government once an enforcement operation begins. Minato described the union's limitations as follows:

> Once ICE comes in, it's very difficult for us because then they're dictating more. So we have to try to get things worked through as much as we can in a positive way before that happens. They don't just come in willy-nilly. That's not normal anymore. Under Bush it was, but not now. (Interview, 26 November 2012)

Her statement indicates the limits of on-the-ground implementation. At some point, larger political projects matter.

Enclaves of Rights

The legal codification of immigration provisions into Unite Here contracts and the protections these provisions materialize for immigrant workers yield "enclaves of rights" at the local level in labor markets such as Los Angeles and Chicago. These rights obtain *in situ*, at the worksite, and are geographically and temporally bound. Immigrants are protected by the union contract when and while they are at work. These rights and protections stop when the immigrant leaves work. Thus, the union contract creates a delimited space in which certain rights not available elsewhere inhere and are actualized. The paradox is that these enclaves of rights contribute to the highly uneven landscape of security for immigrants in the United States.

Because of their influence in local labor markets, their federated (or national) structure (Herod 1991), and role in the broader moral economy, however, unions are positioned to multiply these enclaves as well as extend these rights beyond the confines of the enclave. Because unions such as Unite Here represent workers at multiple worksites, they can create new enclaves of rights through other contracts. Austin Lynch described this process in Los Angeles:

> The best and strongest language goes into the big hotel contract, then kind of bleeds out to the rest of the union, for instance food service contracts, where sometimes we're not able to get the whole version so they may have lesser versions [of the language]. (Interview, 26 November 2012)

As Lynch's quote illustrates, the union's ability to extend these rights depends on its economic power within certain sectors. When robust, that power can facilitate adoption of norms even without a union contract. Lynch described how a new owner of a hotel began using E-Verify without telling the union (its contract with the old company had expired):

> We told the company you can't [use E-Verify] without bargaining with us. We told them (a) your contract has expired and you're going to have to negotiate with us, and (b) we've got E-Verify prohibitions everywhere else [in other contracts]. We had the credibility to say, we're going to get it with you, too. So ultimately they backed off. (Interview, 26 November 2012)

This is how unions can enforce norms, in the absence of a contract, through broad institutional influence.

If its federated structure in the United States limits the power of labor, it also facilitates the heuristics and transfer of tactical innovations. Unite Here locals independently won inclusion of immigration provisions in each of their collective bargaining agreements while coordinating their contract campaigns in message and timing, a strategy facilitated by the International. The hotel contracts in Chicago and Los Angeles—where the immigration contract language is nearly identical—were won in the same year because the expiration of contracts across major U.S. markets had been coordinated in prior rounds of collective bargaining. Finally, the communicative and political power of unions can begin to shift the contours of the moral economy with regard to the treatment, processing, and criminalization of immigrant workers. The contract victories in Chicago and Los Angeles establish localized beachheads from which to expand a broader immigrant rights platform, one designed and mobilized in partnership with immigrant workers.

Conclusion

The clustering of immigrant workers within labor market enclaves, both by type and place of work, has been a pronounced feature of the urban economic landscape (Ellis, Wright, and Parks 2007). Often cast as sites of exploitation, these labor market enclaves can be sites of collective action, leveraging the very networks that connect immigrants to enclave employment for purposes of resistance. For undocumented immigrant workers, the worksite produces both illegality and legal protections—the former under immigration law, the latter under labor and employment law. Unions, as nonstate actors and legal representatives of immigrant workers at the workplace, negotiate the uncertain and often conflictual intersection of these two legal domains. In the case examined in this study, union contracts defend extant rights and innovate new protections for immigrant workers. These protections inhere in place because worker rights (conferred by federal law) and rights attained through social membership (union representation) are triggered, actualized, and defended at the workplace.

These enclaves of rights add a further degree of granularity to geographic research on the local scaling of immigration policy. Research reveals the ways in which other intermediate scales between the border and the workplace, such as state and municipal policy and enforcement, mediate immigrant rights and protections on the ground (Wells 2004). Gleeson (2012), for example, has examined how municipal policy can affect the employment conditions of undocumented immigrant workers. Yet these scales exert less direct influence on the workplace given its unique status relative to federal labor and employment law. Rendering visible the uneven regulatory geography of immigration policy will require continued multiscalar analyses to identify territorially based practices that might generate similar enclaves of rights, while explicating the mechanisms through which these enclaves are produced, defended, and possibly expanded.

Notes

1. Gleeson (2012) identifies *Hoffman Plastic* as indicative of the further tension between the *rights* of undocumented workers and the *remedies* they receive in practice.
2. Collective bargaining agreement between Unite Here Local 1 and signatory Chicago hotel employers, August 2006. Unless otherwise noted, all language is the same in the collective bargaining agreement between Unite Here Local 11 and signatory Los Angeles hotel employers, August 2006. Both contracts are on file with the author.

References

Bennett, B. 2011. Republicans want a return to workplace immigration raids. *Los Angeles Times* 27 January. http://articles.latimes.com/2011/jan/27/nation/la-na-immigration-raids-20110127 (last accessed 8 July 2013).

Bosniak, L. 2000. Universal citizenship and the problem of alienage. *Northwestern University Law Review* 94: 963–82.

Capps, R., R. M. Castañeda, A. Chaudry, and R. Santos. 2007. *Paying the price: The impact of immigrant raids on America's children.* Washington, DC: The Urban Institute.

Coleman, M. 2012. The "local" migration state: The site-specific devolution of immigration enforcement in the U.S. South. *Law & Policy* 34:159–90.

Ellis, M., R. Wright, and V. Parks. 2007. Geography and the immigrant division of labor. *Economic Geography* 83:255–81.

Elmore, R. F. (1979). Backward mapping: Implementation research and policy decisions. *Political Science Quarterly* 94:601–16.

Feere, J. 2012. *An overview of the E-Verify policies at the state level.* Washington, DC: Center for Immigration Studies. http://cis.org/e-verify-at-the-state-level (last accessed 8 July 2013).

Gleeson, S. 2012. *Conflicting commitments: The politics of enforcing worker rights in San Jose and Houston.* Ithaca, NY: Cornell University Press.

Gorman, A. 2008. Mayor chides ICE for workplace raids. *Los Angeles Times* 10 April. http://articles.latimes.com/2008/apr/10/local/me-raids10 (last accessed 8 July 2013).

Greenhouse, S. 2006. Going after migrants, but not employers. *New York Times* 16 April. http://www.nytimes.com/2006/04/16/weekinreview/16greenhouse.html?pagewanted=print (last accessed 8 July 2013).

Hanson, S., and G. Pratt. 1992. Dynamic dependencies: A geographic investigation of local labor markets. *Economic Geography* 68:373–405.

Herod, A. 1991. The production of scale in United States labour relations. *Area* 23:82–88.

Hsu, S. S. 2008. Immigration raid jars small town. *Washington Post* 18 May. http://www.washingtonpost.com/wp-dyn/content/article/2008/05/17/AR2008051702474.html (last accessed 8 July 2013).

Milkman, R. 2000. *Organizing immigrants: The challenge for unions in contemporary California workers and the future of the U.S. labor movement.* Ithaca, NY: ILR/Cornell University Press.

Mountz, A. 2010. *Seeking asylum: Human smuggling and bureaucracy at the border.* Minneapolis: University of Minnesota Press.

Nevins, J. 2010. *Operation Gatekeeper and beyond: The war on illegals and the remaking of the U.S.–Mexico boundary.* 2nd ed. London and New York: Routledge.

Parks, V., and D. T. Warren. 2012. Contesting the racial division of labor from below: Representation and union organizing among African American and immigrant workers. *Du Bois Review* 9:395–417.

Peck, J. 1996. *Work-place: The social regulation of labor markets.* New York: Guilford.

Preston, J. 2007. Immigration raid draws protest from labor officials. *New York Times* 26 January. http://www.nytimes.com/2007/01/26/us/26immig.html (last accessed 8 July 2013).

———. 2010. Illegal workers swept from jobs in "silent raids." *New York Times* 9 July. http://www.nytimes.com/2010/07/10/us/10enforce.html?pagewanted=all&_r=0 (last accessed 8 July 2013).

Ridgley, J. 2008. Cities of refuge: Immigration enforcement, police, and the insurgent genealogies of citizenship in U.S. sanctuary cities. *Urban Geography* 29: 53–77.

Smith, R., and J. Ortega. 2009. ICED out: How immigration enforcement has interfered with workers' rights. Washington, DC: American Rights at Work. http://www.americanrightsatwork.org/dmdocuments/ARAWReports/icedout_report.pdf (last accessed 8 July 2013).

U.S. Department of Homeland Security and U.S. Department of Labor (U.S. DHS and U.S. DOL). 2011. Revised memorandum of understanding between the Departments of Homeland Security and Labor concerning enforcement activities at worksites. http://www.dol.gov/asp/media/reports/DHS-DOL-MOU.pdf (last accessed 8 July 2013).

U.S. Immigration and Customs Enforcement. 2007. *ICE fiscal year 2007 annual report.* Washington, DC: U.S. Immigration and Customs Enforcement. http://www.ailadownloads.org/advo/ICEFY2007AnnualReport.pdf (last accessed 8 July 2013).

———. 2008. *ICE fiscal year 2008 annual report.* Washington, DC: U.S. Immigration and Customs Enforcement. http://cdm16064.contentdm.oclc.org/cdm/singleitem/collection/p266901c0114/id/2170/rec/4 (last accessed 8 July 2013).

Varsanyi, M. W. 2005. The paradox of contemporary immigrant political mobilization: Organized labor, undocumented migrants, and electoral participation in Los Angeles. *Antipode: A Radical Journal of Geography* 37:775–95.

———. 2008. Rescaling the "alien," rescaling personhood: Neoliberalism, immigration and the state. *Annals of the Association of American Geographers* 98:877–96.

———. 2011. Neoliberalism and nativism: Local anti-immigrant policy activism and an emerging politics of scale. *International Journal of Urban and Regional Research* 35:295–311.

Walker, K. E., and H. Leitner. 2011. The variegated landscape of local immigration policies in the United States. *Urban Geography* 32:156–78.

Wells, M. J. 2004. The grassroots reconfiguration of U.S. immigration policy. *International Migration Review* 38:1308–47.

Western, B., and J. Rosenfeld. 2011. Unions, norms, and the rise in U.S. wage inequality. *American Sociological Review* 76:513–37.

On the Work of Urbanization: Migration, Construction Labor, and the Commodity Moment

Michelle Buckley

Department of Human Geography, University of Toronto Scarborough

Construction labor markets are crucial to processes of urbanization, yet they have been largely overlooked as sites for research and theory at the nexus of urban and migration studies. In this article I explore how widespread trends of flexibility have transformed construction labor markets internationally in recent decades and highlight how these trends intersect with the growing incorporation of temporary migrant labor. Employing research conducted on construction labor markets in the city of Dubai, I offer three interventions on theorizing urbanization through the lens of migrant construction work and employment. These include reconceptualizing urbanization as a process of commodity production; highlighting the building process as a site of intersectional politics; and foregrounding how migrant construction work and employment offers a fruitful lens for comparative urban research seeking to draw new connections about the social relations of urbanization across a host of cities internationally. Drawing on feminist migration and postcolonial urban scholarship, I consider how an engagement with the commodified geographies of migrant construction work offers opportunities to reframe and decenter Marxian theories of urbanization.

建筑劳动市场是城市化过程的关键，但却多半不被视为城市与移民研究关係中的研究及理论场域。我将在本文中探讨近数十年来弹性趋势的普及，如何改变了国际的建筑劳动市场，以及这些趋势如何与成长中的临时移工的引进相互交织。我运用在迪拜城市的建筑劳动市场中所进行的研究，透过建筑移民工与聘雇的视角，提供三个介入理论化城市化的方式，包含将城市化再概念化为商品生产的过程；强调营建过程做为多元交织政治的场域；以及凸显建筑移民工及其聘雇，对于寻求连结国际上不同城市的城市化社会关係的比较城市研究而言，如何提供丰硕的研究视角。我援引女性主义迁徙和后殖民城市的学术着作，考量涉入建筑移民工作的商品化地理，如何能够提供做为再框架并去中心化马克思的城市化理论之机会。

Los mercados de trabajo relacionados con labores de construcción son cruciales en los procesos de urbanización, aunque en gran medida hayan sido ignorados como sitios de investigación y teoría en el núcleo de los estudios urbanos y migratorios. En este artículo exploro cómo las tendencias generalizadas de flexibilidad han transformado internacionalmente los mercados laborales de la construcción en décadas recientes, y destaco la manera como estas tendencias se intersectan con la creciente incorporación del trabajo migratorio temporal. Utilizando la investigación llevada a cabo en los mercados laborales de construcción en la ciudad de Dubai, propongo tres instancias para teorizar la urbanización a través de la visión del trabajo y el empleo migratorio en construcción. Estas propuestas incluyen reconceptualizar la urbanización como el proceso de producción de una mercadería; destacar el proceso de construcción como un escenario de políticas interseccionales; y avanzar en la discusión de cómo el trabajo y el empleo migratorio en la construcción ofrecen una perspectiva fructífera para la investigación urbana comparada, buscando delinear nuevas conexiones acerca de las relaciones sociales de la urbanización dentro de un conjunto internacional de ciudades. Basándome en el conocimiento erudito urbano con enfoque feminista y poscolonial, elaboro sobre cómo un compromiso con las geografías comodificadas del trabajo migratorio de la construcción ofrece oportunidades para reformular y descentrar las teorías marxistas de la urbanización.

Over the last three decades, construction work, like so many other occupational sectors, has undergone a profound transformation. Internationally, widespread yet uneven trends such as rising competition by multinational contractors (Chang 2008), the deunionization of construction labor in parts of North America and Europe (Erlich and Grabelsky 2005; Torres et al. 2013), and the pervasive growth of labor subcontracting practices (Wells and Jason 2010) have been accompanied by the rise of insecure, informal, or otherwise precarious forms of employment across many lower-waged occupational segments in the building trades. Meanwhile, geographers, labor rights scholars, and labor historians have argued that labor market flexibility in construction sectors across the Global North and South has been paralleled by a growing reliance on migrant labor[1] (Gibson and Graham 1986; Erlich and Grabelsky 2005; Chang 2008; Theodore et al. 2008).

Migrants' incorporation into construction sectors is occurring to a significant degree in cities (Chang 2008; P. Chan, Clarke, and Dainty 2010), yet these seismic shifts in the contemporary political economies of building work and employment have yet to be substantively integrated into Marxist conceptualizations of capitalist urbanization (e.g., Harvey 1985; Christophers 2011). In this article, I consider the empirical and theoretical potential of exploring urbanization through the lens of migrant construction work and employment. In contrast to migration scholarship that often implicitly delineates the roles that migrants play in urbanization as those relating to their contributions to population change, I am interested here in the relations of urbanization that produce aggregate changes in the physical, materially produced components of the urban landscape, and even more specifically, in forms of urban construction that entail the use of waged labor and the productive circulation of capital. In doing so, I seek to foreground the crucial roles that migrants employed in construction—both formally and informally—play in the capitalist production of urban built environments.

I begin by briefly exploring some engagements by both migration and urban scholarship with migrant construction work and employment. I then draw on research on construction labor markets in the city of Dubai, United Arab Emirates, to foreground two distinct yet interrelated sites that have been peripheral to efforts to theorize urbanization: Gulf cities and construction labor. As one of the most rapidly urbanizing cities of the twenty-first century, Dubai has been almost exclusively reliant on temporary, migrant workers to build the urban landscape. Most national citizens are neither trained nor willing to do construction work; thus, virtually all employees in the sector are nonnatives employed via two- or three-year work visas, and they are predominantly South Asian men employed at the lower end of the industry's occupational spectrum. By 2006, just a few years after the state's launch of a massive urban development program, an estimated 700,000[2] construction migrants lived and worked in Dubai. Many of these workers were living in barrack-style, mass-worker accommodations and were routinely subject to abuses including passport confiscation, breaches of contract, illegal fee extractions, and wage theft by both labor intermediaries and employers.

Although the systemic exploitation of Dubai's migrant builders has been used in recent years to portray the city as a "limit case of contemporary urbanism" (Robinson 2011, 6), the city's construction labor markets are relevant to theorizing the social relations of urbanization internationally. Using analytical frameworks drawn from feminist migration and postcolonial urban scholarship, I employ primary and secondary data on Dubai's migrant construction labor markets to present three interventions in the theorization of urbanization. These include conceptualizing urbanization through the relations of commodity production; foregrounding the building process as a key locus of intersectional politics; and exploring how migrant construction work and employment offers a novel site for comparative research seeking to draw connections about the social relations of urbanization across a host of cities internationally.

Working at the Nexus of Migration and Urban Theory: Migrant Construction Labor and Capitalist Urbanization

The construction trades have always been integral to processes of city-building. Aggregately, the sector is also one of the largest employers worldwide and, similar to other sectors of the economy, such as hospitality or domestic services, it is a major gateway for mobile, noncitizen job seekers. For instance, in 2006, nearly 23 percent of all employed Mexican-born workers in the United States were working in construction-related occupations (Batalova 2006). Construction also tends to be one of the most important sectors for migrant men whose residency or employment status is unauthorized (Swider 2008; Passel and Cohn 2009) and is often one of the few viable job options for less skilled migrant men (Fan 2002; Martin 2012). Moreover, as Erlich and Grabelsky (2005) noted, the seasonal character and spatial immobility of the construction process especially lends itself to the use of migrant labor, as lower labor costs cannot generally be achieved by moving the production process across borders but by capitalizing on cheaper segments of labor in situ.

Recent research points to diverse and extensive processes of deregulation unfolding in construction labor markets on an international scale (Bosch and Philips 2003) that have had important implications for the growing use of migrant workers over the last three decades. For example, Wells and Jason (2010) documented the pervasive growth of subcontracted and informalized construction labor markets in cities across South Africa, Kenya, and Nepal, in which migrants are heavily represented. Chang's (2008) extensive study of construction labor markets across Asia demonstrated that labor subcontracting and wage erosion over the past decade have been accompanied by a significant rise in the use of transnational migrant labor in construction sectors in Malaysia, Thailand, and Indonesia. Meanwhile, construction labor markets in large and small European and North American cities have become increasingly characterized by the rise of heavily subcontracted employment trends (Passel and Cohn 2009; Martin 2012), wage differentials between citizen and (im)migrant workers (Malheiros 1999; Theodore et al. 2008), and the subsequent decline of vocational training and apprenticeship programs for the domestic workforce (K. W. Chan and Dainty 2007). In an industry where long-term and direct employment is no longer profitable (Ruhs and Anderson 2011), the widespread shedding of core workforces by construction firms over the last several decades has been paralleled by the increasing complexity and insecurity of construction employment relationships, including the growth of casualized day labor hiring strategies in the United States (Valenzuela et al. 2006; Theodore et al. 2008) and bogus self-employment practices in the United Kingdom (P. Chan, Clarke, and Dainty 2010).

In light of these findings, it is important to consider the migrant geographies shaping construction work and urbanization processes in the twenty-first century. Labor market segmentation in construction has typically been framed through the sector's portrayal as a quintessential immigrant niche (Rath 2002) in which permanently settling foreigners have undertaken difficult and low-status construction jobs for cheaper wages. However, the increasingly diverse and intersecting conditions of temporariness being produced in rapidly deregulating sectors—including the stipulations of workers' employment contracts, their employment, residency, and citizenship status or their housing accommodations—highlight how construction labor market segmentation, insecurity, and inequality transcend simple citizen–noncitizen binaries (Rogaly 2009, 1976–77). Moreover, following Bailey et al.'s (2002) contention that migration scholarship needs to offer more deeply spatialized conceptualizations of the relations between migrant incorporation and wider processes of political economic restructuring, a focus on construction's evolving divisions of labor offers a way to enrich migration theory by challenging the tendency for migration scholarship to treat cities as mere containers (Çağlar and Glick Schiller 2011) in which migrants work and live and to explore how they are theoretically significant agents in the contemporary production of urban space.

A number of illuminating studies on construction work and employment have begun to do this. For example, Iskander, Lowe, and Riordan (2010) highlighted the significant innovations made by (im)migrant men employed in the residential construction sector in Philadelphia. They contended that the fundamental ways these workers have reshaped the construction process have been both underresearched and undertheorized in part because most workers are informally employed and lack access to institutions (e.g., unions) that make such contributions visible. Global cities-focused research has also emphasized how precarious migrant divisions of labor are a defining facet of London's construction labor market (Wills et al. 2010), while Malecki and Ewers's (2007) research on Arab Gulf cities contextualized migrant labor as a particularly crucial dimension of global city-making strategies specific to the region. Within this literature, however, construction work is largely subsumed within a broad spectrum of insecure and poorly paid migrant occupational sectors that are theorized as invisible yet essential to the inner workings of global cities. Framing the precarious migrant divisions of labor in construction solely as the products of polarized global city economies would be unduly limiting, however, as it would preempt a consideration of the role that migrant labor plays in urbanization processes beyond a small cadre of powerful urban command centers. For example, it would not explain why strikingly similar employment conditions and migrant divisions of labor characterize the construction labor markets of purportedly more ordinary cities such as Austin, Texas (Torres et al. 2013), Lisbon (Malheiros 1999), or Nairobi (Wells and Jason 2010). In what follows, I seek to broaden and build on the insights of this work by briefly delving into the politics and political economies of construction work and employment in Dubai and presenting three avenues for theorizing urbanization through the lens of migrant construction labor.

Cities and the Commodity Moment: De-centering and Recentering the Social Relations of Urbanization

In early 2008, at the height of the postmillennial building boom in Dubai, state officials announced that construction activity had risen to comprise nearly a quarter of the city's gross domestic product.[3] With thousands of active construction sites snarling traffic and dominating the skyline, the size and scope of building activity led to boastful—although highly dubious—claims about the number of construction cranes dotting the landscape and the cubic tons of concrete being consumed by the city. Yet a key relevance of Dubai's construction market to migration and urban theory lies not in its size but in the ways it trains our attention on a set of social relations that are integral to urbanization.

In Marxian terms, these relate to what Harvey (1985, 6) suggested are the original "commodity relations" of the city; that before the material components of urban built environments can be used for production or consumption (e.g., roads used for transporting goods, or buildings rented to tenants) they are initially produced as commodities. Yet the relations of this commodity moment (or in more classically Marxian terms, the commodity form of the built environment) have been largely ignored in recent scholarship on capital accumulation through the built environment. Instead, the productive circulation of capital through urban landscapes has overwhelmingly been examined by urban scholars through the relations of what Harvey (1985) calls the *secondary circuit of capital*: the rent-seeking or productive capital accumulation activities toward which the components of the built environment are employed as already-built fixed capital (e.g., Beauregard 1994; Gottdiener 1994; Gotham 2006; Christophers 2011).[4] Unlike these relations, accumulation through the commodity moment of construction is based on a complex mix of industrial and craft-based relations of material urban production. In Dubai, this has involved hundreds of thousands of waged, migrant workers undertaking activities like concrete forming and curing; steel-fixing; masonry and dry-walling; sand-dredging; and mechanical, electrical, and plumbing system installation. Together, their tasks make up a discrete service of building that, like other commodities, is produced and consumed "within one time period" (Harvey 1985, 4).

If we accept the premise that investment in the material components of urban built environments has been employed internationally as a spatial fix for overaccumulation in the global economy over the past decade (Harvey 2010, 172–74; Christophers 2011), we must consider how these commodity relations and the migrant divisions of labor that often constitute them are centrally implicated in these strategies. For example, against the backdrop of European Union enlargement, Meardi, Martin, and Lozano (2012, 7) argued that the growth of migrant employment in the British and Spanish construction trades has served to create a "hyperflexible buffer" for an increasingly volatile, globalized industry. Martin (2012), meanwhile, noted that unauthorized construction migrants played a key role in keeping labor costs down, thus helping to sustain the U.S. residential construction boom of the mid-2000s (see also Buckley 2012; Torres et al. 2013). Migrant construction labor, in other words, has been implicated internationally in the most recent urban spatial fix, a reality manifesting most starkly in the wake of the 2007–2008 financial crisis when construction markets worldwide not only experienced the most catastrophic job losses of any sector but low-wage, precariously employed migrant workers from Dublin to Shenzhen to Dubai were consistently at the forefront of these retrenchments (Awad 2009; K. W. Chan 2010; Krings et al. 2011; Buckley 2012).

Finally, as a city that in many ways remains "off the map" (Robinson 2002, 535) of sites deemed worthy for the theorization of urbanization, consideration of the wider relevance of Dubai's construction labor markets steers us into new geographies of urban theory. It draws our attention away from the Western, built-up cities that have tended to dominate theory formation in urban studies—Los Angeles, Paris, and Chicago—to the places where the overwhelming amount of material urbanization is and will be taking place over the coming decades: cities and urbanizing spaces across the Global South, where, according to Chang (2008), roughly 70 percent of the world's estimated 111 million construction workers are located. Employing Dubai as a prism for these commodity relations offers one way to diversify the cadre of cities through which urban theory is produced (Robinson 2006; Ehrkamp 2011).

Intersectionalizing Urbanization: Gender/Sexuality, Class, and Ethno-nationality in Urban Construction

A second intervention considers what feminist and intersectional approaches to the study of (im)migrant construction work and employment can bring to research and theory on urbanization. During Dubai's construction boom, the presence of such an immense migrant construction workforce posed a dilemma for the state—how to manage so many working-class men in a city aiming to attract primarily affluent, professional foreign residents and tourists. This prompted the adoption of a host of segregationist state policies aimed heavily, if not exclusively, at low-wage builders. Actions included the rapid construction of an array of large labor camps to house workers on the outskirts of the city or on industrial lands, systematic efforts by municipal officials to evict or prevent ostensibly single South Asian working-class men from being housed together in family-designated subdivisions (Ahmed 2007), and policing efforts to actively discourage these men from occupying tourist spaces such as beaches, malls, and cafe boulevards (Kanna 2007; Elsheshtawy 2008). Meanwhile, construction labor sourcing and hiring practices among contractors and labor-sending consulates have often been based not just on assessments of skills and work experience but on a host of ethno-nationalized stereotypes that workers from some regions and countries are better suited to the city-state's agenda of becoming an affluent enclave of consumer urbanism; migrants from some countries are seen as inherently "calmer" and thus more amenable to practices of spatial segregation,[5] "built" to withstand hot and physically demanding conditions on Dubai's construction sites,[6] or not "prone" to practices such as visiting brothels or ogling women in public.[7]

These embodied spatial politics of policing and segregation aimed at construction migrants prompt us to move beyond narrow notions of class struggle to consider the broader politics of subjectivity that can sustain processes of material urbanization. In contrast to the ways that feminist scholars have demonstrated how the politics of gender and citizenship in the private spaces of the home can shape the social relations of work in migrant occupational sectors like domestic services (Silvey 2004; Raghuram 2005; Fernandez 2011), construction has typically had an acutely public workplace. In Anglo-American cities, the noisy clamor of street-level construction has long been emblematic not only of modernist urban progress but also evocative of a quintessentially masculine (and white) working-class subject. Ironically, it might be precisely due to the overt predominance of men not just in waged construction but in most formal-sector city-building activities—from engineering to architecture to planning and design—that considerations of subjectivity remain underdeveloped in critical research on material urbanization[8]; these sectors are in some locales so hegemonically masculine that their gendered natures have been rendered invisible.[9]

Additionally, Dubai's experiences highlight the unique sociospatial ways in which the politics of urban construction for many migrant builders are often deeply enmeshed with the politics of social reproduction. In a project-based industry where large numbers of nonresident workers might be required for a finite period of time, temporary workers are often housed by their employer in temporary housing on the outskirts of the city or in makeshift accommodations on or next to the construction site (Firman 1991; Olimova and Bosc 2003; Srinivas 2008); this is despite the fact that workers might reside in a city for many years, working on different projects for different companies (Keane and McGeehan 2008). Discursive framings of working-class migrant builders in the city as a problem in need of a solution, meanwhile, can permeate not only neighborhood politics and local news reports (see Rahman 2011; Martinez-Lucio and Connolly 2012) but also scholarly research on urbanization (e.g., Wong, Yeow, and Zhu 2005).

Although being sequestered in large labor camps heavily circumscribed construction migrants' mobility and visibility in Dubai, it was also highly conducive to collective, informal forms of labor organizing for workers seeking unpaid back wages, better treatment from supervisors, and wage increases; these sites of mass social reproduction in fact became spaces enabling tens of thousands of nonunionized migrant men to illegally strike and collectively bargain with both the state and employers. Dozens of migrants' strikes throughout the mid-2000s mounted a real and immediate threat to Dubai's speculative real estate bubble and resulted in a cascade of wage increases and concessions for low-wage migrants across the building trades as well as the institution of city-wide protocols for paying workers on time and responding to worker complaints (Buckley 2013).

The point of attending to questions about the politics of ethnicity and race, citizenship, class, or gender is not to map how such social axes are simply attributes attached to particular bodies participating in the urbanization process but to illuminate how the material production of urban built environments can depend on the parallel production of complex inequalities and intersecting forms of social difference (McDowell 2008). In contrast to some Marxian theorizations of urbanization, which can tend to be constructed through gender-neutral frameworks for understanding relations of capital and class, Dubai's relations of urban production demonstrate the ways in which scales such as the home or the body are important to the day-to-day functioning of capitalist urbanization processes and how gendered, heteronormative, class- and ethnicity-coded practices of migrant segregation, policing, and employment can be integral to the political economies of urbanization. In the final section I build on recent efforts to consider questions about social justice and construction work by scholars such as Theodore et al. (2008) and Torres et al. (2013) by examining how comparative research on migrant construction work and employment offers new opportunities for politically engaged research on urbanization.

Comparing the Commodity Relations of Urbanization: Multisited Research on Construction Migrants' Rights

A report by the organization Human Rights Watch (2006) starkly illuminated the relationships of exploitation endemic within Dubai's construction trades, including systematic wage theft by employers, the confiscation of workers' passports on their arrival in Dubai, and chronic breaches of contractual agreements on wages, overtime pay, and working hours. For thousands of migrant builders, with no access to union representation or state welfare, and often massively indebted to illegal migration moneylenders back home, some scholars argue that together these practices amounted to de facto conditions of forced labor (Keane and McGeehan 2008). Since the mid-2000s, these relations of migrant exploitation have been framed narrowly as a specific feature of Gulf urbanization (e.g., Harvey 2010, 146; see also Davis 2006; Hari 2009), and a focus on migrant agency and governance has been central to explorations of the variegated (Brenner, Peck, and Theodore 2009) character of urban neoliberalism in Dubai (Kanna 2010; Vora 2011; Buckley 2013). Yet to what extent do these narratives of precarious and exploitative migrant construction work tell us something about Dubai or Gulf cities, specifically?

This question arises when we consider that halfway across the world, in the city of Hamilton, Ontario, the largest human trafficking ring in Canadian history was recently uncovered in which migrant men from Hungary were found to be working under conditions of forced labor on an array of construction sites across the province (Morrow 2012). The Canadian state has actively sought to incorporate migrants into local construction labor markets in recent years, in part by designating construction as a priority sector under the country's Temporary Foreign Workers Program. Unbeknownst to the contractors who had hired these workers on a temporary basis, the men had had their passports taken away and had been working for months under constant threat of violence by the Hungarian crime syndicate that had brought them to Canada, while their monthly wages were being confiscated by the syndicate, purportedly to pay off workers' migration debts. Around the same time, accounts of chronic wage theft, passport confiscation, illegal migration fees, and incidents of forced work among migrants employed in construction labor markets of cities as diverse as Beijing, Moscow, Austin, and Washington have also emerged (Human Rights Watch 2008, 2009; Theodore et al. 2008; Torres et al. 2013).

Beyond constructions of regional exceptionalism, the case of Dubai can serve to generate what Roy (2009, 824) calls "travelling theory," in which the labor relations undergirding the city's urbanization might at once retain their historical and geographical specificity but also "cross borders and travel as dynamic vectors of new theoretical conversations and exchanges" on the relationships among migrant construction work, labor rights, and urbanization. The point here is not to develop universalizing theories about the labored landscapes of urbanization in which Dubai provides some paradigmatic case reflecting conditions that are ostensibly "global" (Rankin 2010). Rather, Dubai's relations of urban production merely offer a starting point for more "systematic" (Strauss 2012, 143) yet open-ended analyses in which insights about the challenges to fair and secure work in the sector could be generated by examining different construction labor markets and migrants' situated experiences in relation to each other. For example, an assessment of agreements recently forged between the Dubai government and construction labor–sending states to coordinate the protection of migrant builders could be used to highlight the silences in current policy responses to migrant construction labor exploitation in Hamilton.

Similarly, a comparative lens could also be useful in confronting challenges to fair construction work that arguably extend beyond the borders of any single construction site, city, or nation-state. Alongside the recent labor actions by Dubai's unorganized migrant builders, diverse campaigns are currently being waged by and for migrants in the building trades across a range of sites and scales: Grassroots organizations such as the Tanzania Association of Informal Construction Workers have formed to address issues of employment insecurity, worker abuse, and access to jobs in the sector (Wells and Jason 2010); the Workers' Defense Project in Austin, Texas, is an urban-scaled coalition seeking to improve (im)migrant workers' rights in the city's construction trades (Torres et al. 2013); and Who Builds Your Architecture?, a sector-based alliance of scholars, activists, and educators, is currently campaigning to place accountability on, and leverage the power of, global architects to improve migrants' conditions in construction sectors internationally (WBYA? 2013). Considering these campaigns in relation to one another opens a window to explore what solidarities and scalar tactics could be mobilized in seeking to address these challenges and share insights on labor rights campaigning in construction. More broadly, it requires us to confront larger questions about who, ultimately, is accountable and responsible for shaping the social relations of urbanization.

Conclusions

In this article I have sought to show how comparative and intersectional analyses of migrant construction work and employment offer new ways to theorize urbanization. First, highlighting how the production of urban landscapes can take place through commodified relationships involving the use of insecure waged labor offers a potentially contrasting narrative of urbanization than those revolving around the actions of elite state and capitalist actors. This focus raises a different set of questions about what socially just urbanization might look like by considering the extent to which labor exploitation and insecurity in the construction sector is integral to the contemporary political economies of city building, and adds a new conceptual thread to enduring debates in labor geography on workers' agency in shaping the contours of contemporary capitalism (e.g., Herod 1994). More specifically, this article builds on insightful scholarship on the complex and contradictory interests of construction unions in neoliberal city-building projects (Tufts 2004); by foregrounding the diversity of employment conditions and shifting divisions of migrant labor within the construction sector itself, this lens trains the focus on the interests, agency, and experiences of those workers who are often not part of organized labor.

Exploring the gendered and embodied work involved in producing urban built environments brings a different set of sociospatial power relations into relief than have typically been privileged in Marxian conceptualizations of urbanization. Moreover, it reveals how the commodity moment of urban production is just one form of construction work among many others. In addition to the construction labor embedded in commodity relations of production, nonwaged, noncapitalist, and other uncommodified construction activities are major forces contributing to the building of contemporary cities. Subsistence-based, unauthorized, or illegal construction work in cities often brings together gendered relations of production and social reproduction in very different ways than have been explored in this article, and in some cities forms of bonded (or otherwise unfree) construction labor, often carried out by women, are constituted by vastly different gendered and embodied power relations than those in Dubai.

Finally, focusing a comparative lens on the construction process joins efforts, in Boucher et al.'s (2008, 990) words "to highlight articulations between emplaced yet always mobile voices; voices that speak of specific cities and people, while participating in a transurban language of urbanization." The transurban dimensions of urbanization have been articulated here in two broad ways—first by drawing theoretical and empirical "lines of connection" (989) between the experiences of construction workers in very different cities and, second, by demonstrating how a focus on the migrant dimensions of construction labor markets disaggregates the urban landscape into an array of productive moments with geographies that extend well beyond the material and political borders of the city. In this sense, these interventions are a response to calls in both urban and migration studies to interrogate notions of the city as a legitimate analytical category, a cohering totality or an identifiable spatial unit with clear boundaries (Çağlar and Glick Schiller 2011; Brenner 2013). In these emergent debates in geography, particularly those arguing that the concept of urbanization is due for "systematic reinvention" (Brenner 2013, 101), the frame of commodity relations provides just one means to scratch below the surface of the city in its material form to construct a picture of the diverse labors that produce it.

Acknowledgments

I wish to sincerely thank Jenny Robinson, Kendra Strauss, Mike Ekers, Mark Hunter, and Zack Taylor for their comments on earlier incarnations of this article. Many thanks also to Richard Wright and the reviewers for their extremely helpful feedback on the original manuscript. Responsibility for errors and shortcomings in the article, of course, remains mine alone.

Notes

1. The use of the category *migrant* is always problematic and particularly so considering the wide scope of this article. I use it here, however, to broadly foreground the intersection of diverse temporary or insecure forms of employment in construction with the incorporation of a wide spectrum of both inter- and intranationally mobile workers.
2. Figure provided by a senior researcher for Human Rights Watch (personal communication, September 2008).
3. Figure provided by Sheikha Lubna bint Khalid bin Sultan Al Qasimi, former Minister of Economics and Planning, Arabian World Construction Summit, Abu Dhabi, February 2008.
4. Although, notably, Aalbers (2008) recently broadened these debates through a focus on the built environment and the "quaternary" or financial circuit of capital.
5. Various interviews with construction labor subcontractors, Dubai, 2008.
6. Interview with a labor welfare official from a labor-sending country consulate, Dubai, 2008.
7. Various interviews with labor-sending country consular officials, Dubai, 2008.
8. Although see Datta (2012) and Kern (2010).
9. Though notably see Thiel (2007, 2012).

References

Aalbers, M. 2008. The financialization of home and the mortgage market crisis. *Competition & Change* 12 (2): 148–66.

Ahmed, A. 2007. *Bachelors being evicted from villas*. http://www.gulfnews.com/nation/Society/10145546.html (last accessed 1 May 2013).

Awad, I. 2009. *The global economic crisis and migrant workers: Impact and response*. Geneva, Switzerland: International Migration Programme, International Labour Organization.

Bailey, A., R. Wright, A. Mountz, and I. Miyares. 2002. (Re)producing Salvadoran transnational geographies. *Annals of the Association of American Geographers* 92 (1): 125–44.

Batalova, J. 2006. *Mexican-born persons in the US civilian labor force*. Washington, DC: Migration Policy Institute.

Beauregard, R. 1994. Capital switching and the built environment: United States, 1970–89. *Environment and Planning A* 26 (5): 715–32.

Bosch, G., and P. Philips, eds. 2003. *Building chaos: An international comparison of deregulation in the construction industry*. London and New York: Routledge.

Boucher, N., M. Cavalcanti, S. Kipfer, E. Pieterse, V. Rao, and N. Smith. 2008. Writing the lines of connection: Unveiling the strange language of urbanization. *International Journal of Urban and Regional Research* 32 (4): 989–1024.

Brenner, N. 2013. Theses on urbanization. *Public Culture* 25 (1): 85–114.

Brenner, N., J. Peck, and N. Theodore. 2009. Variegated neoliberalization: Geographies, modalities, pathways. *Global Networks* 10 (2): 1–41.

Buckley, M. 2012. From Kerala to Dubai and back again: Migrant construction workers and the global economic crisis. *Geoforum* 43 (2): 250–59.

———. 2013. Locating neoliberalism in Dubai: Migrant workers and class struggle in the autocratic city. *Antipode* 45 (2): 256–74.

Çağlar, A., and N. Glick Schiller, eds. 2011. *Locating migration: Rescaling cities and migrants*. Ithaca, NY: Cornell University Press.

Chan, K. W. 2010. The global financial crisis and migrant workers in China: "There is no future as a labourer; returning to the village has no meaning." *International Journal of Urban and Regional Research* 34 (3): 659–77.

Chan, K. W., and A. Dainty. 2007. Resolving the UK construction skills crisis: A critical perspective on the research and policy agenda. *Construction Management and Economics* 25 (4): 375–86.

Chan, P., L. Clarke, and A. Dainty. 2010. The dynamics of migrant employment in construction: Can supply of skilled labour ever match demand? In *Who needs migrant workers: Labour shortages, immigration, and public policy*, ed. M. Ruhs and B. Anderson, 225–55. Oxford, UK: Oxford University Press.

Chang, D. 2008. *Global construction and Asian workers: Expansion of TNCs in Asia and implications for labour*. London and Hong Kong: Building and Wood Workers' International and Asia Monitor Resource Centre.

Christophers, B. 2011. Revisiting the urbanization of capital. *Annals of the Association of American Geographers* 101 (6): 1347–64.

Datta, A. 2012. *The illegal city: Space, law and gender in a Delhi squatter settlement*. Burlington, VT: Ashgate.

Davis, M. 2006. Fear and money in Dubai. *New Left Review* 41:47–68.

Ehrkamp, P. 2011. Internationalizing urban theory: Toward collaboration. *Urban Geography* 32 (8): 1122–28.

Elsheshtawy, Y. 2008. Transitory sites: Mapping Dubai's "forgotten" urban spaces. *International Journal of Urban and Regional Research* 32 (4): 968–88.

Erlich, M., and J. Grabelsky. 2005. Standing at a crossroads: The building trades in the twenty-first century. *Labor History* 46 (4): 421–45.

Fan, C. 2002. The elite, the natives, and the outsiders: Migration and labor market segmentation in urban China. *Annals of the Association of American Geographers* 92 (1): 103–24.

Fernandez, B. 2011. Household help? Ethiopian women domestic workers' labor migration to the Gulf Countries. *Asian and Pacific Migration Journal* 20 (3–4): 433–57.

Firman, T. 1991. Rural households, labor flows and the housing construction industry in Bandung, Indonesia. *Tijdschrift voor Economische en Sociale Geografie* 82 (2): 94–105.

Gibson, K., and J. Graham. 1986. Situating migrants in theory: The case of Filipino construction workers. *Capital and Class* 10 (2): 130–49.

Gotham, K. F. 2006. The secondary circuit of capital reconsidered: Globalization and the US real estate sector. *American Journal of Sociology* 112 (1): 231–75.

Gottdiener, M. 1994. *The social production of urban space*. Austin: University of Texas Press.

Hari, J. 2009. The dark side of Dubai. *The Independent*. http://www.independent.co.uk/opinion/commentators/johann-hari/the-dark-side-of-dubai-1664368.html (last accessed 12 September 2012).

Harvey, D. 1985. *The urbanization of capital: Studies in the history and theory of capitalist urbanization*. Baltimore, MD: John Hopkins University Press.

———. 2010. *The enigma of capital and the crises of capitalism*. London: Profile Books.

Herod, A. 1994. On workers' theoretical invisibility in the writing of critical urban geography: A comradely critique. *Urban Geography* 15 (7): 681–93.

Human Rights Watch. 2006. *Building towers, cheating workers: Exploitation of migrant construction workers in the United Arab Emirates*. Beirut, Lebanon: Human Rights Watch, Middle East and North Africa Section.

———. 2008. *One year of my blood: Exploitation of migrant construction workers in Beijing*. New York: Human Rights Watch.

———. 2009. *Are you happy to cheat us? Exploitation of migrant construction workers in Russia*. New York: Human Rights Watch.

Iskander, N., N. Lowe, and C. Riordan. 2010. The rise and fall of a micro-learning region: Mexican immigrants and construction in center-south Philadelphia. *Environment and Planning A* 42 (7): 1595–1612.

Kanna, A. 2007. Dubai in a jagged world. *Middle East Report* 243 (37): 22–30.

———. 2010. Flexible citizenship in Dubai: Neoliberal subjectivity in the emerging "city-corporation." *Cultural Anthropology* 25 (1): 100–29.

Keane, D., and N. McGeehan. 2008. Enforcing migrant workers' rights in the United Arab Emirates. *International Journal on Minority and Group Rights* 15 (1): 81–115.

Kern, L. 2010. Gendering reurbanisation: Women and new-build gentrification in Toronto. *Population, Space and Place* 16 (5): 363–79.

Krings, T., A. Bobek, E. Moriarty, J. Salamonska, and J. Wickham. 2011. From boom to bust: Migrant labour and employers in the Irish construction sector. *Economic and Industrial Democracy* 32 (3): 459–76.

Malecki, E. J., and M. C. Ewers. 2007. Labor migration to world cities: With a research agenda for the Arab Gulf. *Progress in Human Geography* 31 (4): 467–84.

Malheiros, J. M. 1999. Immigration, clandestine work and labour market strategies: The construction sector in the metropolitan region of Lisbon. In *Immigrants and the informal economy in Southern Europe*, ed. M. Baldwin-Edwards and J. Arango, 169–85. London: Frank Cass.

Martin, P. 2012. *Migration and competitiveness in US construction and meatpacking*. Unpublished paper, University of California, Davis, CA.

Martinez-Lucio, M., and H. Connolly. 2012. Transformation and continuities in urban struggles: Urban politics, trade unions and migration in Spain. *Urban Studies* 49 (3): 669–84.

McDowell, L. 2008. Thinking through work: Complex inequalities, constructions of difference and transnational migrants. *Progress in Human Geography* 32 (4): 491–507.

Meardi, G., A. Martin, and M. Lozano. 2012. Constructing uncertainty: Unions and migrant labour in construction in UK and Spain. *Journal of Industrial Relations* 54 (1): 2–21.

Morrow, A. 2012. Human-traffickers treated men on Ontario construction site "like slaves." *The Globe and Mail*. http://www.theglobeandmail.com/news/national/human-traffickers-treated-men-on-ontario-construction-site-like-slaves/article550340/ (last accessed 12 October 2012).

Olimova, S., and I. Bosc. 2003. *Labour migration from Tajikistan*. Dushanbe, Tajikistan: International Organization for Migration.

Passel, J., and D. Cohn. 2009. *A portrait of unauthorized immigrants in the United States*. Washington, DC: Pew Hispanic Research Centre. http://www.pewhispanic.org/files/reports/107.pdf (last accessed 4 May 2013).

Raghuram, P. 2005. The global maid trade: Indian domestic workers in the global market. In *Asian women as transnational domestic workers*, ed. S. Huang, N. A. Rahman, and B. Yeoh, 146–74. Singapore: Marshall Cavendish.

Rahman, M. 2011. Bangladeshi migrant workers in the UAE: Gender-differentiated patterns of migration experiences. *Middle Eastern Studies* 47 (2): 395–411.

Rankin, K. 2010. Reflexivity and post-colonial critique: Toward an ethics of accountability in planning praxis. *Planning Theory* 9 (3): 181–99.

Rath, J. 2002. A quintessential immigrant niche? The non-case of immigrants in the Dutch construction industry. *Entrepreneurship & Regional Development* 14 (4): 355–72.

Robinson, J. 2002. Global and world cities: A view from off the map. *International Journal of Urban and Regional Research* 26 (3): 531–54.

———. 2006. *Ordinary cities: Between modernity and development*. London and New York: Routledge.

———. 2011. Cities in a world of cities: The comparative gesture. *International Journal of Urban and Regional Research* 35 (1): 1–23.

Rogaly, B. 2009. Spaces of work and everyday life: Labour geographies and the agency of unorganised temporary migrant workers. *Geography Compass* 3 (6): 1975–87.

Roy, A. 2009. The 21st-century metropolis: New geographies of theory. *Regional Studies* 43 (6): 819–30.

Ruhs, M., and B. Anderson, eds. 2011. *Who needs migrant workers? Labour shortages, immigration, and public policy*. Oxford, UK: Oxford University Press.

Silvey, R. 2004. Transnational domestication: State power and Indonesian migrant women in Saudi Arabia. *Political Geography* 23 (3): 245–64.

Srinivas, S. 2008. Urban labour markets in the 21st century: Dualism, regulation and the role(s) of the state. *Habitat International* 32 (2): 141–59.

Strauss, K. 2012. Coerced, forced and unfree labour: Geographies of exploitation in contemporary labour markets. *Geography Compass* 6 (3): 137–48.

Swider, S. C. 2008. *Behind great walls: Modes of employment among migrant construction workers in China's informal labor market*. Unpublished paper, University of Wisconsin–Madison, Madison, WI.

Theodore, N., A. Valenzuela, E. Meléndez, and A. L. Gonzalez. 2008. Day labor and workplace abuses in the residential construction industry: Conditions in the Washington, DC region. In *The gloves-off economy: Workplace standards at the bottom of America's labor market*, ed. A. Bernhardt, A. Boushey, L. Dresser, and C. Tilley, 91–110. Champaign, IL: Labor and Employment Relations Association.

Thiel, D. 2007. Class in construction: London building workers, dirty work and physical cultures. *British Journal of Sociology* 58 (2): 227–51.

———. 2012. Builders, bodies, and bifurcations: How London construction workers "learn to labour." *Ethnography* 14 (4): 412–30.

Torres, R., R. Heyman, S. Munoz, L. Apgar, E. Timm, C. Tzintzun, C. R. Hale, J. McKiernan-Gonzalez, S. Speed, and E. Tang. 2013. Building Austin, building justice: Immigrant construction workers, precarious labor regimes and social citizenship. *Geoforum* 45: 145–55.

Tufts, S. 2004. Building the "competitive city": Labour and Toronto's bid to host the Olympics games. *Geoforum* 35:47–58.

Valenzuela, A., N. Theodore, E. Melendez, and A. L. Gonzalez. 2006. *On the corner: Day labor in the United States*. Los Angeles: University of California Los Angeles, Centre for the Study of Urban Poverty.

Vora, N. 2011. From golden frontier to global city: Shifting forms of belonging, "freedom," and governance among Indian businessmen in Dubai. *American Anthropologist* 113 (2): 306–18.

WBYA? 2013. *Who builds your architecture?* http://www.whobuilds.org (last accessed 10 May 2013).

Wells, J., and A. Jason. 2010. Employment relationships and organizing strategies in the informal construction sector. *African Studies Quarterly* 11 (2–3): 107–24.

Wills, J., K. Datta, Y. Evans, J. Herbert, J. May, and C. McIlwaine. 2010. *Global cities at work: New migrant divisions of labor*. New York: Pluto.

Wong, T., M. Yeow, and X. Zhu. 2005. Building a global city: Negotiating the massive influx of floating population in Shanghai. *Journal of Housing and the Built Environment* 20:21–50.

Spaces of Immigrant Advocacy and Liberal Democratic Citizenship

Helga Leitner* and Christopher Strunk†

*Department of Geography, University of California, Los Angeles
†Department of Geography, Augustana College

Challenging representations of immigrants as victims of the biopolitical power of the state, and of citizenship as reinforcing such repression, this article interrogates how immigrant advocacy has been contesting the proliferation of exclusionary immigration policies in the United States and in the course of this pushing against the boundaries of formal liberal democratic citizenship. Using Washington, DC, as a case study, and deploying the concept of insurgent citizenship to theorize these actions, we examine the polyvalence of immigrant advocacy, paying particular attention to the spaces and scales of struggle and flexibly networked mobilizations. Insurgent citizenship creates new political spaces that exceed the national territorial framing of liberal democratic citizenship, that focus on civic action rather than simply relying on the electoral process, and that lay claim to new values and criteria for citizenship.

本文透过挑战移民做为国家生命政治权力的受害者，以及公民权做为强化此一压迫的再现，探讨移民倡议如何与美国激增的排他性移民政策和反对正式自由民主公民权疆界的进程相互竞争。我们以华盛顿特区做为案例研究，并运用反叛公民权的概念，理论化上述行动，以此检视移民倡议政策的多面相关性，并特别关注斗争的空间与尺度，以及网络化的弹性动员。反叛公民权创造出新的政治空间，并超越自由民主公民权中的国家领土框架，其所关注的是公民行动，而非单纯仰赖投票过程，并且主张公民权的新价值与准则。

Cuestionando las representaciones que muestran a los inmigrantes como víctimas del poder biopolítico del estado, y a la ciudadanía como agente que refuerza tal represión, este artículo explora la manera como la defensa del inmigrante ha estado disputando la proliferación de políticas de inmigración selectivas en los Estados Unidos al tiempo que presiona contra los límites de una formal ciudadanía democrática liberal. Adoptando a Washington, DC, como un estudio de caso, y desplegando el concepto de ciudadanía insurgente para teorizar estas acciones, examinamos la polivalencia de la defensa del inmigrante, poniendo particular atención en los espacio y escalas de lucha y movilizaciones flexiblemente encadenadas. La ciudadanía insurgente crea nuevos espacios políticos que sobrepasan el marco territorial nacional de la ciudadanía democrática liberal, más enfocada en la acción cívica que dependiendo simplemente del proceso electoral, y que reclama nuevos valores y criterios de ciudadanía.

> The art of politics teaches men how to bring forth what is great and radiant . . . greatness, therefore, or the specific meaning of each deed, can lie only in the performance itself and neither in its motivation nor its achievement. (Hannah Arendt, as cited in Beltrán 2009, 595)

During the past two decades, immigrants in the United States have encountered an increasingly restrictive political landscape, as changes in federal laws and policies had a sweeping impact on foreigners' access to legal and social services (Miller 2005). The early to mid-2000s also saw the enactment of federal–local immigration enforcement partnership programs such as the 287(g) and Secure Communities programs, both of which enroll local police into federal immigration enforcement.[1] These have been supplemented by a proliferation of exclusionary state and municipal policies designed to deter or expel undocumented immigrants from these territories (Varsanyi 2010; Leitner and Preston 2011; Walker and Leitner 2011; Coleman 2012). The repertoire of such policies ranges from antisolicitation ordinances, crafted to prevent migrant workers from congregating in public space to solicit work, to overcrowding ordinances thwarting extended family living in the same housing unit, to

English-only ordinances and resolutions prohibiting local governments from conducting nonemergency business in other languages. States and localities also have imposed fines on employers and prohibited government contractors from hiring undocumented immigrants. Finally, state-level immigration laws in Arizona and Alabama have sought to require local police to determine the immigration status of persons arrested or detained, if there is reasonable suspicion that they are in the country without authorization.

In response, immigrants and their allies have mobilized to resist exclusionary policies, highlighting such problems as the racial profiling of immigrants of color, the dramatic increase in detention and deportations, violations of workers and civil rights, and the fear that these policies have incited among immigrants. Through legal challenges, protests, and other mobilizations, immigrant advocates have embroiled a number of exclusionary policies in legal battles, causing some to be declared unconstitutional. Crucially, immigrant advocacy alliances also have promoted welcoming policies at a variety of scales. Cities around the country have supported day laborer centers, adopted "sanctuary city" ordinances that prohibit local police from participating in immigration enforcement, and approved resolutions supporting immigrants (Theodore 2007; Ridgley 2008). A smaller number of cities have issued identity cards for immigrants and extended the right to vote in local elections irrespective of formal citizenship status (Varsanyi 2005). The states of California, Illinois, Maryland, and Minnesota have passed versions of the DREAM Act, allowing young undocumented immigrants to pay in-state tuition at local colleges and gain access to state financial support, and Illinois approved a 2013 policy that will permit undocumented immigrants to obtain drivers' licenses (for further details see Mitnik and Halpern-Finnerty 2010).

This article discusses how and to what extent immigrants and their allies push against and exceed the boundaries of liberal democratic citizenship as they contest exclusionary and promote welcoming policies. In *Democracy and the Foreigner*, Honig (2003) reflects on how the foreigner and foreignness are represented in classical and contemporary political thought. In the former, the foreigner has been seen as a threat, to be tightly managed and controlled. Although this thinking persists in contemporary political theory, it is counterposed against postnational positions celebrating both diversity and the foreigner as an agent of change. Honig offers an alternative to these contrasting perspectives by arguing for a switch from asking, "How should we solve the problem of foreignness?" to "What problems does foreignness solve for us?" (Honig 2003, 4).

We take up Honig's question to argue that the demands made through the discourses and actions of immigrants and their allies work on behalf of democracy by stretching the boundaries of liberal democratic citizenship[2] and might lead to "a rethinking of democracy as also a cosmopolitan and not just a nation-centered set of solidarities, practices, and institutions" (Honig 2003, 8). We ask the following questions: What does immigrant advocacy do for liberal democracy and liberal democratic citizenship? What kinds of political spaces are being (re)created and how? Does all immigrant advocacy question established patterns of control and authority? Is immigrant advocacy advancing new values and rationalities beyond those generally associated with liberal democratic citizenship?

It is important to note that immigrants are not acting alone in these struggles. Migrants and migrant organizations are joined by long-term residents and civic organizations, including labor unions, churches and other faith-based groups, community-based organizations, civil rights groups, and student organizations. Immigrant advocacy also is multiethnic and multiracial in character, bringing together immigrants from a variety of countries, native-born whites, African Americans, and Latinos. The examples we draw on are based on fieldwork conducted in Washington, DC. During 2010 and 2011, Strunk conducted twenty-five in-depth interviews with organizational directors and staff of twenty-one different immigrant-serving or advocacy, faith-based, and community-based organizations in the metropolitan area. Interviews were complemented by participant observation at meetings and other organizational activities, supplemented with written documentation of these events. We undertook textual analysis of interview transcripts, field notes, and newspaper and policy reports, examining the political struggles and efficacy of immigrant advocacy groups challenging exclusionary immigration policies (for further details see Leitner and Strunk n.d.). Although not a traditional center of immigrant settlement, Washington, DC, is representative of new immigrant gateways across the United States and provides a valuable case study of immigrant advocacy in the context of restrictive immigration policies. Its foreign-born population is highly diverse in terms of geographic origin and level of education (Price and Singer 2008), and it has been a hotspot of local immigration policy experimentation in recent years. Since 2005, almost half of the cities, counties, and towns in the metropolitan area have passed exclusionary immigration ordinances (Walker and Leitner 2011). In turn, immigrants and their allies have challenged these ordinances while pushing for more welcoming policies, which the District of Columbia and several inner-ring suburbs have adopted or reinforced and reaffirmed in recent years.

Immigrant Advocacy Politics as Insurgent Citizenship

Much of the scholarship on the politics of immigration in the United States has drawn attention to the ubiquitous and often paralyzing biopolitical power of the state, presenting migrants as victims of this power (Coleman 2009; Coleman and Grove 2009; Varsanyi 2010; Mountz 2011). Recent research, however, has emphasized that immigrants—citizens and noncitizens; undocumented, temporary, and permanent residents; labor migrants and refugees—are not simply passive victims (Das Gupta 2006; Lee 2006; McNevin 2006; Laubenthal 2007; Pulido 2007; Antoja, Menjivar, and Magana 2008; Cordero-Guzmán et al. 2008; Shi 2008; Anderson 2010; Ellerman 2010; Campbell 2011; McGregor 2011; Millner 2011; Voss and Bloemraad 2011). This growing body of literature examines everyday acts of resistance to detention and deportation (Ellerman 2010), the history of the sanctuary movement (Coutin 1993; Ridgley 2008), the mass mobilization and street protests in U.S. cities in 2006 (Pallares and Flores-Gonzalez 2010; Voss and Bloemraad 2011), struggles by immigrants and their allies to create day laborer centers (Theodore 2007), recent efforts to expand immigrant residents' access to drivers' licenses and social services (Wells 2004; Varsanyi 2008, 2010), or immigrants' participation in formal politics (Varsanyi 2005). A notable line of work in the United States has advanced the notion of "migrant civil society" (Theodore and Martin 2007; Cordero-Guzmán et al. 2008; Martin 2010), which focuses on the role of nonprofit and religious organizations in providing services and mobilizing immigrants. Other scholars have shown how immigrant and refugee mobilizations in public spaces constructed new understandings of citizenship and sought to claim new rights (McNevin 2006; Beltrán 2009; Squire and Bagelman 2012), often using a "right to the city" framework (Miraftab 2012; Nyers and Rygiel 2012). In the European context, scholars have analyzed the tactics of migrant advocacy groups in refugee camps (Millner 2011; Rygiel 2011) and cities (Gill 2010). This literature theorizes immigrant advocacy in many ways—as civic mobilizations, acts of resistance, constructions of a counterpublic to the dominant public realm, and citizenship practices.

Some scholars have been highly critical of deploying conceptions of citizenship to theorize and make meaningful the political activities and demands of immigrants and their allies. It is argued that even an expanded conception of citizenship is too limiting because it reinforces citizenship as the foremost measure of belonging. Thus, political scientist Nyers (2007, 2) cautions that expanding citizenship always risks reproducing "a kind of conceptual imperialism that effaces other ways of being political . . . for all the innovations in how we conceive of citizenship, the concept remains deeply embedded with practices that divide humanity according to race, ethnicity, gender and geography." Feminist scholars Tyler and Marciniak (2013, 148) are concerned about the deployment of citizenship as a technology of (neo)liberal governance and that "integrationist forms of migrant politics" (for example, through claims to citizenship rights) "risk remaining captured within the existing legal frameworks and prevailing regimes of sovereignty which they contest." They advocate for "migrancy as a creative and productive form of movement which 'escapes' and always 'exceeds' legal and conceptual frameworks of citizenship [cultivating] 'an imaginary and a practical sensibility to what lies after citizenship'" (Papadopoulos and Tsianos 2013, cited in Tyler and Marciniak 2013, 148).

Although acknowledging these dangers associated with the concept of citizenship, we disagree that this necessitates abandoning the concept of citizenship altogether. To the contrary, we suggest that an expanded, agentic conception of citizenship, which dislodges citizenship as located and conferred by the nation-state, offers other insights and solutions (see also Isin 2009). We deploy the conception of insurgent citizenship, which, following Holston and Appadurai (1999) and Holston (2009), we define as discourses and practices that challenge existing laws, policies, and institutions; promote alternative criteria for membership in a polity; and lay claims to and enact new forms of citizenship and rights. Insurgent citizenship moves the focus toward individual practices and organized collective actions. The former include living in a community and engaging with others and individual everyday acts of resistance, such as foot-dragging on the job to resist grueling or unsafe working conditions or immigrants "losing" their documentation to be able to claim refugee status (Lee 2006;

Ellerman 2010). Organized collective actions might include lobbying and testimonies in front of elected government officials, seeking legal remedies against exclusionary policies, performing caravans and protest marches, publicizing counterdiscourses, strikes, and organized intervention into electoral politics. It is important to note that individual and collective acts of resistance are not necessarily directed toward the state but can be struggles against multiple forms of oppression and subordination through claims that do not rely on formal citizenship rights (Das Gupta 2006). Here we focus on collective actions, paying particular attention to polyvalence of immigrant advocacy and its spatial register. We explore whether and how immigrant advocacy is creating new political spaces and advancing new values and rationalities beyond those generally associated with liberal democratic citizenship.

Polyvalence of Immigrant Advocacy

Immigrant advocacy in the United States is highly complex and diverse in terms of the participants, agendas, claims-making, and understandings of citizenship that are mobilized and their spatial register. In short, advocacy and citizenship mean different things to different groups. The most visible groups in formal politics seeking to advance immigrants' rights are national immigrant advocacy and civil rights organizations, such as the American Immigration Lawyers Association, the National Council of La Raza, the National Immigration Forum, the American Civil Liberties Union, the National Immigration Law Center, and the League of United Latin American Citizens (LULAC). Many of these organizations are headquartered in Washington, DC, or other large cities and engage in the formal political process through producing reports, proposing policies, and lobbying elected officials.

Although such national organizations tend to speak for immigrants during immigration reform debates, the base of the immigrant advocacy movement is the plethora of sometimes highly localized, but networked, nonprofit, religious, and civic organizations and groups. These serve and advocate on behalf of immigrants and mobilize citizens and noncitizen migrants as they articulate claims of belonging and make demands on the state and employers. A prominent example was the 2006 immigrant rights protests, when a loosely affiliated network of Spanish-language media, labor unions, churches, and immigrant-serving organizations helped mobilize hundreds of thousands of migrants and nonmigrants to rally in streets across the United States (Voss and Bloemraad 2011).

The landscape of immigrant advocacy is continuously emergent, as new networks and connections form and others dissipate. In recent years, two influential single-issue immigrant rights networks developed that are largely independent of the major national immigrant rights organizations. The first is made up of young, mostly undocumented immigrants who call themselves DREAMers because of their activism in support of the DREAM Act and against continued deportations of unauthorized migrants. Since the latest attempt to pass the federal DREAM Act failed in 2010, DREAMers have engaged in a number of innovative direct action protests that include occupying legislative offices and publicly "outing" themselves as undocumented at press conferences and in social media. DREAM activists played a major role in creating a more positive perception of undocumented immigrants, including successful campaigns to pass or expand state-level versions of the DREAM Act in several states.

A second network of immigrant rights activism has revolved around opposition to the expansion of immigration enforcement and detention. Since the mid-1990s, the federal government has dramatically expanded both policing along the U.S.–Mexico border and enforcement of immigration laws within the U.S. territory (Coleman 2007). Under the Obama administration, almost 400,000 immigrants have been deported annually, with many forced to endure long waits in the labyrinthine network of state- and private-run detention centers. In response to these developments, informal networks of community-based organizations, legal advocacy organizations, and civil rights groups mobilized to oppose the criminalization of immigrants and to directly challenge federal–local immigration enforcement partnership programs such as 287(g) and Secure Communities (Loyd, Mitchelson, and Burridge 2012; for further details see Strunk and Leitner 2013).

These groups and networks differ not only in terms of their geographic scope, agendas, strategies, and tactics but also in how far they extend the boundaries of liberal democratic citizenship. As immigrants and their allies push back against exclusionary immigration policies and propose more inclusive measures, they deploy different discursive framings to express grievances and imaginaries and to make their demands. A common frame, advanced particularly by national immigrant and civil rights organizations, presents immigrants as hard-working and responsible laborers. This representation fits squarely into the conception of the neoliberal citizen subject as a self-reliant and responsible individual. This can be seen as a strategic framing to justify rights claims and migrants' incorporation into and belonging to the national community.

A related but distinct discourse deployed by immigrant advocates highlights the economic contributions of immigrants (and, to a lesser extent, their cultural contributions) to the local and national community—not only through their labor but also as consumers and taxpayers. Although this framing of rights claims based on economic contributions expands the currently dominant political and cultural criteria for the attribution and acquisition of citizenship in liberal democratic states, it also remains state-centered and reaffirms the national community and national citizenship as the dominant measure of belonging. This is not the case with the third, humanitarian frame deployed by immigrant advocates, which argues that undocumented migrants are worthy of inclusion in a community not only because they contribute to the local economy and society but because they are human beings. This is sometimes expressed in the phrase "No Human Being is Illegal" (McNevin 2011). This frame explodes the privileging of the nation-state and national citizenship as the arbiters of decisions about belonging.

Besides stretching the discursive boundaries of liberal democratic citizenship, local and regional advocacy organizations have been active in creating new political spaces, both within and outside of the formal political system. As we show, in seeking to transform the political landscape at a variety of scales, advocates have constructed spaces of encounter and deliberation between migrants and long-term residents of different political persuasions.

Spaces of Insurgent Citizenship

Just as insurgent citizenship exceeds formal citizenship, so do its spatialities exceed the bounded and policed national space of liberal democratic citizenship to encompass the transnational and the local. International migration, particularly migrant transnationalism—the recurrent and continual physical mobility of people, their remittances, goods, and ideas across national borders—challenges conceptions of bounded national citizenship in important ways (Smith and Bakker 2008). Transnational migrants do so by simultaneously engaging with multiple states and public spheres to express their identities, contest exclusion and discrimination, and voice demands to be recognized as members of more than one national community (Leitner and Ehrkamp 2006). For example, hometown associations, migrant organizations composed of members from a single village or town in the sending country, have increasingly mobilized around immigrants' rights in the United States even as they participate from abroad in sending country politics (Smith and Bakker 2008; Sites and Vonderlack-Navarro 2012). Receiving countries often have responded to migrant transnationalism by reinforcing their borders, seeking to more tightly bound and control the national territory and those within it (Staeheli et al. 2012). Border enforcement also has occurred inside national borders through large-scale worksite raids and, more recently, through programs such as 287(g) and Secure Communities.

Immigrants and their allies have resisted heightened immigration enforcement and other exclusionary policies at a variety of different subnational sites, deploying different spatial strategies and tactics. They have created new political spaces of welcome and deliberation. One very visible example of immigrant advocacy in U.S. cities has been the construction of day laborer centers, alongside streets and in shopping centers. In the Washington, DC, metropolitan area, organizations such as CASA de Maryland, the state's largest immigrant-serving organization, operate centers that connect workers with employers, providing English language instruction and job training classes. Yet worker centers have been vulnerable to challenges by anti-immigration activism. In Herndon, a small outer suburb in northern Virginia, a publicly funded worker center operated by a local nonprofit organization was closed in 2007, less than two years after it was opened. Protests and legal challenges by local and national anti-immigration groups resulted in the withdrawal of town council support for day laborers. The controversy in Herndon was not the end of struggles over migrant day laborers in northern Virginia, however. In 2011, after years of debates with community members around the topic of immigration, the Centreville Immigration Forum opened a new worker center in a neighboring part of Fairfax County. The Forum, founded in 2007 by local church leaders, itself is a new political space for deliberation among a diverse set of individuals and groups—including members of immigrant service organizations, churches, politicians, and local businesses.

Indeed, faith-based organizations in collaboration with legal advocacy and human rights groups have taken the lead in creating welcoming spaces for immigrants across the metropolitan area. In another outer ring suburb, for instance, members of several churches serve a weekly lunch to Latino day laborers. Church leaders view the lunch as a space for promoting understanding of cultural differences and disrupting, however temporarily, the uneven power dynamics separating wealthy suburban residents from a largely undocumented day laborer population. As one of the church leaders noted, even though this is a slow way of creating change, interpersonal interactions have the potential to change the minds of voters in northern Virginia and eventually contribute to broader transformation:

> I realize that we are not contributing to immigration reform. I am a big advocate for reform but after seeing what is going on with the national environment I think there are other things that are more important right now. We need to do outreach and build up those networks for when reform does come. I think that our networks are strong and we can mobilize when reform comes. I am convinced that reform will happen but only when it broadens beyond the Latino community. Immigration impacts all of us. (Interview with church leader, May 2011)

Immigrant advocates also have appropriated public spaces in campaigns to convince local authorities to refrain from enacting exclusionary local immigration policies. In the outer northern Virginia suburb of Prince William County, elected officials proposed in 2007 that local law enforcement officers must inquire into the immigration status of anyone suspected of being in the country illegally. The Illegal Immigration Enforcement Ordinance became a model for Arizona's Senate Bill 1070 passed three years later[3] and, as in Arizona, immigrant advocacy groups challenged the ordinance using a variety of tactics. Mexicans without Borders, a local advocacy group, argued that it would harm the county because immigrants would leave, part of the immigrant rights discourse that emphasizes the contributions of all migrants as taxpayers and workers, regardless of formal legal status. Advocates also endorsed a work stoppage and economic boycott to highlight immigrants' economic contributions to the community (interview with community activist, Prince William County, Virginia, 2010).

Mexicans without Borders also organized a series of large-scale public forums at sites across the county to inform residents about the proposed ordinance and mobilize opposition. Participants, mostly Latino immigrants, organized human caravans traversing the county, participated in marches and rallies, and spoke out against the ordinance at county board meetings. Acting in public spaces is a particularly important part of immigrant advocacy in this metropolitan area. Organizations not only protest exclusionary local policies but also frequently are called on to support national-scale mobilizations on the National Mall and outside congressional offices on Capitol Hill (interview with community activist, September 2010). Immigrant advocacy organizations also challenge anti-immigration policies in the courts and lobby political leaders and administrators in state offices.

Acts of insurgent citizenship do not simply involve particular places, of course, but connect them through extralocal relations and flows, in both material and cyberspace. Mobilities, essential to political struggles for a number of reasons, are a strategic tool to gain publicity, register discontent, and promote alternative imaginaries (Leitner, Sheppard, and Sziarto 2008). They also allow people to connect in and across sites and places by exchanging information, learning from one another, and enabling emotional support. Indeed, many groups in the immigrant advocacy movement are actively involved in extralocal networks, spanning a particular metropolitan area or reaching well beyond it, seeing these as an essential toolkit that organizations draw on in responding to restrictive immigration measures and promoting inclusionary policies (cf. Darling [2010] on City Sanctuary networks in the UK). In the Washington, DC, metropolitan area, advocates from well-established immigrant rights organizations in the District of Columbia and the inner suburbs often provided valuable assistance to smaller, volunteer-run organizations in the outer suburbs that support more recent immigrant populations with significantly less funding from local governments. Organizations also drew on broader networks during local campaigns against the Secure Communities program in 2010 and 2011. Immigrant advocates discussed strategies and shared resources on political and legal challenges to the program through conference calls with national immigrant rights networks and advocates in other states (Strunk and Leitner 2013). Immigrant advocacy struggles thus exceed militant particularism, a localism that eschews larger political struggles and entraps citizens in narrowly circumscribed communities (cf. Featherstone 2008). Extralocal networks are not always tension-free, however; advocates in smaller organizations sometimes felt "abandoned" by national immigrant rights organizations continuing to push for comprehensive immigration reform at the expense of struggles around local exclusionary policies (interview with community activist, October 2010).

Finally, in response to particular local political contexts and changing opportunity structures, immigrants and their advocates have devised novel scalar strategies to make their voices heard and potentially enhance their efficacy. For example, after Immigration and Customs Enforcement (ICE) blocked Arlington and other local governments from opting out of the Secure Communities program, activists looked to state capitals, where most Memoranda of Agreement with ICE were being negotiated. In the spring of 2011, advocates successfully lobbied the governors of Illinois, Massachusetts, and New York to opt out of the program. When ICE responded by also blocking state-scale opting out, advocates refocused their attention on local legislation that targeted immigration detainers, a key aspect of local immigration enforcement.[4] Subsequently, Santa Clara, San Francisco, Cook County (Illinois), and the District of Columbia each passed bills limiting the ability of ICE to issue detainers on noncriminal immigrant offenders located in municipal jails. Although these campaigns are decidedly limited in scope, they separate local law enforcement from federal immigration enforcement, challenging the underlying federal–local framework of immigration enforcement partnerships.

Conclusion

The expansion of immigration enforcement has not only created fear among immigrant communities but also has helped mobilize and organize immigrants and their allies against exclusionary policies. Beyond simply opposing these policies, immigrant advocates have both lobbied for comprehensive reform and promoted welcoming measures and spaces in localities and states nationwide. We have examined how and to what extent immigrants and their allies, through these actions, push against and exceed the boundaries of liberal democratic citizenship. Whereas liberal democratic citizenship is associated with a bundle of individual rights and obligations conferred by a single nation-state, insurgent citizenship is an agentic conception of citizenship that involves active participation in civic and political life through discourses and practices that challenge existing laws, policies, and institutions; promote alternative criteria for membership in a polity; and lay claims to and enact new forms of citizenship and rights.

Our research demonstrates that immigrants do not act alone but are joined in their struggles by a broad spectrum of local and national civic organizations and groups forming complex networks advocating on behalf of immigrants. Immigrant advocacy networks are continuously emerging as new networks and connections form and others dissipate. They are also highly polyvalent, using a diverse set of sociospatial discourses and actions to register discontent with existing policies and conditions and to promote alternative criteria for citizenship and belonging. Not all advocacy and immigrant-serving organizations and groups advance and enact discourses that question established patterns of control and authority or promote new values and rationalities beyond those generally associated with liberal democratic citizenship, but many do. Those engaging in insurgent citizenship utilize discursive strategies that stretch the boundaries of liberal democratic citizenship and construct new political spaces at different sites and scales. Through humanitarian frames of belonging and transnational engagement, they exceed the national territorial boundaries of liberal democratic citizenship. Locally, workers' centers, often hotly contested, provide spaces that serve, protect, and welcome immigrants. Last but not least, the appropriation of public space and convening of public forums create new political spaces for deliberation, which are crucial in the formulation of new values and the development of new criteria for citizenship and belonging.

Acknowledgments

We would like to thank two anonymous reviewers, Richard Wright, and Eric Sheppard for their helpful comments on an earlier draft of this article. We are grateful to the immigrant rights advocates in the Washington, DC, metropolitan area for their support throughout the research process. The research for this article was supported by the Graduate Research Partnership Program from the College of Liberal Arts, University of Minnesota, Minneapolis.

Notes

1. Under the 287(g) program, local officials are trained in immigration enforcement and authorized to initiate deportation proceedings. The Secure Communities program is an information-sharing program that integrates criminal and immigration databases, allowing federal immigration officers to identify deportable immigrants when they are stopped, detained, or arrested by local police officers. Similarly, the Criminal Alien Program uses biometric data to scan inmates at all federal and state prisons and hundreds of local jails and then removes unauthorized immigrants from the country.
2. Liberal democratic citizenship is a legal status conferred by the state, defining individual membership in one

nation-state that is associated with a bundle of individual rights and obligations.

3. In 2012, the Supreme Court ruled several parts of Arizona's S.B. 1070 law unconstitutional but upheld the provision that requires police officers to request documentation of the immigration status of any person stopped, detained, or arrested if they are suspected of being in the country without authorization.
4. Immigration detainers are requests from ICE and the Department of Homeland Security to hold potentially deportable individuals in jails or prisons until immigration officials can remove them.

References

Anderson, B. 2010. Mobilizing migrants, making citizens: Migrant domestic workers as political agents. *Ethnic and Racial Studies* 33:60–74.

Antoja, A. D., C. Menjivar, and L. Magana. 2008. The spring marches of 2006: Latinos, immigration, and political mobilization in the 21st century. *American Behavioral Scientists* 52:499–506.

Beltrán, C. 2009. Going public: Hannah Arendt, immigrant action, and the space of appearance. *Political Theory* 37:595–622.

Campbell, K. M. 2011. The road to S.B. 1070: How Arizona became ground zero for the immigrants' rights movement and the continuing struggle for Latino civil rights in America. *Harvard Latino Law Review* 14:1–20.

Coleman, M. 2007. Immigration geopolitics beyond the Mexico–US border. *Antipode* 39:54–76.

———. 2009. What counts as the politics and practice of security, and where? Devolution and immigrant insecurity after 9/11. *Annals of the Association of American Geographers* 99:904–13.

———. 2012. The "local" migration state: The site-specific devolution of immigration enforcement in the U.S. South. *Law & Policy* 34:159–90.

Coleman, M., and K. Grove. 2009. Biopolitics, biopower, and the return of sovereignty. *Environment and Planning D: Society and Space* 27:489–507.

Cordero-Guzmán, H., N. Martin, V. Quiroz-Becerra, and N. Theodore. 2008. Voting with their feet: Nonprofit organizations and immigrant mobilization. *American Behavioral Scientist* 52:598–617.

Coutin, S. B. 1993. *The culture of protest: Religious activism and the U.S. sanctuary movement*. Boulder, CO: Westview.

Darling, J. 2010. A city of sanctuary: The relational re-imagining of Sheffield's asylum politics. *Transactions of the Institute of British Geographers* 35:125–40.

Das Gupta, M. 2006. *Unruly immigrants—Rights, activism, and transnational South Asian politics in the United States*. Durham, NC: Duke University Press.

Ellerman, A. 2010. Undocumented migrants and resistance in the liberal state. *Politics & Society* 38:408–29.

Featherstone, D. 2008. *Resistance, space and political identities—The making of counter-globalization networks*. Oxford, UK: Wiley-Blackwell.

Gill, N. 2010. Tracing imaginations of the state: The spatial consequences of different state concepts among asylum activist organisations. *Antipode* 42:1048–70.

Holston, J. 2009. *Insurgent citizenship: Disjunctions of democracy and modernity in Brazil*. Princeton, NJ: Princeton University Press.

Holston, J., and A. Appadurai. 1999. Cities and citizenship. In *Cities and citizenship*, ed. J. Holston, 1–18. Durham, NC: Duke University Press.

Honig, B. 2003. *Democracy and the foreigner*. Princeton, NJ: Princeton University Press.

Isin, E. 2009. Citizenship in flux: The figure of the activist citizen. *Subjectivity* 29:367–88.

Laubenthal, B. 2007. The emergence of pro-regularization movements in Western Europe. *International Migration* 45:101–33.

Lee, C. T. 2006. Tactical citizenship: Domestic workers, the remainders of home, and undocumented citizen participation in the third space of mimicry. *Theory & Event* 9 (3). doi: 10.1353/tae.2006.0034

Leitner, H., and P. Ehrkamp. 2006. Transnationalism and migrants' imagining of citizenship. *Environment and Planning A* 38:1615–32.

Leitner, H., and V. Preston. 2011. Going local: Canadian and American immigration policy in the new century. In *Immigrant geographies of North American cities*, ed. W. Li, C. Teixeira, and A. Kobayashi, 2–21. Toronto: Oxford University Press Canada.

Leitner, H., E. Sheppard, and K. Sziarto. 2008. The spatialities of contentious politics. *Transactions of the Institute of British Geographers* 33:157–72.

Leitner, H., and C. Strunk. n.d. Assembling insurgent citizenship: Challenging exclusionary local immigration policies in the Washington, DC, metropolitan area. Unpublished manuscript.

Loyd, J., M. Mitchelson, and A. Burridge, eds. 2012. *Beyond walls and cages: Prisons, borders, and global crisis*. Athens: University of Georgia Press.

Martin, N. 2010. The crisis of social reproduction among migrant workers: Interrogating the role of migrant civil society. *Antipode* 42:127–51.

McGregor, J. 2011. Contestations and consequences of deportability: Hunger strikes and the political agency of non-citizens. *Citizenship Studies* 15:597–611.

McNevin, A. 2006. Political belonging in a neoliberal era: The struggle of the sans-papiers. *Citizenship Studies* 10:135–51.

———. 2011. Undocumented citizens? Shifting grounds of citizenship in Los Angeles. In *Citizenship, migrant activism and the politics of movement*, ed. P. Nyers and K. Rygiel, 165–83. London and New York: Routledge.

Miller, T. 2005. Blurring the boundaries between immigration and crime control after September 11. *Boston College Third World Law Journal* 25:81–124.

Millner, N. 2011. From "refugee" to "migrant" in Calais solidarity activism: Re-staging undocumented migration for a future politics of asylum. *Political Geography* 30:320–28.

Miraftab, F. 2012. Right to the city and the quiet appropriations of local space in the heartland. In *Remaking urban citizenship: Organizations, institutions, and the right to the city*, ed. M. P. Smith and M. McQuarrie, 11–35. New Brunswick, NJ: Transaction.

Mitnik, P. A., and J. Halpern-Finnerty. 2010. Immigration and local governments: Inclusionary local policies in the era of state rescaling. In *Taking local control: immigration policy activism in U.S. cities and states*, ed. M. Varsanyi, 51–72. Palo Alto, CA: Stanford University Press.

Mountz, A. 2011. Islands as enforcement archipelago: Haunting, sovereignty and asylum. *Political Geography* 30:118–28.

Nyers, P. 2007. Introduction: Why citizenship studies? *Citizenship Studies* 11 (1): 1–4.

Nyers, P., and K. Rygiel, eds. 2012. *Citizenship, migrant activism and the politics of movement*. London and New York: Routledge.

Pallares, A., and N. Flores-Gonzalez. 2010. *Marcha: Latino Chicago and the immigrant rights movement*. Chicago: University of Illinois Press.

Papadopoulos, D., and V. S. Tsianos. 2013. After citizenship: Autonomy of migration, organizational ontology and mobile commons. *Citizenship Studies* 13 (2): 178–96.

Price, M., and A. Singer. 2008. Edge gateways: Immigrants, suburbs, and the politics of reception in metropolitan Washington. In *Twenty-first century gateways: Immigrant incorporation in suburban America*, ed. A. Singer, S. Hardwick, and C. Brettel, 137–68. Washington, DC: The Brookings Institution.

Pulido, L. 2007. A day without immigrants: The racial and class politics of immigrant exclusion. *Antipode* 39:1–7.

Ridgley, J. 2008. Cities of refuge: Immigration enforcement, police, and the insurgent genealogies of citizenship in US sanctuary cities. *Urban Geography* 21:53–77.

Rygiel, K. 2011. Bordering solidarities: Migrant activism and the politics of movement and camps at Calais. *Citizenship Studies* 15:1–19.

Shi, Y. 2008. Chinese immigrant women workers: Everyday forms of resistance and coagulate politics. *Communication and Critical /Cultural Studies* 5:363–82.

Sites, W., and R. Vonderlack-Navarro. 2012. Tipping the scale: State rescaling and the strange odyssey of Chicago's Mexican hometown associations. In *Remaking urban citizenship: Organizations, institutions, and the right to the city*, ed. M. Smith and M. McQuarrie, 151–69. New Brunswick, NJ: Transaction.

Smith, M. P., and M. Bakker. 2008. *Citizenship across borders: The political transnationalism of el migrante*. Ithaca, NY: Cornell University Press.

Squire, V., and J. Bagelman. 2012. Taking not waiting: Space, temporality and politics in the city of sanctuary movement. In *Citizenship, migrant activism and the politics of movement*, ed. P. Nyers and K. Rygiel, 146–64. London and New York: Routledge.

Staeheli, L., P. Ehrkamp, H. Leitner, and C Nagel. 2012. Dreaming the ordinary: Daily life and the complex geographies of citizenship. *Progress in Human Geography* 36:628–44.

Strunk, C., and H. Leitner. 2013. Resisting federal–local immigration enforcement partnerships: Redefining "Secure Communities" and public safety. *Territory, Politics, Governance* 1:62–85.

Theodore, N. 2007. Closed borders, open markets: Immigrant day laborers' struggle for economic rights. In *Contesting neoliberalism: Urban frontiers*, ed. H. Leitner, J. Peck, and E. Sheppard, 250–65. New York: Guilford.

Theodore, N., and N. Martin. 2007. Migrant civil society: New voices in the struggle over community development. *Journal of Urban Affairs* 29:269–87.

Tyler, I., and Marciniak, K. 2013. Immigrant protest: An introduction. *Citizenship Studies* 17 (2): 143–56.

Varsanyi, M. 2005. The rise and fall (and rise?) of non-citizen voting: Immigration and the shifting scales of citizenship and suffrage in the United States. *Space & Polity* 9:113–34.

———. 2008. Immigration policing through the backdoor: City ordinances, the "Right to the City," and the exclusion of undocumented day laborers. *Urban Geography* 29:29–52.

———, ed. 2010. *Taking local control: Immigration policy activism in U.S. cities and states*. Palo Alto, CA: Stanford University Press.

Voss, K., and I. Bloemraad, eds. 2011. *Rallying for immigrants: The fight for inclusion in 21st century America*. Berkeley: University of California Press.

Walker, K., and H. Leitner. 2011. The variegated landscape of local immigration policies in the United States. *Urban Geography* 32:1–23.

Wells, M. J. 2004. The grassroots reconfiguration of U.S. immigration policy. *International Migration Review* 38:1308–47.

On Distance and the Spatial Dimension in the Definition of Internal Migration

Thomas Niedomysl* and Urban Fransson†

*Department of Human and Economic Geography and the Centre for Innovation, Research and Competence in the Learning Economy, Lund University

†School of Business, Economics and Law, Department of Business Administration, Gothenburg University

Migration is commonly defined by a temporal and a spatial dimension. It is generally agreed that these dimensions are problematic and could seriously distort understanding of migration, but data constraints have effectively obstructed further insights. This article focuses on the spatial dimension where migration is typically defined as movement across administrative borders. Borders usually serve as proxies for migration distance but the validity of such proxies is largely unknown, posing a considerable challenge to migration research. Using data for all internal migrants in Sweden, the only known country where migration distances are available in sufficient detail, we present the first accurate description of actual migration distances and investigate the relationship between actual migration distances and migration-defining boundaries. More specifically, we examine how the volumes of migration and the characteristics of migrants change when migration distances vary and when different types of migration-defining boundaries are employed. The findings show that notable shares of short-distance migrants are included almost regardless of which migration-defining boundary is employed, but migrant differentials are less affected than might be expected.

迁徙一般是由时间与空间的面向定义之。这些面向普遍被认为是有问题的，并可能严重扭曲对于迁徙的理解，但数据的限制,实际上更有效地妨碍了进一步的洞见。本文聚焦的空间面向中，迁徙一般被定义为跨越行政界限的行动。边界经常做为迁徙距离的代替，但此一代替的有效性却多半未知，并对迁徙研究造成相当大的挑战。瑞典是可以取得迁徙距离详尽细节的唯一国家，而我们便运用瑞典国内所有内部移民的数据，呈现第一个有关实际迁徙距离的精确描述，并探讨实际迁徙距离和定义迁徙边界之间的关联。我们特别检视，当迁徙距离有所差异，以及当运用不同的定义迁徙的边界时，迁徙的数量及迁徙者的特徵会如何改变。研究发现显示，无论使用何种定义迁徙的边界，皆包含了可观比例的短程迁徙，但迁徙的差异则是较预期受到较小的影响。

Comúnmente la migración es definida con referencia a una dimensión temporal y a una espacial. Generalmente se acepta que estas dimensiones son problemáticas y que podrían seriamente distorsionar la comprensión del fenómeno migratorio, pero la escasez de datos efectivamente han obstruido mayores avances en estas cuestiones. Este artículo está enfocado en la dimensión espacial, donde la migración se define típicamente como movimiento a través de límites administrativos. Las fronteras usualmente sirven como referentes de la distancia de la migración, aunque la validez de tales referentes es en gran medida desconocida, lo cual plantea considerables retos a las investigaciones sobre migración. Mediante el uso de datos relacionados con todos los migrantes internos de Suecia, único país conocido donde se dispone de distancias de migración con suficiente detalle, presentamos la primera descripción precisa de distancias de migración reales e investigamos la relación entre las distancias de migración reales y los límites definidores de la migración. Más específicamente, examinamos cómo cambian los volúmenes de migración y las características de los migrantes cuando varían las distancias de migración y cuando se emplean diferentes tipos de límites definidores de la migración. Los descubrimientos logrados indican que se incluyen porciones notables de migrantes de corta distancia casi sin importar qué límite definitorio de la migración se emplea, aunque los diferenciales del migrante se ven menos afectados de lo que podría esperarse.

The study of human migration is a topic of significant interest across the social sciences. This is hardly surprising, considering that migration is arguably the most important form of capital mobility and affects the lives of those involved in a most profound way. Although various disciplines tend to specialize in certain areas, they share a common understanding that migration involves two dimensions: time and space. The temporal dimension relates to how long someone has to stay at a new location to be regarded as a migrant, and the spatial dimension relates to how far someone has to move to be counted as a migrant. Migration scholars also share an understanding that each dimension is problematic and potentially constitutes serious obstacles for achieving a better understanding of migration. The problems are not merely related to determining "how long" or "how far"; in fact, they are more fundamental than that. Data constraints have effectively prevented a thorough investigation of these two dimensions, precluding further insights, with the result that the actual scope and consequences of these problems are largely unknown.

This article focuses on the spatial dimension in the definition of internal migration. The key issue here is distance, which is arguably critical for understanding the volumes, causes, and effects of migration for individual migrants and societies. It matters how far someone moves; those who move long distances are more likely to experience significant change and consequences compared to those who move only short distances. In addition, what causes someone to move across a short distance differs from what causes a long-distance move: Housing concerns dominate the former type of movement and employment the latter (Morrison and Clark 2011; Niedomysl 2011). The problem is, however, that data on migration distances are not available from standard censuses. Individuals are conventionally defined as migrants when moving across some administrative border, regardless of the distance of the move. Borders thus serve as proxies for migration distance but are also of functional relevance in the sense that, for example, tax rates and public services usually vary across administrative areas. To add further complexity, administrative areas vary considerably between and within countries in terms of size, shape, and settlement pattern, not to mention that adjacent areas also vary in these regards, making the validity of borders as proxies for distance most uncertain. Such uncertainty, and the risk of confusing short-distance migrants with long-distance migrants, poses a considerable challenge for improving understanding of migration, in particular when considering that these two groups might have little else in common than having changed their address.

This article investigates the differences between migrants who move over certain distances and migrants who move between different administrative areas at different spatial scales. Using longitudinal microdata for all internal migrants in Sweden, the only known country where sufficiently detailed information on migration distances and migrant characteristics is available, we examine how the volumes of migration and the characteristics of migrants change when migration distances vary, as well as when different types of migration-defining boundaries are employed. The use of accurate distance data allows us, first, to determine the risk of confusing short-distance migrants with long-distance migrants under varying types of migration-defining boundaries and, second, to investigate the impact of using proxies for distance instead of actual distance on migrant differentials. Both aspects are fundamental for determining the validity of the spatial dimension in conventional definitions of migration.

Theory and Previous Research

Migration and distance has been of scholarly interest ever since Ravenstein (1885) noted that most migrants proceed only a short distance. He thereby confirmed the opinion of A. Smith (1776 [1981]) that people constitute the most difficult form of capital to be transported over distance. Tellingly, Ravenstein also pointed out one of the key problems of migration research, in that migrants are defined as migrants when they cross administrative borders, but due to the varying size, shape, and situation of the areas, any analysis is fraught with difficulties:

> [A] circumstance likely to lead to misconception, if not error, arises from the very unequal size of the counties. Rutland and Yorkshire are hardly comparable. A journey of 25 miles at the most converts any native of Rutland into a "migrant," whilst a native of Yorkshire to place himself into the same position might have to travel as many as 95 miles. (Ravenstein 1885, 168)

Since then, a number of studies have shown that there is a strong deterrent of distance to migration flows (see, e.g., Makower, Marschak, and Robinson 1938, 1939; Hägerstrand 1957; Olsson 1965; Long, Tucker, and Urton 1988a, 1988b), and because migration is still almost exclusively defined by administrative borders, the problems identified over a century ago are equally valid today. Nonetheless, migration researchers have had to try to cope with these problems, and distance has come to occupy a central place in migration research.

Distance also plays a central role in the development of more formal models of migration. Numerous studies have followed the early works of researchers such as Stouffer (1940), Zipf (1946), and Stewart (1948), in which gravity models were used to explain aggregate migration flows between regions (e.g., Plane 1984; Stillwell and Congdon 1991). In brief, distance is treated as a proxy for variables that are difficult to measure, and by including other variables that are expected to be of importance for migration decision making, estimates can be made regarding their relative impact on attracting or repelling migrants (for some recent examples, see Congdon 2010; Cooke and Boyle 2011; Biagi, Faggian, and McCann 2011; Kalogirou 2012). Spatial interaction modeling of migration flows incorporates a measure of distance, but it tends to be very coarse and is usually given little analytical weight. Exceptions do exist, however, and the work of Gordon (1991) is a good example where distinctive migration flows are separated and investigated to provide further insights. He distinguished three types of flows (local, regional, and national) based on inferred moving distances and found that each type of flow correlated differently with various types of push and pull factors. Nonetheless, insights gained from such exercises likely depend on the accuracy of data on migration distance and the extent to which individual characteristics can be linked to migration flows or, more generally, on the level of data aggregation.

Central to this literature is the modifiable areal unit problem (MAUP) that arises with the aggregation of data into areal units (Openshaw 1984; Wong 2009). Two issues of concern are related to this problem. The scale effect refers to differences in results that might arise simply due to variation in the number of spatial units employed. The zonation effect refers to differences that arise as a result of the numerous ways in which those spatial units could be divided. Because migration data are usually only available at aggregate levels for administrative areas that vary in terms of size, shape, settlement pattern, and situation, the MAUP applies and conclusions drawn based on such aggregate migration data might be very different from conclusions drawn based on data where the exact origins and destinations of individual migrants are available (Rogerson 1990).[1] A related concept is that of ecological fallacy (Robinson 1950), where fallacies might arise when inferences from results of groups are deduced to individuals, a problem that occurs "because areal studies cannot distinguish between spatial associations created by the aggregation of data and real associations possessed by the individual data prior to spatial aggregation" (Openshaw 1984, 14). The MAUP and ecological fallacy might in fact be more serious in migration studies than in other fields because migration can take place across several administrative areas (as already discussed, distance is a key factor distinguishing migration from residential mobility).

Theoretically, the finding that most people move only short distances has mainly been explained with reference to (1) the monetary costs of moving longer distances, (2) the fact that the amount of information on the potential new location decreases with increasing distance, and (3) the psychic costs of moving (e.g., the loss of contacts with family and friends; Greenwood 1975). The notion that monetary costs would constitute a main obstacle, however, was questioned early on and it has been shown that direct transportation costs have a negligible effect, even more so considering the potential economic gains of moving (Lansing and Mueller 1967; Sahota 1968).

Consequently, more credence has been given to lack of information and psychic costs in explaining the distance decay of migration (Ritchey 1976). In an attempt to empirically estimate the relative importance of these two factors, Schwartz (1973) argued that education would offset the dearth of information (the ability to obtain and process information is assumed to increase with education) and that the psychic costs of long-distance migration would increase with age (older individuals have more invested in relations with family and friends). This predicts that if lack of information is the main deterrent, education would be positively related to migration distance, whereas age would have the opposite relation to distance if psychic costs were the main deterrent. Finding the expected relation between education and distance, but not between age and distance, Schwartz (1973, 1153) suggested that "the adverse effect of distance on migration is basically a diminishing-information phenomena."

A quite different approach, the concept of intervening opportunities (Stouffer 1940, 1960), posits that the number of persons going a given distance is directly proportional to the percentage increase in opportunities at that distance. A similar train of thought is found in Zipf (1949), who argued the importance of the principle of least effort (see also T. E. Smith 1978). Zipf's simple explanation is that most people will not go a longer distance than they have to to achieve their goals. This raises a number of questions as to which goals can be achieved over varying migration distances and whether any differences exist across population subgroups.

Although much attention has been paid to examining the characteristics of migrants compared to nonmigrants, less work appears to have focused on how the characteristics change over distance. Rose (1958), however, constitutes one of the first examples. Categorizing migrants by socioeconomic status, he found that high status was positively correlated with migration distance, suggesting that individuals from high-status groups must move longer distances to achieve their goals. In a more recent study, Hunt (2004) suggested that highly educated people are more likely to move longer distances compared to those with low levels of education.

Considering the preceding information, it is not surprising to find variations in self-expressed motives for migration across distance. In one of the first examples of a study that used information from survey data on self-expressed motives for migration, Gleave and Cordey-Hayes (1977) examined motivational variation across migration distance. In a review and empirical extension of this literature, Niedomysl (2011) found considerable variation in migration motives, not only over migration distance but also particularly in relation to migrants' socioeconomic and demographic characteristics. The generalized results showed, however, that housing-related motives were negatively related to increasing migration distance, whereas the opposite was true for employment-related motives (see also Morrison and Clark 2011).

As briefly noted earlier, the migration literature is largely made up of studies examining migration flows between administrative areas where distance is typically included as an independent variable. Findings are often inconclusive, however, sometimes showing insignificant and theoretically implausible results. Arguably, this could very well be a result of using administrative borders as proxies for distance, due to the risk of confusing short-distance migrants with long-distance migrants. Support for such a claim is found in the studies that distinguished between short-distance and long-distance migrations. For example, Jun and Chang (1986) made a distinction between migration flows between contiguous and noncontiguous administrative areas, to sort out short-distance from long-distance migration in the absence of data on actual distance. They concluded that "noncontiguous migration behaves significantly different from contiguous migration, and that noncontiguous migration, rather than total migration, should be used for the study of factors affecting interstate migration" (Jun and Chang 1986, 17). Similarly, White and Mueser (1988) argued that the migration-defining boundary is fundamental in that it distinguishes between different types of social processes and is selective of population characteristics. They found that different kinds of moves are highly dependent on personal characteristics, in concurrence with human capital theories and life cycle theories of local residential mobility.

It is striking how little effect these studies, where distance is taken into more than passing consideration, have had on general migration research. A possible, but not a good, explanation for this neglect is that the reviewed studies have been obstructed by a lack of high-quality data on migration distances. With few exceptions they have been forced to use the administrative borders approach because individuals' migration distances have not been available. In addition, they have relied on data from relatively small samples of migrants. It is possible that this has restricted their impact on migration research. In summary, it is evident that even if migration distance appears to be quite vaguely understood from a theoretical perspective, it plays a crucial role for the fundamental ways migration is analyzed. Regardless of how distance should be understood from a theoretical perspective, for the individual migrant, distance obviously matters when it is no longer possible to see on a regular basis those people and places that used to be part of everyday life (Roseman 1971). For progress to be made, more in-depth data on migration distances have to be employed.

By employing uniquely detailed migration data, this article addresses three specific research questions to provide more accurate descriptions of the importance of the spatial dimension in the definition of migration:

1. What does the distribution of migrants across actual migration distance look like?
2. How does that distribution change when different types of (administrative) migration-defining borders are employed?
3. What do the individual characteristics of migrants across actual migration distance, and across different types of (administrative) migration-defining borders, look like?

Data

Identifying Migrants and Inferring Moving Distances

In Sweden, all individuals must, according to law, be registered as living in unique real property units. The centroids of all properties have known exact coordinates collected by the Swedish mapping, cadastral, and land registration authority. For confidentiality reasons, the coordinates of the properties are truncated into 100 × 100 m squares when transferred to Statistics Sweden and made available for research. As the exact coordinate is not known to researchers, the centroids of the coordinate squares constitute the starting point for estimations of moving distances in this article. Because Sweden is a sparsely populated country, a vast majority of squares do not contain any properties at all, whereas in the more populated parts of Sweden, a square can,

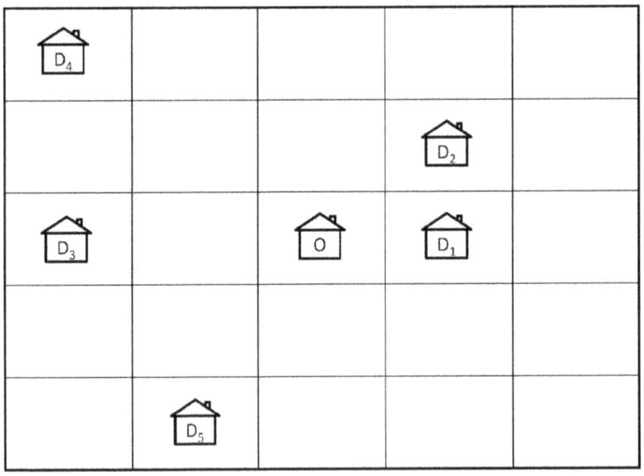

Figure 1. Illustration of coordinate squares used for calculating migration distances in Sweden. *Note:* O = origin; D = destination. Each square represents 100 × 100 m.

at an extreme, contain up to fifty-seven properties. In addition, some large properties cover areas larger than the 100 × 100 m square grid.

In 2008, a total of 1,241,519 individuals changed their 100 × 100 m square of residence. When limiting the population to those eighteen years or older, 1,000,363 individuals had changed their residence. Because the whole of Sweden is covered by the 100 × 100 m grid, Euclidian moving distances between the squares can be calculated with considerable accuracy (the maximum error is only ±142 m) and at this level of precision the MAUP is not a concern.[2]

A more detailed presentation with illustrations of inferred Euclidian moving distances among six different properties is provided in Figure 1 and Table 1. We assume that each migration takes place over a distance of at least 1 m and that all properties are located within one square (that they do not extend across the borders of the 100 × 100 m squares). Hence, the distance between the centroids for a migrant moving from his or her origin to destination D_1 is 100 m (inferred distance in Table 1). Because the exact location of the properties O and D_1 are not known, however, the actual moving distance can range from 1 m to 224 m (if the exact location of the origin and destination is very close or very distant to the border).[3] Similarly, if moving from O to D_2, the distance between the centroids is 141 m but could in theory be 1 m or 282 m. Furthermore, Table 1 shows that the maximum error (regardless of actual migration distance) does not exceed 142 m.

A problem might arise if properties are extended (e.g., if someone buys a neighboring property and the latter property obtains the same property code as the original property). In such cases, the coordinates of the property might change so that a property located in one 100 × 100 m square in 2007 might appear in an adjacent square in 2008. Hence, even if no one actually moved, the coordinates could change, making it falsely appear as if someone had moved. Conversely, the same problem might arise if a property is parceled out (e.g., if someone sells part of their property and the new property obtains a different property code). To minimize the risk that no such "false" movers were included, we made two restrictions. First, we set the minimum inferred moving distance to be greater than 100 m, thus blocking out adjacent 100 × 100 m squares as shown in Figure 2. This restriction reduced the number of movers to 858,584 individuals and sets the minimum inferred moving distance to 141 m (see Table 1), although it is still possible to move 1 m and end up in a destination square located diagonally from the square of origin. Second, for those remaining individuals who had the same property code in 2008 as in 2007 but had changed their 100 × 100 m square between the two years, we added a further restriction that they would have to move a distance greater than 1,000 m. This further reduced the number of movers to 846,320. These measures also remove an unknown number of "true" movers. Some "true" movers are not possible to detect in these data, because some properties cover more than one 100 ×

Table 1. Distance calculations of the examples given in Figure 1

Inferred distance (between centroids)	Minimum distance	Maximum distance	Maximum error
O to D_1 = 100 m	O to D_1 = 1 m	O to D_1 = 224 m	O to D_1 = 124 m
O to D_2 = 141 m	O to D_2 = 1 m	O to D_2 = 282 m	O to D_2 = 141 m
O to D_3 = 200 m	O to D_3 = 101 m	O to D_3 = 316 m	O to D_3 = 116 m
O to D_4 = 282 m	O to D_4 = 142 m	O to D_4 = 424 m	O to D_4 = 142 m
O to D_5 = 223 m	O to D_5 = 101 m	O to D_5 = 360 m	O to D_5 = 137 m

Note: O = origin; D = destination.

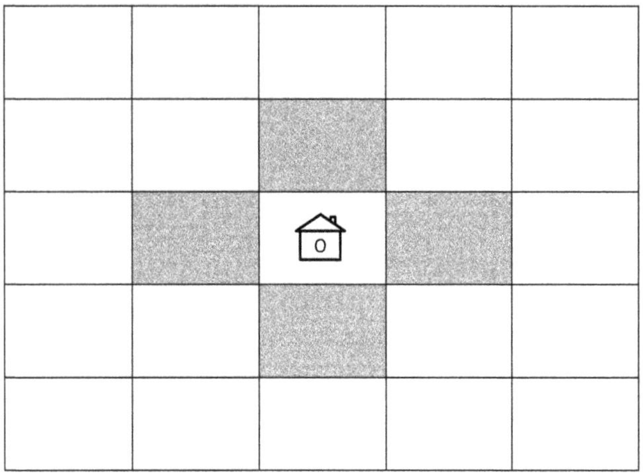

Figure 2. Setting a restriction that inferred distances have to be greater than 100 m. *Note:* O = origin. Each square represents 100 × 100 m.

100 m square and might contain many residents. In such cases, an individual could indeed move but still be registered as a resident at the same property code and thus not appear as a mover.

To identify movers across administrative borders (discussed later), layers of the administrative units are simply merged onto the 100 × 100 m square grid, so that each individual square is identifiable within specific administrative units. Movers can then be identified from the square of their origin, across any administrative border, to the square of their destination. Although Euclidian migration distances can be measured with a maximum error margin of ±142 m, for simplicity we rounded all reported distances to the nearest kilometer.

Although the Swedish register data are of exceptionally high quality for migration research purposes, they are not without flaws (Amcoff 2009). In addition to the risk of including "false" movers, three general issues need to be mentioned. First, only one migration event is recorded per year. Also, the exact date of the event is unavailable. Third, although people are required to report a "permanent" change of residence, some individuals fail to do so. These issues lead to an undercount of migration volumes (presumably short-distance migration), but the extent of these problems is not known.

We distinguish between short-distance (residential mobility) and long-distance migration based on whether a migrant changes his or her location of workplace when moving (see Zax 1994). Because there is no exact way to determine a distance at which a combined change of residence and workplace is "optimal," we examined the proportion of employed movers that changed the location of their workplace by distance of move. The results (available from the authors) showed that the increase in the proportion that changed their workplace location leveled off at an approximate moving distance of 100 km and we thus define migration as a move exceeding 100 km. This is a definition that also resonates with general notions on the distinction between short-distance and long-distance migration, discussed earlier, as a move exceeding 100 km is most likely to involve a significant change in the daily life of the individual concerned (Hägerstrand 1957; Roseman 1971).

The Individual Characteristics of Migrants

Information on the individual characteristics of migrants is drawn from the Swedish register data by Statistics Sweden. The term *register data* refers to data collected on individuals by various Swedish government agencies for administrative purposes and updated on a continuous basis (and some of these data are transferred to Statistics Sweden). In Sweden, all legal inhabitants obtain a unique social security code from birth or, for immigrants, when a permanent residence permit is granted. Using each individual's social security number as a unique key, Statistics Sweden can link the various pieces of data concerning each legal inhabitant of Sweden over time. Researchers seeking access to these data must specify their demands and Statistics Sweden will, after a confidentiality assessment, provide a database in which the individuals' social security numbers have been replaced by randomly generated unique identification codes. This provides researchers with access to a wealth of data on socioeconomic and demographic characteristics for all individuals (including the location of their properties described previously). For a detailed overview of the Swedish register data, see Statistics Sweden (2011).

The register data we employ here cover the entire adult Swedish population in 2008. Table 2 shows the individual characteristics we selected, and these are standard characteristics reported in the migration literature. Table 2 shows both migrants and nonmigrants. Three demographic variables—sex, age, and whether children are present in the household—were included. Sex and children are self-explanatory, but it might be worth mentioning that the four age classes were based on a rough estimate of different life phases that could influence migratory behavior.

Four socioeconomic variables were also included. The level of education was divided into four classes.

Table 2. Descriptive characteristics

Category	Nonmigrants		Migrants	
	N	%	*N*	%
Gender				
Male	3,014,239	49.2	420,128	49.6
Female	3,111,884	50.8	426,192	50.4
Age				
18–25	602,762	9.8	284,124	33.6
26–40	1,439,696	23.5	305,751	36.1
41–64	2,659,424	43.4	187,691	22.2
65+	1,424,241	23.2	68,754	8.1
Children living at home				
Yes	2,545,456	41.6	357,599	42.3
No	3,580,650	58.4	488,721	57.7
Education				
11 years or less	3,328,507	54.3	335,379	39.6
12 years	1,079,900	17.6	242,933	28.7
13–16 years	1,608,858	26.3	256,541	30.3
17 or more years	108,857	1.8	11,467	1.4
Income				
First quartile	1,054,866	17.2	232,796	27.5
Second quartile	1,394,021	22.8	192,204	22.7
Third quartile	1,270,829	20.7	169,225	20.0
Fourth quartile	2,406,406	39.3	252,095	29.8
Job change				
No change	4,986,149	81.4	554,559	65.5
Job change	1,139,974	18.6	291,761	34.5
Occupation				
Employed	3,808,784	62.2	589,989	69.7
Unemployed	82,468	1.3	99,741	11.8
Student	207,294	3.4	80,635	9.5
Other	2,027,577	33.1	75,955	9.0
Moving distance (km)				
0–9	—	—	514,058	60.7
10–19	—	—	98,417	11.6
20–39	—	—	62,804	7.4
40–69	—	—	35,761	4.2
70–109	—	—	24,657	2.9
110–159	—	—	20,185	2.4
160–229	—	—	24,601	2.9
230–329	—	—	22,518	2.7
330+	—	—	43,319	5.1
Total	6,126,123		846,320	

Compulsory school in Sweden currently encompasses nine years; older people might have two additional years of schooling, but this is no longer the case. Upper secondary school (twelve years) is now the standard and this explains why twelve years constitute a class of its own, before it is possible to embark on a university education. Those with university degrees were divided into those with degrees completed in less than four years and those with degrees completed in four or more years.

To measure income, we divided disposable incomes into quartiles (the first quartile had less than 99,000 SEK per year and the fourth quartile had more than 200,000 SEK).[4] Information on whether the migrant changed job in connection with migration was also included, as was information on occupation. The latter we divided into four groups: employed (including self-employed), students, unemployed, and an "other" category that mainly included retirees and those on long-term sick leave. Unlike the other characteristics, change of job and occupation are measured on 1 November. This produces an undercount (two months) of those who change jobs. Finally, included in Table 1 is the key factor of interest, migration distance, which was divided into nine categories of incrementally increasing length.

The differences between migrants and nonmigrants in Table 2 are as expected from the literature on migrant differentials. Clearly, these two groups are different in many respects, as only gender and presence of children in the household have an equal balance. Regarding the other characteristics, in comparison to nonmigrants, migrants tend to be much younger, with 34 percent in the age group between eighteen and twenty-five years, compared to 10 percent in the corresponding age group among nonmigrants. Some differences are not as clear; there is a slight tendency for migrants to have a higher level of education. Further, migrants are considerably more inclined to have changed jobs during the last ten months, more likely to have been unemployed or a student prior to moving, and to have lower incomes (the latter in part due to their younger age and in part due to the greater risk of having been unemployed).

Administrative Regions

We used six different types of administrative regions for our analysis (number of units in brackets): parish (1,785), municipality (290), LA region (local labor market regions, 78), county (21), Nomenclature des Unités Territoriales Statistiques (NUTS) 2 (8), and NUTS 1 (3). The *parish* is an administrative level mainly used by the Swedish church (the Swedish church is nowadays separated from the Swedish state), but many municipalities still use it for analytical purposes. (With few exceptions, parishes merge into municipalities and Statistics Sweden presents annual reports on the number of inhabitants for all parishes.) *Municipalities* constitute the most important administrative level in Sweden. Municipalities are independent bodies with an elected local government, taxation rights, and various responsibilities such as running child care, schools, and health services. Further, *LA regions* are based on

Table 3. Descriptive statistics for the administrative areas, 2008

	Parish	Municipality	LA region	County	NUTS 2	NUTS 1
Number of areal units	1,785	290	78	21	8	3
Area (mean) km^2	N/A	1,522	5,809	21,024	55,188	147,169
Area (median) km^2	N/A	692	4,201	11,422	38,953	80,823
Area (*SD*) km^2	N/A	2,668	5,554	24,639	50,532	143,186
Population (mean)	5,186	31,918	121,792	440,771	1,157,024	3,085,412
Population (median)	2,074	15,258	37,111	273,382	1,096,085	3,526,807
Population (*SD*)	7,928	62,477	310,926	499,434	612,935	1,223,094
Population density (mean) km^2	N/A	131	22	42	66	42
Population density (median) km^2	N/A	25	15	30	30	50
Population density (*SD*) km^2	N/A	458	29	62	96	34

Note: The total area of Sweden is 441,506 km^2. No up-to-date maps at the parish level were available. LA region = local labor market region; NUTS = Nomenclature des Unités Territoriales Statistiques.

commuting patterns and constructed from municipalities but do not merge into counties and have no administrative functions. *Counties* constitute the second type of administrative regions that, along with municipalities, has real administrative functions—partly via the "Landsting," which overlaps counties—such as taxation rights and some responsibilities for public services (municipalities are merged into counties). The *NUTS 2* and *NUTS 1* regions are determined by the European Union for comparisons across member countries but are rarely used for internal analyses in Sweden (counties are merged into NUTS regions).

Two maps of Sweden, showing the municipality and the NUTS 2 levels as examples, along with a table that presents statistics on area and population for the administrative levels are provided in Figure 3 and Table 3.

Method

We use the following logistic regression model to calculate the odds ratio for an individual with certain characteristics to move across varying types of administrative borders or distance: $\text{odds}(Y = 1) = e^{\ln[\text{odds}(Y = 1)]} = e^{(\alpha+\beta_1 X_1 + \beta_2 X_2 + \dots \beta_k X_k)}$ where Y is the dependent variable (moving across a specific migration distance or a specific administrative border), $X_1, \dots, X_k$ constitute the independent variables (migrant characteristics), and $\beta_1, \dots, \beta_k$, are the regression parameters that determine the relationship between the odds ratio and the independent variables. When reporting the results of the regressions we present the relative odds, exp (β), and the Nagelkerke R^2, which is a "pseudo" R^2 that can be used for comparisons across regressions (higher values suggest a greater fit of the model). Furthermore, for all regressions we report the number of migrants, also including those in the reference group (the numbers might seem large but this is because most migrants fall into the reference category). The shortest migration distance and smallest administrative areas are only used as reference categories in the regressions.

Unlike actual distance, administrative borders, as proxies for distance, cannot be divided into comparable classes to fit a multinomial model (a migrant might cross different types of borders in a single move). Therefore, binary logistic regressions were employed to investigate how the characteristics of migrants vary across actual distances and administrative borders. (Nine distance classes or models are used; fewer classes were also tested but the division of distance into nine classes presented the most distinct results and one regression for each type of administrative border, using movers within parish borders as the base for all regressions.)

This approach, where a migrant can cross additional borders in a spatial administrative hierarchy, means that moves over the longest distances will be present in all administrative border regressions. Only shares of the shortest distances are gradually excluded when the spatial resolution decreases. Even though this might disguise the effect of distance, the approach mimics conventional use of administrative borders as proxies for distance where, regardless of which administrative border is employed as a migration-defining boundary, the longest distances are always present. As far as we are aware, it is not possible to make a direct statistical test in one statistical model between actual migration distance and administrative borders as proxies for distance for the simple reason that a large variation of actual migration distances are present in the administrative border regressions. Hence, only comparisons between the models

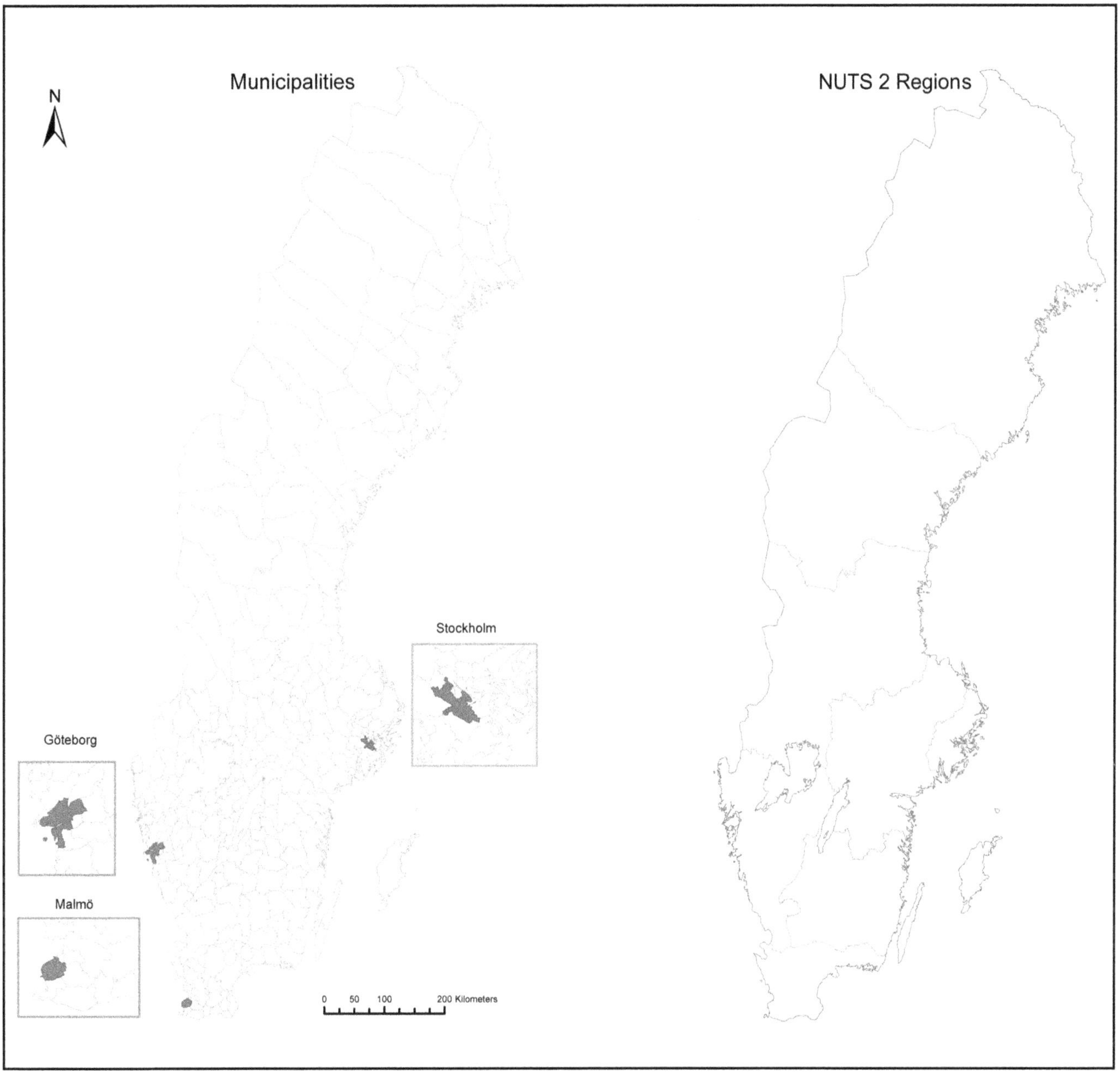

Figure 3. Municipalities (left) and Nomenclature des Unités Territoriales Statistiques (NUTS) 2 regions (right) of Sweden. The three metropolitan municipalities and their surroundings inserted. (Color figure available online.)

are feasible. To test the robustness of the regressions, all independent variables were included in a stepwise manner and changes in the coefficients were analyzed. This resulted in a reduction of the independent variables and classes originally included. A 5 percent sample was also tested, but the results were clearly unsatisfactory. This is quite expected considering the highly skewed distribution of migration distances whereby too few migrants moving longer distances are included when a small random sample is used. In countries where only samples of migration data are available, it is thus clearly advisable to use a sample of migrants stratified by migration distance. Regressions were also run where the island of Gotland (the largest island located off the southeast coast; see Figure 3) was excluded because anyone migrating from the island will have to move a distance of at least 100 km to reach the mainland (Gotland is a municipality, a local labor market region, and a county even though it holds only approximately 60,000 inhabitants). Excluding Gotland did not have an effect on the outcome of the models, however, so migrants to and from Gotland were included.

Results

Descriptive Results

Table 4 presents the first detailed description of migration distances for a whole country in which all movers moving a distance longer than 141 m are included. In 2008, more than 840,000 people changed residence, distributed over 190,000 areal units. The mean distance moved was 53 km, but the median was only 5 km. The mode was zero, revealing that the most common move is shorter than 500 m (all distances were rounded to the nearest kilometer). It is further notable that the minimum distance is zero (again, due to rounding) regardless of administrative border employed, even when Sweden is divided into only three regions (column 7). Evidently this means that someone moved a distance shorter than 500 m but happened to cross a NUTS 1 region border (the largest administrative area), showing that what might appear as merely a theoretical problem (i.e., including short-distance movers when employing large administrative areas) also exists in reality.

The number of migrants is reduced by almost 300,000 when parish borders are used as the migration-defining boundaries and the median distance increases to 14 km. When the municipality border is used as the migration-defining boundary—as is occasionally done in Swedish migration studies, in part because of data availability and in part because of the importance of municipalities as administrative units—distance increases further to a median of 55 km. As illustrated more clearly in Figure 4, a large share of migrants across municipality borders are not even close to being considered "long-distance" movers. Evidently, researchers should be wary of using municipality borders as migration-defining boundaries, and these results cast doubt on studies that have done so, unaware of the large share of short-distance movers.

Fortunately, as most policy implications drawn from research on internal migration in Sweden implicitly assume a focus on long-distance migration, it has become more common to use local labor market regions (LA regions) as migration-defining boundaries. Figure 4 shows that there is a difference in migration distance compared with municipalities where the median distance is 190 km for LA regions compared to only 55 km for municipalities. Even so, large shares of those who cross LA regional borders do not move very far. Approximately 20 percent move distances less than 81 km, arguably within commuting range in the most populous parts of Sweden with well-developed infrastructure.

The number of units decreases from seventy-eight to twenty-one when county borders are employed instead of LA regions, but migration distances increase only modestly and there is no longer a drastic reduction in the number of migrants. Now, the 20 percent of migrants moving the shortest distance reach 92 km, likely beyond commuting distance and close to the definition of migration as a move exceeding 100 km employed in this research.

Table 4. Migration distances in kilometers across different administrative borders, 2008

	All movers	Parish	Municipality	LA region	County	NUTS 2	NUTS 1
Number of areal units	190,092	1,785	290	78	21	8	3
Number of migrants	846,320	555,955	301,453	156,311	141,050	119,613	74,233
Mean	53	80	141	250	268	297	380
Median	5	14	55	190	212	245	349
Mode	0	2	16	66	62	62	396
SD	131	155	189	209	213	216	222
Minimum	0	0	0	0	0	0	0
10th percentile	0	2	9	51	58	68	141
20th percentile	1	3	15	81	92	116	197
30th percentile	1	6	22	112	132	163	247
40th percentile	3	9	33	151	171	201	294
60th percentile	9	21	90	240	261	299	397
70th percentile	17	42	160	307	335	376	453
80th percentile	41	108	247	398	410	447	513
90th percentile	173	265	401	513	519	549	687
Maximum	1,498	1,498	1,498	1,498	1,498	1,498	1,498

Note: Except for the two first rows, all values are rounded to the nearest kilometer. For further details see text. LA region = local labor market region; NUTS = Nomenclature des Unités Territoriales Statistiques.

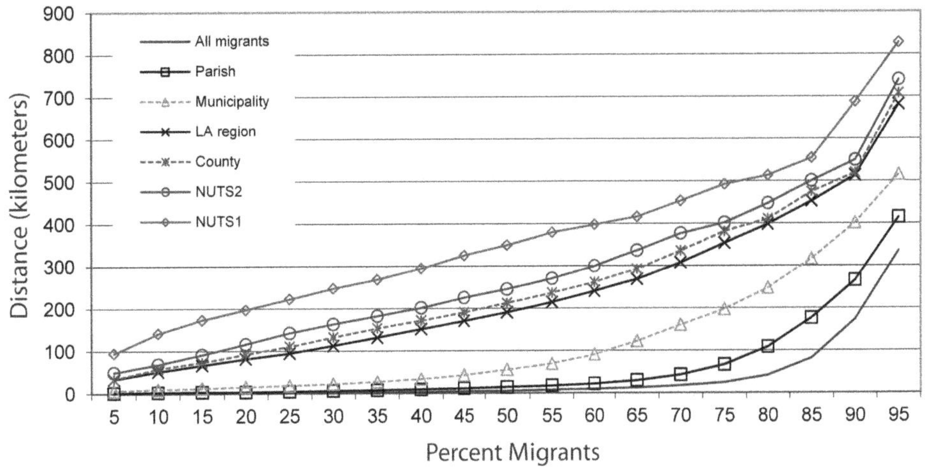

Figure 4. Distribution of migrants across distance and administrative borders, 2008. (Color figure available online.)

The division of Sweden into eight or three regions (NUTS 2 and NUTS 1, respectively) further increases migration distances, consolidating the mean and median values. For researchers concerned with not including a significant number of short-distance migrants, the NUTS 1 regional division is the most feasible choice. The 10 percent of migrants who cross this type of border and move the shortest distances reach 141 km, and even the 5 percent are in the vicinity of 100 km.

Despite some obvious expectations (i.e., fewer administrative units produce fewer movers who move longer than average distances), the picture is more complicated. For example, comparing LA regions (seventy-eight units) with counties (twenty-one units), we might expect substantial differences in the number of movers and their moving distances (as was the case when comparing municipalities [290 units] with LA regions [78 units]). Only marginal differences arise, however, which calls into doubt straightforward expectations about the relationship between number of administrative units and moving distances.

These descriptions clearly demonstrate that the administrative borders approach for defining migration is problematic. A large share of movers travel only across short distances, a pattern present almost regardless of border type. Hence, there is a risk that migration scholars will confuse long-distance migrants with short-distance ones. Although such evidence has not been available until now, it is far from evident what the implications of these findings are.

Regression Results

Table 5 presents the results of the logistic regressions. Women are somewhat less likely to move across longer distances than men. The greatest difference in this regard is seen for migration over 160 to 229 km, where women are 1.11 (1/0.90) times less likely to move. Recalling Table 2, where descriptive characteristics of migrants and nonmigrants were presented, it is interesting to note that women actually showed a higher propensity of moving versus staying than men.

Young people generally move the most, and this is evident also across migration distances. As a rule, age differences become more pronounced over longer distances, mainly longer than 110 km, but it is notable that the age group from forty-one to sixty-four years old has about the same probability of moving over short distances (less than 70 km) as the youngest age group. It is also not surprising to find that people who have children living in their household have a lower propensity to move longer distances. This does not mean that families with children are immobile: In fact, they are 1.1 times (1/0.91) more likely to move over 10 to 19 km compared to those without children, but this relationship changes as distance increases. Presumably, this is related to parents' reluctance to uproot their children, which is inevitably the case when moving longer distances (although it could be argued that for small children in particular, even moving a few kilometers is likely to be a significant change). There appears, however, to be a threshold effect of distance; migration distances exceeding 40 km do not change the estimates to any noteworthy extent (with the exception of the longest distances).

The results for education are mixed even though the general pattern is clear: The higher the level of education, the higher the probability of moving over longer distances. For the shortest moves, however, the highly educated display a lower probability of moving.

Table 5. Results of binary logistic regressions for migration across distance (migration across 0–9 km is reference category in all regressions)

	Migration distance							
Category	10–19 km Exp(B) Sig.	20–39 km Exp(B) Sig.	40–69 km Exp(B) Sig.	70–109 km Exp(B) Sig.	110–159 km Exp(B) Sig.	160–229 km Exp(B) Sig.	230–329 km Exp(B) Sig.	330+ km Exp(B) Sig.
Sex (reference = male)								
Female	0.99	0.99	0.96***	0.94***	0.92***	0.90***	0.94***	0.93***
Age (reference = 18–25)								
26–40	0.95***	0.88***	0.75***	0.68***	0.63***	0.65***	0.64***	0.76***
41–64	1.01	1.06***	0.98	0.87***	0.69***	0.66***	0.65***	0.67***
65+	0.74***	0.62***	0.59***	0.50***	0.52***	0.48***	0.44***	0.45***
Children at home (reference = yes)								
No	0.91***	1.03**	1.12***	1.11***	1.05**	1.08***	1.06***	1.17***
Education (reference = 11 years)								
12 years	1.03**	1.02	1.08***	1.14***	1.32***	1.39***	1.50***	1.50***
13–16 years	0.93***	0.85***	1.22***	1.40***	1.62***	1.87***	1.93***	2.17***
≥ 17 years	0.86***	0.84***	1.65***	2.09***	2.28***	2.99***	3.20***	3.75***
Income (reference = first quartile)								
Second quartile	0.90***	0.90***	0.85***	0.82***	0.86***	0.84***	0.82***	0.79***
Third quartile	0.90***	0.90***	0.77***	0.71***	0.70***	0.67***	0.64***	0.62***
Fourth quartile	0.94***	0.95***	0.81***	0.65***	0.60***	0.59***	0.57***	0.58***
Job change (reference = no job change)								
Job change	1.09***	1.21***	1.52***	1.87***	2.12***	2.32***	2.34***	2.56***
Occupation (reference = employed)								
Unemployed	0.87***	0.96**	1.07***	1.15***	1.18***	1.10***	1.14***	1.14***
Student	0.80***	0.88***	1.24***	1.40***	1.54***	1.53***	1.52***	1.48***
Other	0.98	1.10**	1.31***	1.43***	1.35***	1.31***	1.52***	1.47***
Constant (B)	–1.465	–1.992	–2.703	–3.105	–3.366	–3.263	–3.395	–2.935
N	612,475	576,862	549,819	538,715	534,243	538,659	536,576	557,377
Nagelkerke R^2	0.005	0.006	0.018	0.039	0.058	0.074	0.076	0.100

$^{*}p < 0.05$.
$^{**}p < 0.01$.
$^{***}p < 0.001$.

For those with a university degree, the probability of moving is 1.65 times higher than for those with only a compulsory education (over 40–69 km) and increases to 3.75 times over 330 km. That this group stands out so markedly over the longest distances is clearly interesting and could possibly be explained by the uneven geography of universities in Sweden, although further research would be required to establish this.

Whereas the results for education are largely in line with the literature, the results for income are not. There is a lower probability of moving with increasing income. For example, migrants in the highest income quartile are approximately 1.7 to 2 times less likely to move distances longer than 75 km compared to migrants in the lowest income quartile. It should be kept in mind, however, that the income variable does not measure change of income.

As expected, people are more likely to change jobs the farther they move. It is also expected that people who were unemployed prior to moving have higher odds of moving longer distances compared to those who were employed prior to moving. In this case it is interesting to note that this effect does not increase with a linear progression for the longest distances. Students have somewhat higher odds compared to the unemployed.

Turning to Table 6, where administrative borders are employed as proxies for distance, the results appear at first to be very similar to those of Table 5. On closer

Table 6. Results of binary logistic regressions for migration across administrative borders (migrants within parish is reference category in all regressions)

	Migration across type of border					
Category	Parish Exp(B) Sig.	Municipality Exp(B) Sig.	LA region Exp(B) Sig.	County Exp(B) Sig.	NUTS 2 Exp(B) Sig.	NUTS 1 Exp(B) Sig.
Sex (reference = male)						
Female	0.96***	0.95***	0.92***	0.93***	0.93***	0.93***
Age (reference = 18–25)						
26–40	0.83***	0.78***	0.66***	0.67***	0.68***	0.69***
41–64	0.77***	0.73***	0.65***	0.62***	0.61***	0.62***
65+	0.51***	0.40***	0.38***	0.36***	0.35***	0.37***
Children at home (reference = yes)						
No	1.32***	1.33***	1.36***	1.35***	1.37***	1.36***
Education (reference = 11 years)						
12 years	1.18***	1.24***	1.35***	1.40***	1.42***	1.48***
13–16 years	1.65***	1.81***	2.01***	2.27***	2.37***	2.44***
≥ 17 years	2.05***	2.44***	3.46***	3.84***	4.17***	4.44***
Income (reference = first quartile)						
Second quartile	0.85***	0.79***	0.78***	0.77***	0.76***	0.75***
Third quartile	0.77***	0.69***	0.62***	0.60***	0.59***	0.58***
Fourth quartile	0.86***	0.80***	0.60***	0.62***	0.60***	0.58***
Job change (reference = no job change)						
Job change	1.42***	1.74***	2.30***	2.34***	2.43***	2.58***
Occupation (reference = employed)						
Unemployed	1.00	1.02**	1.18***	1.14***	1.15***	1.15***
Student	1.18***	1.29***	1.55***	1.57***	1.57***	1.57***
Other	1.02	1.11***	1.30***	1.34***	1.37***	1.40**
Constant (B)	0.507	–0.153	–0.901	–1.057	–1.265	–1.790
N	846,320	591,818	446,676	431,415	409,978	364,598
Nagelkerke R^2	0.051	0.092	0.158	0.168	0.173	0.165

Note: LA region = local labor market region; NUTS = Nomenclature des Unités Territoriales Statistiques.
*$p < 0.05$.
**$p < 0.01$.
***$p < 0.0$.

inspection, however, it is evident that there is less dispersion for these estimates. Even though this is to be expected because there are fewer models in Table 6, the smaller dispersion is not random. Overall, the results are in agreement with the estimates gained from actual distance, but it should be remembered that the distance categories in Table 5 and the administrative borders categories in Table 6 are not directly comparable (the latter categories cover a wide range of distances).

Although women in this portion of the analysis have a lower propensity to move over long distance (about the same odds as in Table 5), we find a larger age effect across administrative borders. For example, comparing the odds of the oldest age group in Table 5 with the equivalent in Table 6, the odds range from 0.74 to 0.44 in Table 5 and from 0.51 to 0.35 in Table 6. Moreover, in Table 6, not having children living at home increases the probability of moving by 1.3 times, regardless of the type of administrative border. When actual distances were measured in Table 5, the odds were notably lower, and in the case of moving 10 to 19 km, even negative.

For education, the higher the level of education, the higher the probability of moving longer distances. For example, having spent seventeen years or more in school increases the probability of moving across a NUTS 1 border by 4.4 times compared to those with only compulsory education. Recalling Table 5, the odds became positive only for those moving distances longer than 40 km. Regarding income, the results are very similar when comparing Table 6 with Table 5.

Turning to employment and occupation, those moving across a parish border are 1.4 times more likely to change jobs compared to those moving within parish borders. Here, it is important to keep in mind that the median migration distance is only 14 km for those moving across parish borders, hardly a distance where job change is obvious. Similarly, the odds (1.7) of changing jobs when moving across a municipality border are higher than might be expected, considering that large shares of this group move very short actual distances. The overall impression is that the administrative borders approach shown in Table 6 produces higher odds for job change than might be expected given the relatively large shares of short-distance migrants present for all types of areal units. Finally, only minor differences are detectable for the last variable, occupation, when the odds in Table 6 are compared to those in Table 5.

Concluding Discussion

It has been 128 years since Ravenstein (1885) first identified the key problems with the spatial aspect of conventional definitions of migration. Meanwhile, migration scholars have had little choice but to hope that these problems are not too serious. This article has cast new light on these issues by investigating the relationship between actual migration distances and migration-defining boundaries. Using unusually detailed data on migration distances, this article has provided the first large-scale analysis of how, first, the volumes of migration and, second, the characteristics of migrants change when migration distances vary as well as when different types of migration-defining boundaries are employed. Two main findings stand out.

First, a significant share of migrants move only short distances, even when migration-defining boundaries of relatively large administrative regions are employed. This finding confirms long-held qualms and is clearly a cause for validity concerns. There is a risk of confusing short-distance migrants with long-distance ones, increasing the potential danger of drawing wrong conclusions about many aspects of migration. Even if the data employed in this article are unique, there are no obvious reasons to believe that the findings would not be generalizable to other developed countries (at least those with similar population geography; i.e., a high degree of urbanization and a relative concentration of population to coastal areas). The evidence presented in this article thus suggests that much previous migration research could be significantly hampered due to failure to account for actual migration distance. For example, if a study employs a random sample population and defines migrants as those who have changed the location of their dwelling, only a tiny fraction of those who changed their dwelling will be long-distance migrants. If the study population is not the intended one, the ability to draw valid conclusions is compromised.

A counterstrategy occasionally employed in previous research to mitigate the risk of including short-distance migrants has been to exclude migrants moving to adjacent administrative areas (usually referred to in the literature as noncontiguous migration). This strategy is rarely employed, however, possibly because little is known about the migration population included or excluded by such measures. More common are approaches that have tried to improve distance estimates, typically by measuring the centroids of administrative regions, sometimes weighted by population (Boyle and Flowerdew 1997), thus going beyond the standard approach of simply assuming that movers across certain administrative region types are long-distance migrants. Rogerson (1990) used geometrical probability to estimate migration distances under various assumptions related to the MAUP but the extent to which such measures correspond with actual migration distances remains to be explored. An appropriate next research step would therefore be to investigate the extent to which such measures are suitable for providing a more reasonable estimation of migrants. This is important because in the vast majority of countries, only migration flow data between relatively large administrative areas are available, making it virtually impossible to use actual distance as a migration-defining criterion (see Bell et al. 2002).

Another interesting finding was the nonlinear relationship between the number of administrative units and the number of migrants. A reduction in the number of administrative units does not automatically correspond with an expected reduction of migrants. This finding implies that at a certain point the different region types pick up similar population processes in spite of considerable areal differences.

Second, and perhaps somewhat surprising, the comparative analyses of how migrants' characteristics change across actual distance and administrative borders were more consistent than might be anticipated considering the above and previous research (Jun and Chang 1986; White and Mueser 1988; Biagi, Faggian, and McCann 2011). To be sure, differences do exist between the estimates gained from the two approaches, but they do not appear to be of such magnitude as to be cause for great concern. Hence, one interpretation of these findings is that even though there are obvious validity issues with the spatial definition of migration, previous research results on individuals' migration differentials appear not to have been affected by the inclusion of short-distance migrants. An alternative interpretation based on the fact that all migrants of an entire country were used and that a 5 percent sample failed to produce satisfactory results suggests that sample data on migration unstratified by distance will be less easily interpreted.

In addition, the article has advanced the literature by showing that migrant differentials exist also across distance and administrative borders, not only when investigating differences between migrants and nonmigrants. The generalized results showed that women, being older, having children, or high income are characteristics that decrease the propensity to move across longer distances. Conversely, being highly educated, having changed jobs, being unemployed, or being a student increased the propensity to move across longer distances. Again, there is little to suggest that these findings would not be generalizable to other countries, although caution is required because migrant differentials are more likely to be influenced by country-specific characteristics such as differences in welfare systems.

In essence, then, although this study has confirmed validity concerns held since the late nineteenth century, the question of whether these concerns are significant enough to produce invalid conclusions on migration is still open for debate and empirical investigation. For further research, important extensions include incorporating distance in analyses of the Swedish migration system where the explicit role of types of regional settings and the directionality of moves are explored. Finally, however, it needs to be emphasized that only differences in terms of migrant characteristics have been investigated here, not other types of migration determinants, or the effects on individual migrants or the societal consequences of migration. The relatively large shares of short-distance migrants present, almost regardless of the administrative migration-defining boundary employed, will have serious consequences in such cases.

Acknowledgments

The authors would like to thank reviewers for valuable comments on previous drafts.

Notes

1. Another partly related issue pertains to changes in statistical boundaries that make it problematic to compare migration over time (see, e.g., Blake, Bell, and Rees 2000; Amcoff 2006).
2. For readers who are used to working with origin–destination migration matrixes, instead of the data employed here, it might be instructive to know that these data potentially provide a migration matrix with more than 3.6 billion inhabited cells.
3. Note that we use the term *actual migration distance* simply to denote Euclidian distance measured with considerable accuracy. It should not be taken to imply that we know anything about the routes of transportation.
4. 1 SEK = 0.15 USD (17 June 2013).

References

Amcoff, J. 2006. Rural population growth in Sweden in the 1990s: Unexpected reality or spatial–statistical chimera? *Population, Space and Place* 12 (3): 171–85.

———. 2009. Rapid regional enlargement in Sweden: A phenomenon missing an explanation. *Geografiska Annaler: Series B, Human Geography* 91 (3): 275–87.

Bell, M., M. Blake, P. Boyle, O. Duke-Williams, P. Rees, J. Stillwell, and G. Hugo. 2002. Cross-national comparison of internal migration: Issues and measures. *Journal of the Royal Statistical Society: Series A (Statistics in Society)* 165 (3): 435–64.

Biagi, B., A. Faggian, and P. McCann. 2011. Long and short distance migration in Italy: The role of economic, social and environmental characteristics. *Spatial Economic Analysis* 6 (1): 111–31.

Blake, M., M. Bell, and P. Rees. 2000. Creating a temporally consistent spatial framework for the analysis of interregional migration in Australia. *International Journal of Population Geography* 6 (2): 155–74.

Boyle, P. J., and R. Flowerdew. 1997. Improving distance estimates between areal units in migration models. *Geographical Analysis* 29 (2): 93–107.

Congdon, P. 2010. Random-effects models for migration attractivity and retentivity: A Bayesian methodology. *Journal of the Royal Statistical Society: Series A (Statistics in Society)* 173 (4): 755–74.

Cooke, T. J., and P. Boyle. 2011. The migration of high school graduates to college. *Educational Evaluation and Policy Analysis* 32 (2): 202–13.

Gleave, D., and M. Cordey-Hayes. 1977. Migration dynamics and labour market turnover. *Progress in Planning* 8 (1): 1–95.

Gordon, I. 1991. Multi-stream migration modeling. In *Migration models: Macro and micro approaches*, ed. J. Stillwell and P. Congdon, 73–91. London: Belhaven.

Greenwood, M. J. 1975. Research on internal migration in the United States: A survey. *Journal of Economic Literature* 13 (2): 397–433.

Hägerstrand, T. 1957. Migration and area. In *Migration in Sweden: A symposium*, ed. D. Hannerberg, T. Hägerstrand, and B. Odeving, 27–158. Lund, Sweden: C.W.K. Gleerup.

Hunt, J. 2004. Are migrants more skilled than non-migrants? Repeat, return, and same-employer migrants. *Canadian Journal of Economics* 37 (4): 830–49.

Jun, I. S., and H. S. Chang. 1986. Functional forms and the relevance of contiguous migration in the study of migration and employment growth. *The Annals of Regional Science* 20 (2): 17–27.

Kalogirou, S. 2012. Testing local versions of correlation coefficients. *Jahrbuch für regionalwissenschaft* 32 (1): 45–61.

Lansing, J. B., and E. Mueller. 1967. *The geographic mobility of labor*. Ann Arbor: University of Michigan, Institute for Social Research.

Long, L., C. J. Tucker, and W. L. Urton. 1988a. Measuring migration distances: Self-reporting and indirect methods. *Journal of the American Statistical Association* 83 (403): 674–78.

———. 1988b. Migration distances: An international comparison. *Demography* 25 (4): 633–40.

Makower, H., J. Marschak, and H. W. Robinson. 1938. Studies in mobility of labour: A tentative statistical measure. *Oxford Economic Papers* 1 (1): 83–123.

———. 1939. Studies in mobility of labour: Analysis for Great Britain, Part I. *Oxford Economic Papers* 2:70–79.

Morrison, P. S., and W. A. V. Clark. 2011. Internal migration and employment: Macro flows and micro motives. *Environment and Planning A* 43 (8): 1948–64.

Niedomysl, T. 2011. How migration motives change over migration distance: Evidence on variations across socioeconomic and demographic groups. *Regional Studies* 45 (6): 843–55.

Olsson, G. 1965. *Distance and human interaction: A review and bibliography*. Philadelphia: Regional Science Research Institute.

Openshaw, S. 1984. *The modifiable areal unit problem*. Norwich, UK: Geobooks.

Plane, D. A. 1984. Migration space: Doubly constrained gravity model mapping of relative interstate separation. *Annals of the Association of American Geographers* 74 (2): 244–56.

Ravenstein, E. G. 1885. The laws of migration. *Journal of the Royal Statistical Society* 48 (2): 167–235.

Ritchey, N. P. 1976. Explanations of migration. *Annual Review of Sociology* 2: 363–404.

Robinson, W. S. 1950. Ecological correlations and the behavior of individuals. *American Sociological Review* 15 (3): 351–57.

Rogerson, P. A. 1990. Buffon's needle and the estimation of migration distances. *Mathematical Population Studies: An International Journal of Mathematical Demography* 2 (3): 229–38.

Rose, A. M. 1958. Distance of migration and socioeconomic status of migrants. *American Sociological Review* 23 (4): 420–23.

Roseman, C. C. 1971. Migration as a spatial and temporal process. *Annals of the Association of American Geographers* 61 (3): 589–98.

Sahota, G. S. 1968. An economic analysis of internal migration in Brazil. *Journal of Political Economy* 76 (2): 218–45.

Schwartz, A. 1973. Interpreting the effect of distance on migration. *The Journal of Political Economy* 81 (5): 1153–69.

Smith, A. 1776 [1981]. *An inquiry into the nature and causes of the wealth of nations*. Chicago: Phoenix Books.

Smith, T. E. 1978. A cost-efficiency principle of spatial interaction behavior. *Regional Science and Urban Economics* 8 (4): 313–37.

Statistics Sweden. 2011. Integrated database for labour market research: 1990–2009. *Background Facts, Labour and Education Statistics* 2011:4.

Stewart, J. Q. 1948. Demographic gravitation: Evidence and applications. *Sociometry* 11 (1–2): 31–58.

Stillwell, J. C. H., and P. Congdon. 1991. *Migration models: Macro and micro approaches*. London: Belhaven.

Stouffer, S. A. 1940. Intervening opportunities: A theory relating mobility and distance. *American Sociological Review* 5 (6): 845–67.

———. 1960. Intervening opportunities and competing migrants. *Journal of Regional Science* 2 (1): 1–26.

White, M. J., and P. R. Mueser. 1988. Implications of boundary choice for the measurement of residential mobility. *Demography* 25 (3): 443–59.

Wong, D. 2009. The modifiable areal unit problem (MAUP). In *The Sage handbook of spatial analysis*, ed. A. S. Fotheringham and P. A. Rogerson, 105–24. Los Angeles: Sage.

Zax, J. S. 1994. When is a move a migration? *Regional Science and Urban Economics* 24 (3): 341–60.

Zipf, G. K. 1946. The PP/D hypothesis: On the intercity movement of persons. *American Sociological Review* 11:677–86.

———. 1949. *Human behavior and the principle of least effort*. Reading, MA: Addison-Wesley.

The Tactics of Asylum and Irregular Migrant Support Groups: Disrupting Bodily, Technological, and Neoliberal Strategies of Control

Nick Gill,* Deirdre Conlon,† Imogen Tyler,‡ and Ceri Oeppen§

*Department of Geography, Exeter University, UK
†School of Geography, Leeds University, UK
‡Sociology Department, Lancaster University, UK
§Department of Geography, Sussex University, UK

States are exercising an increasing array of spatial strategies of migration control, including in the area of asylum migration. Drawing on interview data with thirty-five British and American irregular migrant and asylum support groups (MASGs), this article explores the spatial "tactics" (De Certeau 1984) employed by MASGs in response to strategies of migration control. We consider their infiltration of highly securitized physical spaces like detention centers and courts. We analyze their appropriation of control technologies and discuss their exploitation of inconsistencies within the neoliberalization of controls. These tactics highlight the importance of resistive actions that are carried out "within enemy territory" (De Certeau 1984, 37). As such, they represent a complementary set of actions to more radical forms of protest and consequently enrich our understanding of the diversity of forms of resistance.

国家正逐渐增加运用空间策略配置进行移民控管，包含难民迁徙的区域。本文运用三十五个英国与美国的非正式迁徙与难民支持团体 (MASGs) 的访谈资料，探讨这些团体回应移民控管策略的空间 "战术" (De Certeau 1984) 。我们考量这些团体对于诸如拘留中心和法院等高度安全化的实体空间的渗透。我们分析他们如何挪用控管技术，并探讨他们如何利用新自由主义化控制中的不一致性。这些战术凸显出 "在敌人的领域上" 从事抵抗行动的重要性 (De Certeau 1984, 37) 。据此，它们代表着对于更为激进的抗议形式的补充行动，并因而丰富了我们对于多样化的抵抗形式之理解。

Para controlar la migración, los estados están echando mano del creciente número de estrategias espaciales asociadas a la migración por asilo. Con base en los datos de entrevistas de treinta y cinco grupos de apoyo británicos y norteamericanos al migrante irregular y a los asilados (MASGs), este artículo explora las "tácticas" espaciales (De Certeau 1984) empleadas por los MASG como respuesta a las estrategias utilizadas por el control migratorio. Consideramos la infiltración de espacios físicos altamente seguros por estos grupos, como son los centros de detención y los juzgados o cortes. Analizamos su apropiación de tecnologías de control y discutimos su explotación de las inconsistencias dentro de la neoliberalización de los controles. Tales tácticas relievan la importancia de acciones de resistencia efectuadas "dentro de territorio enemigo" (De Certeau 1984, 37). En esa condición esas tácticas representan un conjunto de acciones complementarias a formas más radicales de protesta y que en consecuencia enriquecen nuestro entendimiento de la diversidad que caracteriza las formas de resistencia.

This article examines what can be done, and what is being done, in response to the increasingly harsh and exclusionary asylum systems that countries such as the United States and United Kingdom maintain. The asylum systems of most developed countries have been transformed since the 1980s through parallel processes of devolution, international cooperation, and privatization of migration controls. Developed states have deployed various geographical measures as part of these broad trends and innovations. The differential slowing down and speeding up of procedures and apparatuses of control to suit governmental logics illustrates a power-laden use of pacing and timing of border control processes (Bailey et al. 2002; Conlon 2011; Hyndman and Giles 2011). The exteriorization and devolution of migrant controls away from national territories through processes of offshore, extraterritorial interception and the use of remote, largely invisible detention and processing facilities constitutes a set of powerful and exclusionary spatial measures (Darling 2009; Mountz 2010; Coleman 2012).

It would be understandable to see these developments as a coherent and immutable thicket of strategies against which resistance is all but impossible. Drawing on De Certeau's (1984) concept of "tactics," however, this article discusses some of the struggles irregular migrant and asylum support groups (MASGs) carry out against this dominant constellation of strategies from a marginal and dominated position. The article therefore presents a rare view of counterhegemonic efforts to undermine exclusionary border control practices by operating within or close to, but against, systems of subjugation and exclusion. There is a significant body of work that examines radical activist actions against border controls that seek their wholesale replacement with a new system of rights and obligations, such as "No Borders" groups that advocate the outright abolition of border controls (see, for example, Nicholls 2011; Loyd, Mitchelson, and Burridge 2012; Oliveri 2012; Rigby and Schlembach 2013), but less work has addressed the place of "non-reformist reforms" (Apple 1995, 120) within immigration control that seek nontrivial improvements and alterations to existing practices and procedures.

MASGs are a compelling object of analysis because they are currently under considerably more pressure than before the financial crisis of the late 2000s. Reductions in funds available from both public and private sources have precipitated closures or downsizing of MASGs and heightened competition among them for remaining funds in both the United States and the UK (Gill et al. 2012). More financial resources have simultaneously been funneled toward border enforcement activities by states. The increasing regulation and criminalization of support for irregular migrants, with more frequent prosecution of those housing, transporting, or employing overstayers, for example, as well as stricter regulation of legal advice, have added to the challenges that MASGs encounter (Fekete 2009). Their dwindling room for maneuver is a worrying barometer of the health of counterhegemonic practices in the age of global neoliberalism.

The article draws on primary research with thirty-five MASGs in the United Kingdom and United States conducted in 2010 and 2011. MASGs are heterogeneous, varying by size, motivation, and scale of operation. Given the paucity of research on this sector, our sample was purposefully inclusive. Half of the groups sampled employed less than ten people, one third self-identified as faith motivated, one fifth self-identified as politically motivated (not necessarily different groups), and over two thirds cited concern for human rights as a primary motivation. Few had connections to other MASGs in different countries. Another important fault line among MASGs is the degree to which they are financed by public funds or are prepared to work with public or private organizations involved in securing borders. We focus on groups that have developed close—yet critical—relationships with either state organizations or private immigration enforcement operations. Our interest is in the motivations and ramifications of this geographical "proximity" to established nodes and networks of subjugating institutional power.

First, we briefly review why De Certeau (1984) differentiates "strategies" from "tactics." We then draw on the analysis of our data, which involved systematic coding of interview transcripts, to present three on-the-ground tactics used by many of our respondents. Although we make no claims that these tactics—which relate to the body, technological innovations, and exploiting the contradictions within the neoliberalization of migrant controls—are exhaustive of the tactical approaches employed by MASGs, they are significant because they were repeatedly discussed by respondents as effective in interrupting the effects of exclusionary constellations of border control.[1]

Strategies and Tactics in Migration Control

The asylum and migration enforcement systems in the United Kingdom and United States enact forms of spatial violence. They lay claim to a place, the "United Kingdom" or "United States," whose borders operate by identifying individuals who are conceived as "out of place" (Cresswell 1996). For individuals subjected to the consequent controls, the result is often "disorder and hardship" (Coleman 2009, 912). De Certeau describes how subjugating social forces maintain influence through "strategies." A strategy assumes that a space can be circumscribed as proper, thus implicitly defining an "improper" space, in which hegemonic groups can isolate the subject of their will and power (De Certeau 1984, xix).

Yet, for De Certeau, certain *tactics* remain available despite the strategies of dominating groups. These tactics are "the procedures of everyday creativity" (De Certeau 1984, xiv) and are at least as important as the strategies of the powerful. They are the "clever tricks" and "knowing how to get away with things" (xix) that present themselves in the course of even the most subjugated of everyday lives. Although they are marginal, De Certeau urges us not to underestimate them.

A question arises as to what forms of power tactics are effective against. Foucault's distinction between sovereign and disciplinary power proves useful here. Tactics that rework particular configurations within existing systems of control risk strengthening sovereign power by implicitly shoring up the legitimacy of the sovereign. This is their principal flaw; they contest the way that strategies are effected rather than contesting the very right to effect a strategy. This can result in petitioning the sovereign power, which performs its authority.

If tactics risk strengthening sovereign power, however, De Certeau (1984) is clear that they address disciplinary power more effectively. In reference to Foucault he writes:

> If it is true that the grid of "discipline" is everywhere becoming clearer and more extensive it is all the more urgent to discover how an entire society resists being reduced to it, what popular procedures . . . manipulate the mechanisms of discipline and conform to them only in order to evade them? (xiv)

From this we suggest that De Certeau's tactics align with Foucault's conception of power as a productive and relational force rather than one that is only understood as top-down and always oppressive.[2]

For De Certeau (1984) then, the strength of tactics is that they are able to "manipulate the mechanisms" (xiv) or, as Katz (2004) has put it, *rework* power relations "on the very grounds on which they are cast" (247). Hence, tactics operate within disciplining systems, turning formations of disciplinary power against themselves. "Although they are composed within the vocabularies of established languages," they "trace out the ruses of other interests and desires that are neither determined nor captured by the systems in which they develop" (De Certeau 1984, xviii).

On the one hand, then, tactics seem ill fated in the face of the dominance they react against. They are only ever "manipulations of imposed spaces, tactics relative to particular situations" (De Certeau 1984, 24). On the other hand, tactics can be strikingly effective. They involve "victories of the 'weak' over the 'strong'" (xix) on the "home ground" of the established order (26). In the next section we draw on MASGs' experiences to outline three types of on-the-ground tactics.

The Body as a Site of Tactics

The emergence of asylum as an abjectifying technology of governance constitutes populations without formal rights and protections. It would be a mistake, however, "to assume that all we are left with is a wasteland of abject subjectivities, discarded and jettisoned from political life" (Nyers 2004, 207). On the contrary, "with each act of desubjectification comes another attempt of resubjectification by both the state and political subjects themselves" (Nyers 2004, 207; Tyler 2013). The body is a key site through which counterhegemonic tactics can be enacted, often deep within subjugating spaces. Detainee hunger strikes and lip sewing are material and symbolic acts of defiance that can jolt public sensibilities in powerful ways (Perera 2002). They articulate an uncontainable excess at the level of the body that subjugating strategies cannot control (Wolfram and Minahan 2004; McGregor 2011; Conlon 2013; Tyler 2013).

Similarly, some of the tactics used by MASGs combine the times and spaces of immigration enforcement with bodily actions. Our interviewees were involved in supporting asylum seekers and irregular migrants through various corporeal activities. For example, one U.S. faith-based MASG, deeply disturbed by the buses used to deport undocumented and irregular migrants from a county jail, decided to focus on gaining access to the buses to offer pastoral care and spiritual support to migrants whose deportations were imminent. Although predominantly Christian, the group operates as an interfaith coalition providing spiritual support to detained migrants of all religions. One group member described some of the struggles the group faced. They combined "getting on the buses," "laying down in front of the buses," "a lot of publicity," and "peaceful and consistent" requests to the jail's management and to Immigration and Customs Enforcement (ICE) officials regionally and nationally. They persisted for five years and eventually gained access to the jail and the buses, enabling members of the group to pray with and offer comfort to the detainees and deportees. Reflecting on their tactics, one group member remarked, "We couldn't take no for an answer . . . I think with us they [prison authorities] kept saying 'Well if we just ignore 'em they'll finally give up' and [then] they found out we wouldn't!"

What was important for this activist group was physical, bodily presence in the same place, on the buses, with the detainees and deportees. Presence is used to communicate solidarity and compassion, as a way to monitor the treatment of detainees, and to enable spiritual and emotional connections with them. Such bodily tactics resonate with what Askins (forthcoming) describes as the "quiet politics" of "meaningful encounters" that have become increasingly significant as the "proper space" of refuge for irregular migrants shrinks.

The same group describes the importance of their physical presence at court hearings determining the fate of the migrants they support. In response to the unpredictable behavior of local immigration judges, the group initiated a program of court observations. This involved organizing supportive members of the public to sit in court galleries when detention and deportation cases were being heard so that the judges and clerks "know you're there, that you care about those people and that you're seeing what's going on." Although some of our respondents claimed that "judges don't embarrass," others were convinced that such tactics make a difference. They have seen improvements in the relationship among judges, lawyers, and applicants; a less hostile court atmosphere; and fewer instances of intimidating behavior on the part of judges. These court observations are now a large-scale form of activism, drawing in student volunteers from nearby universities. This activity is termed *court-watching*, a nomenclature that is significant because the work of watching the state is an important form of "sousveillance" (inverse surveillance) that reflects the disciplining thrust of panoptic power back on state actors.

These forms of activism mobilize a specific activist body involved in processes of blocking and watching. Neither of these tactics represents a challenge to the existing system of asylum seeker management likely to cause its wholesale failure. Indeed, in both cases the demands were relatively modest: The supporters of the soon-to-be-deported asylum seekers merely wanted to offer spiritual and emotional support and the court watchers demanded a more transparent and consistent system rather than an end to detention and deportation or a wholesale overhaul of the judicial arbitration of these enforcement regimes. In this sense they resemble De Certeau's maneuvers within a system of power and are distinct from, although arguably complementary to, more radical approaches like those used by No Borders and the sanctuary movement, who advocate a complete rethinking of the allocation of rights and responsibilities pertaining to borders. They nevertheless represent responses that draw attention to the practices of state actors, like judges and detention facility guards, and expose migrant governance to scrutiny.

Technological Tactics

There are echoes of what has been termed "political jiu jitsu" (Sharp 1973) in these migrant activist tactics: the use of an adversary's energy against them to unbalance them. As De Certeau (1984) writes:

> A tactic . . . must play on and with a terrain imposed on it and organized by the law of a foreign power. It does not have the means to *keep to itself*, at a distance, in a position of withdrawal, foresight, and self-collection; it is a manoeuver "within the enemy's field of vision." (36–37)

States use technological innovations in the governance of asylum seekers and irregular migrants. Dean (1999, 212) defines technologies as "forms of notation [and] ways of collecting, transporting and storing information." Technologies are central to the strategy of marking out territory, and the intensification of the use of digital and other surveillance technologies at the border has been well documented (Amoore and Hall 2009; Martin 2010). As part of this inscription, accumulation, and standardization there has been a "veritable 'avalanche' of printed numbers" (Rose 1999, 211), from censuses to registers, that allows those who govern to assume centrality at the hub of an informational infrastructure.

Technological records, however, do not only facilitate state activities but also hold them in check by providing an opportunity to increase accountability. One of our interviewees, Jasmin,[3] was supporting a trauma survivor, Lucy. After being refused asylum in the United Kingdom and given a date for her deportation, Lucy had made multiple suicide attempts and was not fit to travel. Despite repeatedly trying to present the Home Office with new information about her case in writing, she had been instructed to attend a meeting in London in person, several hours' journey away. Jasmin recalled:

> This was someone who couldn't get out of bed, who was in the fetal position twenty-four hours a day in her bedroom. . . . Totally disengaged, just believing that [this was] the end of her life.

After navigating a difficult journey and long queues, Lucy and Jasmin attended the meeting. Jasmin described how on arrival they received a numbered ticket, indicating their queue position, their location, and the date. After reaching the end of the queue they arrived at the desk only to be told that the application should have been made in writing. Jasmin explained:

> I can't swear but what I'm convinced happened is that she [the official] looked on the computer and saw that

> [Lucy] was going to be picked up next morning because it was next morning that they tried to pick her up [to be deported]. If she had managed to make her application they couldn't do that. So [the official] deliberately lied to us and refused to take the application.

To compound the situation, the Home Office official refused to give her name to Lucy and Jasmin, thus denying them information that could have helped them in an objection to this situation. Furthermore, after returning from the restroom, Jasmin discovered that, in her absence, Lucy had been asked to return the numbered ticket they had earlier received. Lucy, however, had realized that the ticket was the only proof that they had been present in London on that date and had put it in Jasmin's briefcase earlier. The ticket became a key piece of evidence in their official complaint, which led to a judicial review, the reopening of the case and, eventually, to Lucy receiving refugee status.

Although Lucy was deeply subjugated throughout this ordeal, her experience shows that the mundane technologies of the state, often used to cement the link between government and information through the "numericization of the population" (Rose 1999, 210), can sometimes be subverted and set against their own original purposes. In other words, the paper trail that the state generates to numerically document, order, and govern can be usurped and put to use holding bureaucratic procedures to account. Thus, even deep within subjugating systems there are opportunities for "putting one over" on the established order (De Certeau 1984, 25). As De Certeau observes, "The actual order of things is precisely what 'popular' tactics turn to their own ends, without any illusion that it will change any time soon. . . . Into the institutions to be served are thus insinuated styles of social exchange, technical invention, and moral resistance" (26).

MASGs' responses to video linking in courts are another tactic that turns subjugating technologies to alternative ends. Governments in both the United States and United Kingdom increasingly provide a two-way on-screen facility that allows applicants to be seen in the court and vice versa, but from a distance. Video links are intended to facilitate efficiency; more cases can be heard, as judges do not have to travel to detention centers. They can reduce time spent in detention, help clear court backlogs (which can be very lengthy in the United States; see Transactional Records Access Clearing House 2013), and, ultimately, speed up the time between arrival and possible deportation.

Almost all of our respondents were dismayed at how video-linking undermines detainees' cases, which echoes concerns voiced previously by pressure groups (Bail for Immigration Detainees and the Refugee Council 2008; Campaign to Close Campsfield 2011). In addition to describing various technical drawbacks associated with video-linking, interviewees detailed how difficult it is for lawyers to communicate with video-linked clients if there is a need for private discussion while the case is being heard. Alternatively, sometimes a lawyer will be with a client during the video-linking, but this means that the lawyer is not present in the courtroom, which has its own disadvantages. In this manner, detained migrants and legal representatives are relegated to the inauspicious space of the detention facility while immigration judges and state officials govern from a distance.

Nevertheless, notwithstanding these disadvantages, one respondent's organization explicitly exploited the technology to strengthen migrants' cases. Health professionals are regularly asked to give expert testimony in courts, but accessing detainees and attending immigration hearings, frequently in remote locations, often means giving up a day's work. Offering doctors and psychologists the opportunity to use video-linking strengthened the ability of our respondent to recruit expert testimonies in support of migrants' accounts. As one of the organization's coordinators put it:

> Our . . . position is that [video-linking] might be a good thing. . . . The in-court testimony can be difficult because [health professionals] may end up using their whole day because, of course, you have a certain time that it's supposed to happen but you can end up waiting around in court. And a lot of the doctors really don't like it. Particularly if it means they're giving up a vacation day or giving up time when that could be seeing paying patients.

To the extent that video-linking technology has been paid for by the state and put in place to expedite legal procedures that typically lead to deportation, this usurping of its intended purpose represents the possibility of "divert[ing] resources . . . [i]n the very place where the machine . . . reigns supreme" (De Certeau 1984, 25).

Exploiting the Contradictions Within Neoliberal Migrant Controls

Both the formal privatization of border controls and the more insipid roll-out of logics of territorial exclusion away from the state toward actors as diverse as teachers, lawyers, doctors, and the "citizen-detective" (Vaughan-Williams 2008) have been well-documented. In the United States this has been associated with a large-scale "delegation program" (Coleman 2009, 907) that results in heightened responsibilities of an increasing number of local actors for a proliferating range of migration enforcement issues. For Coleman, many immigration policies in the United States aim to render undocumented migrants "more legible" (911) to immigration authorities and consequently enroll various organizations in practices of detection and control, including driver's license issuers, tax authorities, real estate agents, hospitals, and schools.

In the United Kingdom, too, devolution of enforcement has picked up pace with plans to require landlords to monitor the immigration status of tenants (Travis 2013). In addition, the contracting of private security companies such as Serco, Mitie, and G4S has not only underpinned the expansion of the immigration detention system from a capacity of 250 in 1993 to more than ten times that number by 2005, but has also seen the development of payment-by-results contracts to detect and deport overstayers (British Broadcasting Corporation 2012) and the involvement of these same security firms in the provision of regular, supposedly noncarceral, housing for asylum seekers (Grayson 2012).

The neoliberalization of immigration enforcement in both countries, including the privatization, delegation, and devolution of controls, allows the state to deflect criticism of migrant control policies onto privately contracted companies and local deputies. It can also provide a mechanism of cooptation of third-sector organizations that purport to support migrants, who work with security companies in the delivery of "services" to migrant groups. The involvement of Barnado's, the prominent UK children's charity, in a new secure center for migrant families awaiting deportation highlights this risk.

Yet even in this bleak neoliberal landscape there are opportunities to tactically resist. Our respondents identified a number of ways that they were able to exploit the internal contradictory logics within the neoliberal project of border control and enforcement. The high cost of detaining irregular migrants in the United States and United Kingdom is a case in point. No matter whether detention exists as a reaction to migrant inflows (McBride 1999), as a deterrence strategy (Martin 2012), or as part of migrants' initiation into neoliberal society (Conlon and Gill 2013), its presence announces a performative and expanding state. Today, with states under pressure to reduce spending, MASGs can use economic costs and the neoliberal mantra of austerity to their advantage. Where arguments against detention that draw on human rights discourse have limited impact, calls for reduced state expenditure have much more traction. UK-based research with 188 long-term detainees conducted in 2007–2008 demonstrated that only 18 percent were eventually deported (Detention Action 2009). Given that detention cost £68,000 ($110,000) per detainee for one year in 2005–2006, each individual deportation costs the UK taxpayer almost £700,000 ($1.13 million; Detention Action 2009).[4] In the United States, the federal budget for ICE, which controls migrant detention, increased from $3.1 billion to $5.9 billion (an 87 percent increase) between 2005 and 2012 (Meissner, Kerwin, Chisti, and Bergeron 2013).[5] With such large outlays, heightened concern about government debt and expenditure "makes . . . it easier to say, 'Look you shouldn't be detaining this many people, we don't have the funds as a nation. This isn't a good use of our money or our taxpayer dollars'" (U.S.-based respondent). Utilizing a neoliberal lexicon in this way bolsters arguments against detention. Our respondent continued, "We've said all along that we think this nation overdetains immigrants. Now we can say we overdetain immigrants and it's really expensive and we can't afford it."

By way of illustration, recent sequester cuts have prompted the release of several hundred immigrants from detention centers across the United States (Semple 2013), demonstrating the value of these counter-hegemonic tactics.

Other tactical advantages arise as the overextension of the delegation of border controls causes a communication breakdown between the central state and locally contracted or subcontracted agents. Our interviewees revealed ways in which they were able to play organizations and departments that operated at different scales off each other, exploiting weaknesses in communication between scales of government and private security firms, and working together to share information faster than migration control organizations to strengthen advocacy. One advocate commented of the state, "One wonders if there is a strategic plan in place. It seems rather chaotic the way that the system is managed, right? I think it's because [the government] is so large."

If the government seems chaotic, however, this disarray contrasts with increasingly coordinated horizontal interdependence between MASGs. Although financial austerity has put pressure on many organizations, it has also prompted them to find ways to work together in an attempt to overcome the damaging competition for funds that austerity promotes. As one UK-based organization put it, "Oddly enough now there is no money, we're finding less to argue about." Our research revealed that a high proportion of our participants are linked to other groups as part of larger umbrella organizations that fund programs and activities, as well as providing other resources. Others form working group coalitions that are issue specific, including developing a coordinated response to recent funding cuts in the United Kingdom, responding to government policies such as detention, coordinating around benefits issues, and preparing funding bids.

One of the advantages of this horizontal interdependence is that it allows the chaotic inconsistencies between layers of state administration to be quickly identified, communicated, and exploited. One organization involved in national-level advocacy work in the United States reflected on how helpful it is to have local organizations feed issues to them:

> It's hard for our organization to see what it looks like on the ground. But we also have much more of an inside voice and recognition within the federal government, and so we're able to advocate. Then the local organizations . . . have such great information and experience because they're working so much more closely with the clients themselves and with the impacted population. [So, we] recognize where I sit versus where the people on the ground sit and recognizing the strengths and the barriers to both of those positions.

This interdependent relationship also allows MASGs to feed information to national groups discretely, which might be important for local groups that "cannot say something. So for example they need to stay in the good graces of the [local officials] at the end of the day . . . they need their clients not to be punished for the actions that they are doing . . . [but] they can tell us things that we can take back to headquarters." These relationships render information agile in a way that bureaucratic and cumbersome, multilayered government structures struggle to emulate. In the United States, the federal government has introduced a public advocate to liaise with immigrant support groups. Although the position has been criticized by conservatives as playing into the hands of undocumented migrants and by immigration advocates as little more than a public relations stunt, the fact that groups can now demand the "transparency and accessibility" (Foley 2012) that accord with the public advocate's role highlights how governments must respond to the increasingly organized and adroit tactics employed by MASGs. Even though groups operate within a deeply subjugating system, their tactics exemplify a deftness, which allows MASGs to seize "the possibilities that offer themselves at any given moment" (De Certeau 1984, 37).

Conclusion

Either by necessity or because of their political motivations, most MASGs in the United States and United Kingdom seek to improve conditions for asylum seekers and irregular migrants by dealing with the state on its own terms through what we can conceive of in geographical terms as tactical closeness and proximity to institutionalized centers of power. This means that MASGs often do not explicitly challenge the political ideologies and systems of border controls that have led to the abjection of asylum seekers and migrants, even while they contest specific governmental policies and practices. As such, they "alter the organization but not the polarization of power relations" (Sparke 2008, 2).

This approach risks the possibility of being coopted by states. For example, providing welfare to destitute migrants arguably enables governments to diminish welfare support. These kinds of "humanitarian" tactics arguably reduce "protest to a contest over 'the possible' which can only ever mean, at a fundamental level, a conservative acceptance of the existing framework for grasping problems and their solutions" (Rigby and Schlembach 2013, 166; Tyler 2013). Nevertheless the kinds of tactics described in this article do strive toward system change, even if this change is only gradual or incremental. As De Certeau (1984) writes, tactics can be likened to "the snowy waves of the sea slipping in among the rocks and defiles of an established order . . . that it in fact gradually erodes and displaces" (34). The closeness of MASGs to the state also allows the possibility of using tactics to disrupt migration enforcement "within enemy territory" (De Certeau 1984, 36–37), whether that is the physical territory of a detention center or predeportation bus, or the virtual territory of a technological paper trail. Although most of our respondents were not explicitly challenging the underlying systems and structures that shape migration enforcement in the United States and United Kingdom, the tactics they use represent a powerful and increasingly common means by which exclusionary migrant controls are consistently troubled on the ground. Rather than seeing tactics as a poor alternative to radical activism, then, they might be seen to provide an important complement to more radical work.

Acknowledgments

We gratefully acknowledge the financial support of the Economic and Social Research Council, grant number RES-000-22-3928-A. Thanks also to Richard Wright for editorial assistance and to the anonymous referees for their helpful suggestions. All errors remain our own.

Notes

1. Among several other tactics detailed were the development of collaborations and networks across the sector to counteract the effects of funding cuts and fragmentation. For a detailed account of responses by MASGs to austerity and government funding cuts see Gill et al. (2012).
2. It is beyond the scope of the article, however, to give a full analysis of the relationship between De Certeau's and Foucault's conceptions of power.
3. All such names are pseudonyms.
4. Based on 2005–2006 costs. Exchange rate: £1.00 = $1.617 (January 2007).
5. Equivalent to an increase from £2 billion to £3.8 billion. Exchange rate £1.00 = $1.517 (May 2013).

References

Amoore, L., and A. Hall. 2009. Taking people apart: Digitised dissection and the body at the border. *Environment and Planning D: Society and Space* 27:444–64.

Apple, M. W. 1995. *Education and power*. London and New York: Routledge.

Askins, K. Forthcoming. Being together: Everyday geographies and the quiet politics of befriending. *ACME: An International E-Journal for Critical Geographies*.

Bail for Immigration Detainees and the Refugee Council. 2008. Immigration bail hearings by video link: A monitoring exercise. http://www.refugeecouncil.org.uk/OneStopCMS/Core/CrawlerResourceServer.aspx?resource=A9E70BD8-C72A-4013-A31C-7247FADAFD2C&mode=link&guid=fe44150bf7e940af87ea01b9c60cf0b5 (last accessed 8 October 2012).

Bailey, A., R. Wright, A. Mountz, and I. Miyares. 2002. (Re)producing Salvadoran transnational geographies. *Annals of the Association of American Geographers* 92: 125–44.

British Broadcasting Corporation. 2012. Capita gets contract to find 174,000 illegal immigrants. http://www.bbc.co.uk/news/uk-politics-19637409 (last accessed 8 October 2012).

Campaign to Close Campsfield. 2011. Immigration bail hearings: A travesty of justice? Observations from the public gallery. http://closecampsfield.files.wordpress.com/2011/03/ccc-bop-report-low-res.pdf (last accessed 8 October 2012).

Coleman, M. 2009. What counts as the politics and practice of security, and where? Devolution and immigrant insecurity after 9/11. *Annals of the Association of American Geographers* 99:904–13.

———. 2012. The "local" migration state: The site-specific devolution of immigration enforcement in the U.S. South. *Law and Policy* 34 (2): 159–90.

Conlon, D. 2011. Waiting: Feminist perspectives on the spacings/timings of migrant (im)mobility. *Gender, Place & Culture: A Journal of Feminist Geography* 18:353–60.

———. 2013. Hungering for freedom: Asylum seekers' hunger strikes—Rethinking resistance as counter-conduct. In *Carceral spaces: Mobility and agency in imprisonment and migrant detention*, ed. D. Moran, N. Gill, and D. Conlon, 133–48. FarnhamUK: Ashgate.

Conlon, D., and N. Gill. 2013. Gagging orders: Asylum seekers and paradoxes of freedom and protest in liberal society. *Citizenship Studies* 17 (2): 241–59.

Cresswell, T. 1996. *In place-out of place: Geography, ideology, and transgression*. Minneapolis: University of Minnesota Press.

Darling, J. 2009. Becoming bare life: Asylum, hospitality, and the politics of encampment. *Environment and Planning D: Society and Space* 27:649–65.

Dean, M. 1999. *Governmentality: Power and rule in modern society*. London: Sage.

De Certeau, M. 1984. *The practice of everyday life*. Berkeley: University of California Press.

Detention Action. 2009. Detained lives: The real cost of indefinite immigration detention. http://detentionaction.org.uk/wordpress/wp-content/uploads/2011/10/Detained-Lives-report.pdf (last accessed 2 October 2012).

Fekete, L. 2009. Europe: Crimes of solidarity. *Race & Class* 50:83–97.

Foley, E. 2012. Immigrant advocates skeptical on new DHS advocate position. *The Huffington Post* 7 February. http://www.huffingtonpost.com (last accessed 17 May 2013).

Gill, N., D. Conlon, C. Oeppen, and I. Tyler. 2012. *Networks of asylum support in the UK and USA: A handbook of ideas, strategies and best practice for asylum support groups in a challenging social and economic climate*. Exeter, UK: Exeter University. http://steedee.files.wordpress.com/2012/03/networks-of-asylum-support-print2.pdf (last accessed 14 November 2013).

Grayson, J. 2012. *Another shambles as G4S is entrusted with running asylum seeker housing*. http://www.opendemocracy.net/ourkingdom/john-grayson/another-shambles-as-g4s-is-entrusted-with-running-asylum-seeker-housing (last accessed 1 October 2012).

Hyndman, J., and W. Giles. 2011. Waiting for what? The feminization of asylum in protracted situations. *Gender, Place and Culture* 18:361–80.

Katz, C. 2004. *Growing up global: Economic restructuring and children's everyday lives*. Minneapolis: University of Minnesota Press.

Loyd, J., M. Mitchelson, and A. Burridge, eds. 2012. *Beyond walls and cages: Prisons, borders and global crisis*. Athens: The University of Georgia Press.

Martin, L. 2010. Bombs, bodies, and biopolitics: Securitizing the subject at the airport security checkpoint. *Social & Cultural Geography* 11:17–34.

———. 2012. "Catch and remove": Detention, deterrence, and discipline in US noncitizen family detention. *Geopolitics* 17 (2): 312–34.

McBride, M. 1999. Migrants and asylum seekers: Policy responses in the United States to immigrants and refugees from Central America and the Caribbean. *International Migration* 37 (1): 289–317.

McGregor, J. 2011. Contestations and consequences of deportability: Hunger strikes and the political agency of non-citizens. *Citizenship Studies* 15:597–612.

Meissner, D., D. Kerwin, M. Chishti, and C. Bergeron. 2013. *Immigration enforcement in the United States: The rise of a formidable machinery*. Washington, DC: Migration Policy Institute.

Mountz, A. 2010. *Seeking asylum: Human smuggling and bureaucracy at the border*. Minneapolis: University of Minnesota Press.

Nicholls, W. 2011. Cities and the unevenness of social movement space: The case of France's immigrant rights movement. *Environment and Planning A* 43:1655–73.

Nyers, P. 2004. What's left of citizenship? *Citizenship Studies* 8:203–15.

Oliveri, F. 2012. Migrants as activist citizens in Italy: Understanding the new cycle of struggles. *Citizenship Studies* 16:793–806.

Perera, S. 2002. What is a camp . . . ? *Borderlands* 1:1.

Rigby, J., and R. Schlembach. 2013. Impossible protest: NoBorders in Calais. *Citizenship Studies* 17:157–72.

Rose, N. 1999. *Powers of freedom: Reframing political thought*. Cambridge, UK: Cambridge University Press.

Semple, K. 2013. Mass release of immigrants is tied to impending cuts. *New York Times* 26 February:A1. www.nytimes.com (last accessed 17 May 2013).

Sharp, G. 1973. *The politics of nonviolent action*. Boston: Porter Sargent.

Sparke, M. 2008. Political geography: Political geographies of globalization III—Resistance. *Progress in Human Geography* 32:1–18.

Transactional Records Access Clearing House. 2013. Immigration data tools and application. http://www.trac.syr.edu/immigration/tools (last accessed 17 May 2013).

Travis, A. 2013 Immigration bill—The devil's in the detail. *The Guardian* 8 May. http://www.guardian.co.uk (last accessed 17 May 2013).

Tyler, I. 2013. *Revolting subjects: Social abjection, resistance and dissent in contemporary Britain*. London: Zed Books.

Vaughan-Williams, N. 2008. Borderwork beyond inside/outside? Frontex, the citizen-detective and the war on terror. *Space and Polity* 12:63–79.

Wolfram, J., and S. Minahan. 2004. Unravelling Woomera: Lip sewing, morphology and dystopia. *Journal of Organizational Change Management* 17:292–301.

Chaos and Crisis: Dissecting the Spatiotemporal Logics of Contemporary Migrations and State Practices

Alison Mountz* and Nancy Hiemstra†

*Balsillie School of International Affairs, Wilfrid Laurier University
†Department of Cultural Analysis and Theory, State University of New York at Stony Brook

Discussions of chaos and crisis feature prominently in contemporary literature on the nation-state, globalization, security, and neoliberalism. In scholarship, policy, and public discourse on global migration, references to chaos and crisis abound and yet have not received sustained critical inquiry. In this article, we examine why these alarmist terms emerge so frequently, scrutinizing in particular what is relayed about the contemporary "state" of migration and its new geographical expressions. We take the recurrent discourse of chaos and crisis as our starting point, with the goal of delving more deeply into the logics driving their mobilization. We analyze when and where these discourses become prominent and trace key moments when migrants become securitized (the border crossing, the detention center, the deportation). This examination of the spatiotemporal logics of chaos and crisis advances understandings of the often contradictory efforts by nation-states to facilitate mobility and containment. We argue that states mobilize constructions of chaos and crisis to create exceptional moments in which sovereign reach and geopolitical influence are expanded. By analyzing the use of these discourses as tools, the article provides insight to the relationship between migration and state power.

当代有关国族国家、全球化、安全与新自由主义的文献, 以探讨混乱与危机为主要的特徵。在有关全球迁徙的学术着作、政策及公共论述中, 大量存在着混乱与危机的指涉, 但却尚未充分受到批判性的探问。我们将在本文中, 检视这些危言耸听的词彙为何经常出现, 并特别探究何者接替了当代的迁徙 "状态" 及其新的地理展现。我们将不断出现的混乱与危机论述当作出发点, 旨在更进一步探究驱动上述论述的动员逻辑。我们分析这些论述在何时、何地变得显着, 并追溯迁徙变得安全化的主要时间点 (跨越边界、拘留中心, 以及驱逐出境) 。检视混乱与危机的时空逻辑, 增进了理解国族国家经常以矛盾的施为来促进流动与进行吓阻。我们主张, 国家动员了混乱与危机的建构以创造例外时刻, 其中主权所及范围和地缘政治的影响得以扩张。透过分析这些论述做为工具般的使用, 本文对于迁徙和国家力量之间的关係提供了洞见。

La discusión sobre caos y crisis es notable en la literatura contemporánea, en especial la referida al estado-nación, la globalización, la seguridad y el neoliberalismo. En la erudición, políticas y en el discurso público dedicados a la migración global abundan las referencias al caos y a la crisis, pero estos temas en general todavía no reciben una atención crítica sostenida. En este artículo, examinamos por qué estos términos alarmistas aparecen con tanta frecuencia, concentrándonos en particular en lo que se dice sobre el "estado" contemporáneo de la migración y sus nuevas expresiones geográficas. Como punto de partida tomamos el recurrente discurso sobre caos y crisis, con la intención de ahondar mucho más en las lógicas que alientan su movilización. Analizamos cuándo y dónde alcanzaron prominencia estos discursos y marcamos los momentos claves cuando el peso de la seguridad es aplicada a los migrantes (el cruce de la frontera, el centro de detención, la deportación). Este examen de las lógicas espacio-temporales del caos y la crisis hace avanzar los entendimientos de los esfuerzos a menudo contradictorios por las naciones-estado que buscan facilitar tanto movilidad como contención. Argüimos que los estados movilizan construcciones del caos y la crisis para crear momentos excepcionales cuando se expandan el alcance de la soberanía y la influencia geopolítica. Al analizar la utilización de estos discursos como herramientas, el artículo brinda mucha mayor comprensión a la relación existente entre la migración y el poder del estado.

Discussions of chaos and crisis figure prominently in contemporary scholarship and public discourse about the nation-state, globalization, security, migration, and neoliberalism. In the arena of global migration, references to chaos and crisis abound and yet have not received sustained critical inquiry. Why do these terms emerge so frequently in discourses surrounding migration? What do they tell us about the contemporary state of migration and its geographical expressions? We take the recurrence of chaos and crisis as our starting point, with the goal of delving more deeply into the logics driving the discourse. We argue that examination of the spatiotemporal logics of these terms advances understandings of the often-contradictory efforts by nation-states to facilitate mobility and containment through geographical extensions of sovereign power. By analyzing the use of chaos and crisis in recent work on sovereignty, state, and the securitization of migration, the article lends insight to the relationship between migration and state power.

Politicians, bureaucrats, policymakers, journalists, scholars, and petty sovereigns (Butler 2004) all prove fluent in the language of chaos and crisis. Drawing on the *Oxford English Dictionary*, we understand chaos to be the appearance of confusion and disorder and crisis as a condition of sudden change that causes alarm. Although each emerges at particular moments, both hold in common the projection of danger, instability, panic, and dramatic upheaval, and both surface frequently at the nexus of human mobility with state power. In a neoliberal era where bureaucrats must work within frames of risk assessment and the bottom line, this language is first and foremost a discourse of states.[1] It is used by government employees when they discuss human smuggling, spontaneous arrivals, and border enforcement (Mountz 2010), and it is circulated by media in the front-page sensationalism of large migration events. Whether narrating the crossing of Syrians into Turkey, Mexicans into the United States, or Eritreans into Italy, each conjures notions of humanitarian disaster and images of disorder.

Scholarship on migration engages and responds to these uses of chaos and crisis in ways that simultaneously mirror and challenge the trend. Many researchers have observed that state responses to migration and border enforcement practices are steeped in rationales of crisis and chaos. Some scholars reference these themes explicitly, whether writing about spontaneous mass arrivals (Mountz 2010), border enforcement (Andreas 2000; Brown 2010; Nevins 2010), or building on Agamben's (2005) arguments about states of emergency (Coleman 2007; Gregory 2007; Secor 2007). Still others conduct analyses couched in language that conveys parallel underlying logics. We contribute a more sustained consideration of the spaces of securitization created through the contemporary freefall into chaos and crisis. By tracing key moments and sites wherein migrants become securitized (the border crossing, detention, deportation), we identify times and spaces in which expressions of disorder and alarm become prominent.

This language is often invoked regarding zones of transition identified by Agamben (1998) as thresholds. Walters (2008) suggests that migration scholars pay more attention to the thresholds where land meets sea. We build on his analysis to show that crises repeatedly erupt where migrants attempt to cross between regions of great economic disparity (which are often places where land meets sea). We focus on thresholds because they are sites of crossing narrated as exceptional, overwhelming, and threatening regardless of the actual number of migrants involved. Investigations of these sites frequently lean on Agamben's mobilization of zones of exception. At thresholds of human mobility, we argue, sovereign power is reconfigured through new geographies of border enforcement.

Human migration often involves crossing territorial borders imbued with ambiguity around jurisdiction and legality. The crisis that now stands in for crossings in the sites discussed here, however, links migration to fundamental changes in political geography. States mobilize the language of chaos to affect lasting reconfigurations of geographies of sovereignty. In the very zones of border crossings where crises are constantly produced, states use this discourse to expand sovereign claims and powers. This expansion involves the extension of territorial claims, such as states' assertion of the right to enforce at the source of threats to their state, wherever that might be (Mountz 2010; Mountz and Hiemstra 2012). This is particularly evident in the externalization of asylum processing (Collyer 2007; Hyndman and Mountz 2008), the offshore interception and processing of migrants (Mountz 2011; Mountz and Hiemstra 2012), and the internal expansion of policing power (Coleman 2007). Like Agamben, we argue that discourses of chaos and crisis are tied intimately to geographical assertions of sovereign power. The exceptionalism of crisis cloaks expansion of sovereign reach and geopolitical influence in migration governance (Mountz 2010). Therefore, deeper understandings of the logic underlying much discourse, policy, and research on migration also offer potential to analyze sovereign powers and practices.

We do not concern ourselves here with policy-driven distinctions between economic migrants and asylum seekers but instead observe that securitization blurs these categories, rendering them less intelligible in processes of enforcement and exclusion. Furthermore, we move intentionally among geographical regions to demonstrate that the logic and language of crisis do not exist only in one region where seemingly exceptional crossings occur but instead operate in a similar fashion across times and places. Particular kinds of bodies are repeatedly recast as security threats and associated with chaos throughout transnational fields of travel and enforcement as the forces we detail next work to criminalize, securitize, and exclude.

We proceed by first discussing in more depth how we understand chaos and crisis to be working. Our third section follows the routes where transnational migrants experience these logics in their journeys across borders and between states. In our conclusions, we address some of the implications of these discourses.

Conceptualizing Chaos and Crisis in Scholarship on Migration

In this section, we analyze the scholarship on migration that highlights narratives of chaos, crisis, emergency, and exception. We trace their antecedents, recurrence, and implications for political geographic analyses of migration and sovereignty. Economic disparity among regions often produces fraught sites of crossing and policing, and therefore crisis. This reality quickly becomes apparent when comparing narratives of immigration at particular national borders. For example, the U.S.–Mexico border, where the disparity in standard of living is starkly evident, is the source of more attention and panic regarding foreign (and now terrorist) invasion than the U.S.–Canada border. Similarly, the borders separating European Union (EU) and non-EU states inspire seemingly endless public discussion of political doom and economic gloom in Europe, regardless of the risks posed by intra-EU mobility (Huysmans 2000; Carr 2012). Significant gaps in economic resources between migrants and citizens (whether presumed or real) propel narratives of threat, risk, and danger in destination countries. These discourses portray migrants as greedy, self-serving foreigners motivated by financial gain (Greenberg and Heir 2001; Nevins 2010). Such depictions spring from fears that resources are scarce and immigrants take jobs from citizens (Honig 1998). They are typically linked to nativists insisting on distinctions between "us" and "them" and depicting immigrants as threats to the national fabric (Higham 2002; Carr 2012; Mainwaring 2012). Likewise, economic disparity is inscribed onto individual bodies through racial and cultural markers, so that individuals in transit, at borders, and inside particular national spaces are regarded differently based on presumed economic status according to appearance (De Genova 2005; Bhandar 2008; Nevins 2010).

Processes of racialization and criminalization simultaneously propel narratives of chaos and crisis and mask their effects. Belief in the supremacy of whites underlies Western liberal ideals, even in states espousing multiculturalism (Hage 2000). Discourses of disorder and danger to the nation act as a cover for racist imaginaries targeting particular groups who appear to threaten this supremacy (Greenberg and Hier 2001; Nevins 2010). In the past decade, panic in the wake of terrorist attacks in the United States, Spain, London, India, and elsewhere has driven a dramatic intensification of racialization and criminalization, as well as a decreased need to mask racist assumptions regarding immigrant criminality (Coleman 2007; Staeheli and Nagel 2008). Immigrants who are racialized as non-white are more easily narrated as criminals who endanger the security of citizens. Additionally, as Nevins (2010, 148) pointed out, attacks on immigrants are often simultaneously attacks on the poor and women (a fact often obscured with simple charges of racism). Media images of punished and confined bodies act recursively as performative spectacle, confirming fears that the targeted population is, indeed, dangerous (Mainwaring 2012). In Malta, for example, the government's mobilization of a crisis narrative reinforces processes of criminalization and marginalization and, consequently, the Maltese have increasingly demonstrated racism and xenophobia in reactions to migrant arrivals (Mainwaring 2012).

As states feel threatened by processes related to globalization, moments of apparent chaos surrounding migration provide opportunities through which nation-states reassert themselves. Scholars commonly argue that ongoing changes resulting from growing global interconnectedness generate instability and contradiction at multiple scales. Sassen (1996, xvi), for instance, points to problems in the existing economic, political, and social order generated by the "existence of two different regimes for the circulation of capital and the circulation of immigrants." Cuttitta (2006, 28) identifies a "crisis of territoriality" caused by technological developments altering distances between peoples and countries. This time–space compression

(Harvey 1989) in turn throws the traditional idea of the fixed physical border—so central to the concept of the sovereign state—into crisis. Amid neoliberal capitalist order, issues related to human mobility—not the mobility of capital—become a focus of the resulting anxiety about sovereignty (Bigo 2002). For example, the building of border walls to keep out unwanted bodies and goods represents attempts by nation-states to reaffirm sovereignty, despite the increased mobility of elite bodies, capital, and goods (Brown 2010).

In the neoliberal era, narratives of crisis are mobilized by both public and private interests who gain financially from hardening immigration enforcement regimes. Proponents of free-market economics have repeatedly taken advantage of disasters to make reforms that ultimately advance neoliberal policies, what Klein (2007) calls "disaster capitalism." Klein's hypothesis bears out in the realm of immigration enforcement, where some nonstate actors (e.g., companies that run detention facilities or sell surveillance technologies) profit from punitive immigration laws (Heyman 2012).[2] Bigo (2002) points to the pivotal role of nongovernment "security professionals" in presenting immigrants as threats to the nation in ways that legitimate and advance securitization. Additionally, governments' increasing adoption of neoliberal capitalist metrics for measuring success and ensuring their own continued existence means that state actors (e.g., agencies involved in border patrol) also benefit from chaos and crisis-driven interpretations of human migration (Heyman 2012).

Similarly, the privatization of immigration detention means that corporations potentially gain through criminalization (Flynn and Cannon 2009). In multiple places, private interests adeptly employ discourses of chaos to promote passage of laws expanding the detention and deportation net (Hiemstra 2013). Without adequate regulation and oversight, few mechanisms counter the machinations of those who gain from intensified pursuit of immigrants.

Finally, the contemporary negotiation of national and individual identity in zones between regions relies on narratives of chaos and crisis. The idea of fixed territory anchoring the state is the foundation of the modern Westphalian system. Borders thus function as physical perimeters and parameters for establishing and shaping national identity, economic and ontological security, and particular racial and ethnic imaginaries (Nevins 2010). Images of people crossing boundaries generate feelings of insecurity that, in turn, justify intensified processes of bordering (Walters 2004). Therefore, rhetoric that escalates fear of immigrant invaders and borders out of control—in combination with the terrorist threat—propels the expansion of immigration and border control (Andreas 2000; Nevins 2010) and the proliferation of walls between states (Brown 2010; Jones 2012). Narratives of disorder and alarm catalyze the expansion of sites where bordering occurs, as states assert the right to police places and persons perceived as a threat both within and outside of territorial boundaries (Coleman 2007; Walters 2008; Mountz 2011; Mountz and Hiemstra 2012).

This overview points to significant themes. First, fear and insecurity emerge as driving forces. Fear is a powerful emotion that can profoundly shape everyday life (Hyndman 2007; Pain and Smith 2008) and has dramatic effects on immigration regimes when harnessed by state and private actors. Particularly in the aftermath of 11 September 2001, personal and national security and belonging are fundamentally defined in opposition to the figure of the immigrant (Kaplan 2003; Walters 2004). Therefore, whatever the particular reason cited for increased immigration securitization—national identity, sanctity of borders, or risk—fear and insecurity feature centrally (Amoore and de Goede 2008). Second, processes of securitization often lead to discourses and practices of exceptionalism regarding immigration. As such, many scholars find Agamben's (1998, 2005) exceptionalized *homo sacer* violently included in the juridical order only by virtue of her exclusion, in the figure of the everyday migrant and refugee (Perera 2002; Pratt 2005; Coleman 2007). Third, these eruptions of chaos and crisis reconfigure geographies of sovereignty. Immigrant-receiving states effectively expand claims of sovereign power while whittling away at states' traditional role of protecting citizens (Walters 2004). New techniques of migration governance are evolving as dispersed assertions of sovereign power merge with increasingly biopolitical modes of controlling individual mobility.

Transnational Journeys: From Migration to Chaos and Crisis

State actors and mainstream media tend to depict particular thresholds as chaotic and crisis driven, exceptional in nature. These places are points of crossing, stasis, and limbo where states deploy chaos to suspend migrants in time and space for prolonged periods. They are often sites that involve ambiguity around jurisdiction and legal status (Mountz 2011). They might be nodes in global networks such as ports of entry that serve what Heyman (2004) identifies as a sorting function.

Here, we trace these global networks to understand the transnational migration journey that tends to be illicit and circuitous, rather than documented and direct, and to outline how chaos and crisis are broadcast and woven into the production of these sites for consumption by large publics. The routes, we show, function cyclically, rather than the more linear time–space trajectories that once dominated assimilationist understandings of immigration. Human rights monitors often refer to these cyclical movements as *revolving doors* of perpetual rejection and repeated attempts (Human Rights Watch 2008; *Médecins sans Frontières* 2011). We now discuss a few of these moments (border crossing, detention, deportation) to examine embedded discourses of disorder and alarm, emphasizing that these are not sequential but circular in form. Examination of these moments also reveals how they are leveraged to broaden state power.

When undocumented migrants are en route to their intended destination, specters of chaotic border crossings by land and sea often take hold. In the United States, images abound of migrants at the U.S.–Mexico border scaling walls, running across highways, and hiding in vehicle compartments. In addition to this idea of the steady flow of arrivals (see Ellis and Wright 1998), perceived hysteria around the mass spontaneous arrival contributes to ideas of the border under threat. Since high-profile arrivals of the 1990s in Australia, Europe, and North America, the boat has become a symbol of chaotic rather than orderly migration. Although all foreign arrivals wishing to settle Britain's colonies and the "new world" came by boat, contemporary boat arrivals in those same places signal illicit behavior and lack of control construed in such a way as to detract from rather than advance nation-building projects (Mountz 2010). Some of these arrivals are so iconic that their mere mention conjures fears of disorder and bodies out of control and place (Cresswell 1997): the Mariel boatlifts that brought Cubans to the United States in 1980 and the interception of overcrowded, unsafe boats in waters between North Africa and Spain. When the *Golden Venture* was grounded in the New York harbor in 1993, media images of Chinese migrants making dangerous and desperate leaps into frigid water to reach shore exacerbated public perceptions of chaos. In response, then-President Bill Clinton created an interdepartmental task force on human smuggling and increased resources devoted to interception and detention. As Kwong (1997) shows, such government initiatives do little to decrease smuggling, but they increase vulnerability for migrant laborers after their arrival.

Illicit journeys by land, too, evoke narratives of borders under threat, as the media projects images of migrants crowded into trucks, treks through wastelands between countries, and brutalized abandoned bodies. Whether journeys are made by sea or land, the figure of the nefarious smuggler or snakehead often rises to the fore in public discourse. The illegality of smuggling is linked to the racialization and criminalization of migrants and fears of terrorism.

The sense of chaos that accompanies undocumented journeys and out-of-control borders produces perceived threats to national identity, economy, and sovereignty (Honig 1998; Hage 2000). These constructions of disorder become gateways to exceptionalism and opportunities for states to legitimate enforcement activities in ways that result in the expansion of sovereign reach. The idea of exceptionalism signals the justification of measures that previously would have been considered extreme and unjust. For example, the building of physical border walls is a popular state response to perceptions of nations under threat from foreigners (Brown 2010; Dear 2013). High walls surround Spain's small external enclaves of Ceuta and Melilla in Morocco, promulgating feelings of crisis as migrants continue to attempt to scale the walls (Human Rights Watch 2002). Walls constructed by Israel have involved territorial expansion, deepened segregation, and immobility (Stewart 2012). India has worked feverishly to wall its borders with Pakistan and Bangladesh, particularly since 9/11 (Jones 2012).

The reactions to boat arrivals also provide key examples of exceptional responses. In Australia, Canada, and the United States, the numbers of migrants who arrive by boat are relatively small yet elicit disproportionately large response from the public, politicians, and policymakers alike. Discourses often target smugglers as purveyors of transnational criminal networks. Their clients, depicted with metaphors of disease and images of crowded boats, become moving targets for racialized offshore policing (Mountz 2010). Along with this intensification of enforcement comes increased precarity as migrants and smugglers take greater risks and more dangerous routes. For example, along the southern frontier of the EU, Italy and Malta join forces with Libya to police the waters separating southern Europe from northern Africa. These enforcement practices move farther southward, expanding into detention and policing in northern Africa. In these desert crossings, it is not uncommon for travelers to encounter the bodies of those who died en route. In 2011, detention facilities along Libyan and Tunisian borders included people who had encountered border authorities and been returned repeatedly (*Médecins sans Frontières* 2011). The exceptional logic of spatially expanding border enforcement is asserted as the EU struggles to protect its interests by managing migration offshore, farther away and earlier on in transnational journeys.

The perceived chaos surrounding boat arrivals is extended with alarming frequency to make broader changes in national immigration policies that affect much larger populations. In 2001, the Norwegian freighter the *MV Tampa* rescued 433 asylum seekers traveling on a boat in distress not far from Australia's Christmas Island. In response, Prime Minister John Howard—engaged in a tough reelection campaign—enacted the Pacific Solution, which restricted access to asylum by detaining asylum seekers offshore and involved the retroactive declaration of parts of sovereign territory to no longer be part of Australia for the purposes of migration (referred to as "the power of excision"; Perera 2002). Although this policy was ended by subsequent Prime Minister Kevin Rudd in 2007 and declared illegal by Australia's high court in 2011, it was reinstated in practice in 2012 with the reopening and expansion of offshore detention facilities on Nauru and Papua New Guinea's Manus Island (ABC Radio 2012).

Recent Canadian legislation also offers a remarkable example of the leveraging of so-called chaos to profoundly alter the legislative environment for all refugee claimants based on sparse boat arrivals. The Canadian government intercepted only two boats in the last ten years and only six boats in the last twenty years, all off the coast of British Columbia. After the most recent two boats were intercepted in 2009 and 2010, carrying a combined total of fewer than 500 Sri Lankan Tamils (most of whom made refugee claims), new legislation (Bill C-31) dramatically altered the landscape for all refugees, refugee claimants, and immigrants on other paths to Canadian residency and citizenship (Canadian Council for Refugees 2012). These changes will expand the landscape of detention in Canada, concentrate decision-making powers with the Minister of Citizenship and Immigration, reduce rights to family reunification, and introduce the right to strip refugees of permanent residence (Canadian Council for Refugees 2012).

In Italy, Prime Minister Silvio Berlusconi twice called states of emergency after boats from northern Africa arrived on Lampedusa in 2008 and 2011. These measures produced a series of expansions of, and intense international attention to, the island's detention facility. The state of emergency then extended to legislation that changed national immigration policies. Hepworth (2011) shows how Italy's emergency legislation was aimed at all racialized migrants who subsequently became targets of attacks on the streets of northern industrial cities.

As these examples suggest, the detention of migrants becomes another moment in the migration journey constructed as chaotic and crisis ridden. The same countries that specialize in expanding immigration policing also tend to invest heavily in the construction of detention facilities. Israel recently constructed the world's largest detention center in the Negev desert to contain up to 8,000 African asylum seekers attempting to enter Israel through Egypt (Stewart 2012). Prime Minister Benjamin Netanyahu expressed a need "to stop this growing influx that threatens Israelis' jobs and changes the character of the state" (Stewart 2012). Once a haven for those displaced, Israel is mobilizing the crisis of geographic proximity to poverty and political instability as a force that requires containment in a constant state of emergency. The United States detained approximately 430,000 foreign nationals in 2011 (Department of Homeland Security Office of Immigration Statistics 2012) in some 250 facilities (Detention Watch Network 2013). Both private corporations and county governments compete for immigration detention contracts as sources of revenue and approximately two thirds of detainees in the United States are held in sites not operated directly by the federal government (Detention Watch Network 2013). Narratives of chaos again enable the overlap of financial enterprise with anti-immigrant imaginaries.

As detention centers proliferate close to borders where migrants enter and from hence they are deported, they are simultaneously proliferating offshore along migration routes in transit countries. The Australian government has invested heavily in detention on Pacific islands. Asylum seekers trying to reach Australia to make claims now find themselves detained on Christmas Island, Nauru, and Manus Island and in some twelve facilities across Indonesia. There and elsewhere (e.g., Cuba, Malta, Lampedusa), islands increasingly serve as sites where images and sentiments of chaos and crisis unfold (Mountz 2011). Islands have become focal points for border enforcement to reduce anxieties over invasion access to rights gained on sovereign territory. Importantly, whether within or outside of the country orchestrating immigrants' detention, the "outsourcing" of detention to other entities—whether private companies, lower level government agencies, or other countries' governments—also allows nation-states to sidestep charges of human rights abuses by blaming subcontractors (Gordon 2002).

Experiences of chaos and crisis are prolonged through processes of deportation. At this point in the migration journey, narratives of instability and danger are mobilized to deport people in ways that make the very experience of exit itself chaotic. The United States offers an extreme example, deporting nearly 400,000 people in 2011 (Department of Homeland Security Office of Immigration Statistics 2012). In spite of the extensive infrastructure needed to detain and deport at historically high rates, neither deportees nor their lawyers have much access to information about their deportation. Detainees often find themselves transferred multiple times through the U.S. detention system prior to a deportation flight (Hiemstra 2013). The criminalization of this population and the mobilization of crisis and chaos preclude chances for adequate legal representation and appeal prior to deportation (Human Rights Watch 2009). As a result, even U.S. citizens unable to prove identity and secure legal representation have been mistakenly deported to "home" countries where they have never set foot (Stevens 2011).

The EU, too, now deports hundreds of thousands of migrants per year; Collyer (2012) identifies Europe's "deportation turn" as occurring in 2003. Carr (2012, 127) notes, "For many governments, deportation has become a kind of political theater in which crowd-pleasing statistics and deportation quotas are presented as evidence of toughness and implacability in the fight against illegal immigration." In recent years, for example, French police have increased ID checks, and Spanish police were ordered to target Moroccan immigrants to fill deportation quotas (Carr 2012).

Deportation's visible show of force elicits public and political approval by seemingly restoring order to chaos. In the United States, workplace raids ending in deportation typically receive sustained attention, both from the agencies carrying them out and politicians applauding them. Such raids, however, account for a mere fraction of the total number of deportations (De Genova 2005). Most immigrants deported from the United States are caught through spatially ubiquitous, routine policing, but raids occupy greater space in popular media due to their place specificity and conspicuously large numbers.

Governments also prove willing to circumvent national and international law to elevate deportation numbers. Italian authorities, for example, bypassed national laws requiring individual hearings for potential deportees by reclassifying immigrants detained on Lampedusa (Carr 2012). The Canadian government devised the idea of the "long tunnel" to restrict Chinese asylum seekers' access to rights in ways that made their deportation more likely (Mountz 2010). In the rush to perform state power amidst crisis, governments portray such machinations as victories.

When conceptualized as part of a broader pattern, these moments prove far from extraordinary but rather routine; yet an exceptionalized event alters practice and immigration policy writ large. Crisis in each site is leveraged into something larger. Small numbers of foreign bodies are represented in terms of racialized crowding (Hage 2000), criminalized, and leveraged to impact on larger populations. The mobilization of chaos and crisis thus functions as a blunt tool of migration governance, one that dispels distinctions between forced–nonforced and political–economic migrations.

Conclusions

This article builds on scholarship within migration studies that engages themes of chaos, crisis, and exceptionalism. We advance this literature by tracing the political–geographic work accomplished therein. We have explored multiple geographic locations of these discourses, the role that they play in changing national immigration, asylum, and refugee policies and the significant shifts invoked in sovereign power. Our analysis opens the way for further research pertaining to these discourses.

As Agamben (1998, 2005) argues, sovereign power emerges from the ability to categorize and spatially enclose people within and outside juridical law in the state of emergency. Our scrutiny of spectacles of chaos and crisis around processes of migration reveals thresholds of human mobility as sites for the assertion of sovereign power. For example, sovereign power expands geographically through ever more creative techniques of governance to move the border, shrink spaces of asylum (Hyndman and Mountz 2008), and conceal the power of the immigration and detention industrial complex (Hiemstra 2013), all restrictions intended to limit human mobility. The often contradictory efforts by nation-states to facilitate mobility and containment can thus be understood as opportunities for the exercise of state power and the expansion of territorial claims.

The shifts in enforcement portrayed in this article have dire consequences. As the border moves offshore, for example, threats of migration crises violently foreclose on paths to asylum. The criminalization of those on the move means that a blanket exceptionalism surrounds asylum, such that the asylum seeker viewed as legitimate has become a rarity. In so doing, they are chipping away at other legal categories, paths, and rights associated with citizenship through a proliferating series of changes (e.g., Stevens 2011; Canadian Council for Refugees 2012; Howard-Hassmann 2013). As a result, the significance of state borders is increasing, rather than decreasing.

Notes

1. Although we use the discourse of "the state," we do not understand states to be monolithic entities. Instead, the state is composed of individual actors with personal goals, backgrounds, and imaginaries; agencies with diverse (and at times contradictory) purposes; and legislative and judicial institutions driven by different mandates, constituencies, and funding sources.
2. There are also private interests that lose money due to immigration enforcement activities, such as companies involved in cross-border trade and industry that lose due to underdeveloped port of entry operations. Such contradictions demonstrate the inherently contradictory and complex nature of capitalism (Heyman 2012).

References

ABC Radio. 2012. Manus Island detention centre "better" than Nauru. 26 November 2012. http://www.radioaustralia.net.au/international/radio/program/pacific-beat/manus-island-detention-centre-better-than-nauru/1051524 (last accessed 30 November 2012).

Agamben, G. 1998. *Homo sacer: Sovereign power and bare life*. Stanford, CA: Stanford University Press.

———. 2005. *State of exception*. Chicago: University of Chicago Press.

Amoore, L., and M. de Goede, eds. 2008. *Risk and the war on terror*. London and New York: Taylor & Francis.

Andreas, P. 2000. *Border games: Policing the U.S.–Mexico divide*. Ithaca, NY: Cornell University Press.

Bhandar, D. 2008. Resistance, detainment, asylum: The onto-political limits of border crossing in North America. In *War, citizenship, territory*, ed. D. Cowen and E. Gilbert, 281–302. London and New York: Routledge.

Bigo, D. 2002. Security and immigration: Toward a critique of the governmentality of unease. *Alternatives* 27:63–92.

Brown, W. 2010. *Walled states, waning sovereignty*. New York: Zone Books.

Butler, J. 2004. *Precarious life: The powers of mourning and violence*. New York: Verso.

Canadian Council for Refugees. 2012. Protect refugees from Bill C-31: Joint statement. http://ccrweb.ca/en/protect-refugees-c31-statement (last accessed 29 November 2012).

Carr, M. 2012. *Fortress Europe: Dispatches from a gated continent*. New York: New York Press.

Coleman, M. 2007. Immigration geopolitics beyond the Mexico–US border. *Antipode* 39 (1): 54–76.

Collyer, M. 2007. In-between places: Trans-Saharan transit migrants in Morocco and the fragmented journey to Europe. *Antipode* 39 (4): 668–90.

———. 2012. Deportation and the micropolitics of exclusion: The rise of removals from the UK to Sri Lanka. *Geopolitics* 17 (2): 276–92.

Cresswell, T. 1997. Weeds, plagues, and bodily secretions: A geographical interpretation of metaphors of displacement. *Annals of the Association of American Geographers* 87 (2): 330–45.

Cuttitta, P. 2006. Points and lines: A topography of borders in the global space. *Ephemera* 6 (1): 27–39.

Dear, M. 2013. *Why walls won't work: Repairing the US–Mexico divide*. Oxford, UK: Oxford University Press.

De Genova, N. 2005. *Working the boundaries: Race, space, and "illegality" in Mexican Chicago*. Durham, NC: Duke University Press.

Department of Homeland Security Office of Immigration Statistics. 2012. Immigration enforcement actions: 2011. http://www.dhs.gov/sites/default/files/publications/immigration-statistics/enforcement_ar_2011.pdf (last accessed 15 November 2013).

Detention Watch Network. 2013. About the U.S. detention and deportation system. http://www.detentionwatchnetwork.org/resources (last accessed 15 November 2013).

Ellis, M., and R. Wright. 1998. The Balkanization metaphor in the analysis of U.S. immigration. *Annals of the Association of American Geographers* 88 (4): 686–98.

Flynn, M., and C. Cannon. 2009. The privatization of immigration detention: Towards a global view. A Global Detention Project Working Paper. http://www.globaldetentionproject.org/fileadmin/docs/GDP_PrivatizationPaper_Final5.pdf (last accessed 15 November 2013).

Gordon, N. 2002. Outsourcing violations: The Israeli case. *Journal of Human Rights* 1 (3): 321–37.

Greenberg, J., and S. Hier. 2001. Crisis, mobilization and collective problematization: "Illegal" Chinese migrants and the Canadian news media. *Journalism Studies* 2 (4): 563–83.

Gregory, D. 2007. Vanishing points: Law, violence, and exception in the global war prison. In *Violent geographies: Fear, terror, and political violence*, ed. D. Gregory and A. Pred, 205–36. London and New York: Routledge.

Hage, G. 2000. *White nation: Fantasies of white supremacy in a multicultural society*. London and New York: Routledge.

Harvey, D. 1989. *The condition of postmodernity: An enquiry into the origins of cultural change*. Oxford, UK: Basil Blackwell.

Hepworth, K. 2011. Encounters with the clandestino/a and the nomad in Milan: Securitisation, irregularisation and illegitimate outsiders through the Italian "security package." Doctoral dissertation, University of Technology, Sydney, Australia.

Heyman, J. 2004. Ports of entry as nodes in the world system. *Identities: Global Studies in Culture and Power* 11:303–27.

———. 2012. Capitalism and US policy at the Mexican border. *Dialectical Anthropology* 36 (3–4): 263–77.

Hiemstra, N. 2013. "You don't even know where you are": Chaotic geographies of U.S. migrant detention and deportation. In *Carceral spaces: Mobility and agency in imprisonment and migrant detention*, ed. D. Moran, N. Gill, and D. Conlon, 57–75. Farnham, UK: Ashgate.

Higham, J. 2002. *Strangers in the land: Patterns of American nativism, 1860–1925*. Rutgers, NJ: Rutgers University Press.

Honig, B. 1998. Immigrant America? How foreignness "solves" democracy's problems. *Social Text* 16:1–27.

Howard-Hassmann, R. 2013. Slippery citizenship. In *Slippery citizenship*, ed. R. Howard-Hassmann and M. Walton-Roberts, 1–27. Philadelphia: University of Pennsylvania Press.

Human Rights Watch. 2002. Discretion without bounds: The arbitrary application of Spanish immigration law. http://www.hrw.org/reports/2002/spain2/Spain0602.pdf (last accessed 26 November 2012).

———. 2008. Stuck in a revolving door: Iraqis and other asylum seekers and migrants at the Greece/Turkey entrance to the European Union. http://www.hrw.org/sites/default/files/reports/greeceturkey1108web_0.pdf (last accessed 19 July 2012).

———. 2009. Locked up far away: The transfer of immigrants to remote detention centers in the United States. http://www.hrw.org/en/reports/2009/12/02/locked-far-away (last accessed 26 November 2012).

Huysmans, J. 2000. The European Union and the securitization of migration. *Journal of Common Market Studies* 38 (5): 751–77.

Hyndman, J. 2007. The securitization of fear in post-tsunami Sri Lanka. *Annals of the Association of American Geographers* 97 (2): 361–72.

Hyndman, J., and A. Mountz. 2008. Another brick in the wall? Neo-refoulement and the externalization of asylum in Australia and Europe. *Government and Opposition* 43 (2): 249–69.

Jones, R. 2012. *Border walls: Security and the war on terror in the United States, India, and Israel*. New York: Zed Books.

Kaplan, A. 2003. Homeland insecurities: Reflections on language and space. *Radical History Review* 85:82–93.

Klein, N. 2007. *The shock doctrine: The rise of disaster capitalism*. New York: Picador.

Kwong, P. 1997. *Forbidden workers: Illegal Chinese immigrants and American labor*. New York: New Press.

Mainwaring, C. 2012. Constructing a crisis: The role of immigration detention in Malta. *Population, Space and Place* 18:687–700.

Médecins Sans Frontières. 2011. *Torture, exploitation and abuse of migrants in North Africa*. Geneva, Switzerland: Médecins Sans Frontières.

Mountz, A. 2010. *Seeking asylum: Human smuggling and bureaucracy at the border*. Minneapolis: University of Minnesota Press.

———. 2011. The enforcement archipelago: Detention, haunting, and asylum on islands. *Political Geography* 30 (3): 118–28.

Mountz, A., and N. Hiemstra. 2012. Spatial strategies for rebordering human migration at sea. In *A companion to border studies*, ed. T. M. Wilson and H. Donnan, 455–72. Malden, MA: Blackwell.

Nevins, J. 2010. *Operation Gatekeeper and beyond: The rise of the "illegal alien" and the making of the U.S.–Mexico boundary*. 2nd ed. London and New York: Routledge.

Pain, R., and S. J. Smith. 2008. Fear: Critical geopolitics and everyday life. In *Fear: Critical geopolitics and everyday life*, ed. R. Pain and S. J. Smith, 1–19. Burlington, VT: Ashgate.

Perera, S. 2002. What is a camp . . . ? *Borderlands E-Journal* 1 (1): 1–10.

Pratt, G. 2005. Abandoned women and spaces of the exception. *Antipode* 37 (5): 1052–78.

Sassen, S. 1996. *Losing control? Sovereignty in an age of globalization*. New York: Columbia University Press.

Secor, A. J. 2007. Between longing and despair: State, space, and subjectivity in Turkey. *Environment and Planning D: Society and Space* 25:33–52.

Staeheli, L. A., and C. R. Nagel. 2008. Rethinking security: Perspectives from Arab-American and British Arab activists. *Antipode* 40 (5): 780–801.

Stevens, J. 2011. U.S. government unlawfully detaining and deporting U.S. citizens as aliens. *Virginia Journal of Social Policy and the Law* 18 (3): 606–720.

Stewart, C. 2012. Israelis build the world's biggest detention center. *The Independent* 10 March. http://www.independent.co.uk/news/world/middle-east/israelis-build-the-worlds-biggest-detention-centre-7547401.html (last accessed 12 November 2012).

Walters, W. 2004. Secure borders, safe haven, domopolitics. *Citizenship Studies* 8 (3): 237–60.

———. 2008. Bordering the sea: Shipping industries and the policing of stowaways. *Borderlands E-Journal* 7 (3): 1–25. http://www.borderlands.net.au/vol7no3_2008/walters_bordering.pdf (last accessed 15 November 2013).

Living the Way the World Does: Global Indians in the Remaking of Kolkata

Pablo S. Bose

Department of Geography, University of Vermont

The transnational activities of migrants continue to gain attention from scholars, national governments, international institutions, and social movements. The migration–development nexus is of special interest and includes the various ways in which both recent and more established diasporas engage in reshaping their former homes. This article considers the proliferation of gated communities and ethnic enclaves meant to house (primarily elite) return migrants in several cities of the Global South using the case of luxury condominium developments in the Indian city of Kolkata. These buildings are marketed equally to overseas Indians as second homes and to locals who wish to inhabit a transnational, cosmopolitan lifestyle in these spaces. The analysis interrogates the figure of the global Indian—a character central to the design and marketing of these developments—in terms of how it has been mythologized, monopolized, and deployed to help reshape the city. Interviews with apartment owners—both local and nonresident—from one of the most prominent of these international-style projects in Kolkata highlight the disjuncture between the assumed lifestyle of transnational migrants and the buildings meant to house them. The findings suggest that we need to think more carefully and critically about what kinds of changes are made possible by either the involvement or the invocation of wealthy overseas communities in development within their former countries.

移民的跨国活动, 持续受到学者、国家政府、跨国机构以及社会运动的关注。"移民—发展" 的关係特别令人感兴趣, 包括新近和已建立基础的海外离散者参与重构他们之前的家的各种管道。本文运用印度加尔各达城市中的豪华大厦开发案例, 考量本为全球南方数个城市中的 (以菁英为主的) 返乡移民提供住所的门禁社区与族裔飞地之增长。这些建筑不但行销给海外印度人做为第二个家, 亦行销给期望住在此般具有跨国寰宇主义生活风格空间的在地人。研究分析探问全球印度人的角色——一个设计与行销上述建案的核心特徵——如何被神秘化、垄断, 并用来协助城市的再造。本文访谈这些位于加尔各达的国际风格建案的其中一个最为突出的公寓所有者们——包含在地居民与非住民, 凸显出预设的跨国移民生活风格, 与原本旨在提供他们住所的建筑两者之间的断裂。研究发现主张, 我们必须更谨慎且批判性地思考, 富裕的海外社群透过参与或祈愿, 有可能会为他们的母国发展带来什麽样的改变。

Las actividades transnacionales de migrantes siguen atrayendo la atención de eruditos, gobiernos nacionales, instituciones internacionales y movimientos sociales. La conexión migración–desarrollo reviste especial interés e incluye las diferentes maneras como diásporas recientes y bien establecidas se involucran en reconfigurar sus anteriores patrias. Este artículo estudia la proliferación de conjuntos residenciales cerrados y enclaves étnicos construidos para dar albergue a migrantes de retorno (primariamente del tipo élite) en varias ciudades del Sur Global; se toma como caso ilustrativo el desarrollo de condominios lujosos en la ciudad de Kolkata, India. Estas unidades residenciales se venden por igual a hindúes residentes en el extranjero, a manera de una segunda casa, y a residentes locales que desean disfrutar en estos espacios de un estilo de vida transnacional y cosmopolita. El análisis examina la figura del hindú global—un carácter central tomado en cuenta para el diseño y mercadeo de estos desarrollos—en términos de cómo ese modelo figurado ha sido mitificado, monopolizado y desplegado para ayudar en la reconfiguración de la ciudad. Las entrevistas con propietarios de los apartamentos—tanto locales como no residentes—de uno de los proyectos más representativos del estilo internacional construido en Kolkata resaltan el desfase entre el pretendido estilo de vida de los migrantes transnacionales y los edificios que se supone deben albergarlos. Los descubrimientos sugieren que deberíamos pensar más cuidadosa y críticamente acerca de qué tipo de cambios pueden sobrevenir gracias al involucramiento o la convocatoria de comunidades extranjeras adineradas alrededor de proyectos de desarrollo dentro de sus anteriores países.

The connections between émigré communities and their former homes has long been a source of interest for many in sending and receiving countries, but a number of factors have intensified the contemporary focus on the transnational practices of newer and older migrant groups. These include geopolitical and security concerns, the integration of immigrants into new host communities, and the economic impact of remittances (Levitt 2001; Demmers 2002; Mohan 2006). The potential for diasporas or overseas workers to help build or rebuild their home countries has spawned a flurry of activity by international institutions, national governments, and local authorities alike (De Haas 2010; Hugo 2012). The fact that remittances—money sent back by workers to their home countries—rank second only to oil exports globally and far outpace both foreign direct investment and aid to developing countries makes it clear why the migration–development nexus has become such an integral growth strategy for countries like Haiti, Lesotho, and Tajikistan, among many others (World Bank 2012). Some scholars, however, have questioned the long-term viability of such strategies given the vulnerability of migrant income to external pressures—natural disasters, economic downturns, and political conflicts, to name but a few (Fix et al. 2009). Others have suggested that the scramble for diasporic resources by various actors leaves the actual nature of development and the roles played by the diaspora relatively unscrutinized (Mohan 2008; Raghuram 2009). As the U.S.-based transnational solidarity network Association for India's Development (2004) has asked, "What kinds of developments in India are Indians in the US (and other countries) making possible?"

To answer such a question, this article examines one particular type of diaspora-influenced development—new housing construction in the Indian city of Kolkata. Although much of the migration and development literature has focused on remittances and investments, it is equally important to look at "less visible, quantifiable and tangible forms ... [which may] have a more critical impact than their pecuniary counterpart" (Kapur 2005, 357). In Kolkata, money remitted by migrant workers is not remaking the city. Instead, it is the proliferation of new housing projects marketed as "global" to local elites who wish to live cosmopolitan lifestyles and to diasporic Indians seeking a familiar and comfortable second home that is transforming the landscape. Such projects have been undertaken by private developers—some of them nonresident Indians themselves—or through public–private partnerships (Roy 2011). A particular narrative of migration and transnational identity—aspirational and idealized—is central to the appeal of these buildings. This is the story of what I call the global Indians: successful, educated professionals, connected to international networks via work, family, or their own overseas experiences and desires to return or remain connected to their homeland. Such developments thus become a home place for at least two distinct populations—diasporic or overseas Indians and local nonmigrants who wish to emulate or approximate these mythologized migrant lifestyles.

Kolkata is not the only place in which we see diasporas and locals sold on the allure of global—usually gated—communities. Suburban villas on the outskirts of Beijing have been sold to both wealthy locals and to overseas Chinese living in Canada, the United States, and Australia (King 2004). Gated condominium complexes in Beirut and Riyadh are similarly marketed equally to overseas communities across the world and to local elites (Glasze and Alkhayyal 2002). Baecker's (2013) examination of postconflict reconstruction in Beirut notes the importance of connecting the refashioned city center to global networks through new luxury apartment buildings built explicitly to entice the bodies and resources of the Lebanese diaspora. Perera (2011) makes a similar observation about the emergence of new apartment towers in Colombo, Sri Lanka, designed to attract and house rich locals and transnational migrants alike. In Kolkata, too, these new building projects offer potential buyers the opportunity to "live the way the world does,"[1] while suggesting to planners, politicians, and local communities that such developments are a key signifier of achievement for cities that wish for global prominence (Bose 2007).

What does the life-space of the world look like, as reproduced in a luxury, gated apartment designed for transnational subjects and aspiring locals? I interrogate the notions of global living as embodied in the figure of the global Indian—a mythologized set of migrant experiences and preferences at the heart of urban transformations in cities like Kolkata. The return of the global Indian and the housing that he or she inhabits gives evidence of Kolkata's resurgence from a site of urban decay through much of the twentieth century (Roy 2003; Chattopadhyay 2005) to one of revitalization and growth in recent years. Global Indians are not the reason for the adoption of neoliberal urbanization strategies by governments of all stripes in Kolkata, but their presence—both physical and ideological—helps to materialize the city's connections to global circuits of capital, labor, and ideologies (Roy

2011). It is they—as well as locals also employed in similar professional fields—who are meant to populate the industrial parks and high-tech suburbs on the fringes of Kolkata, built to attract large corporations and multinationals (Dey 2011; Samaddar 2012).

This article explores these dynamics in three parts. The first examines the mythologies that produce the figure of the global Indian as an ambiguous concept that simultaneously blurs the line between forms and histories of migration and yet becomes monolithic and focused on certain types of trajectories at the expense of others. The second examines the changing landscapes of Kolkata and the increasing importance of housing developments that embody the assumed lifestyle and preferences of the global Indian. The third focuses on the case of the South City Towers, a prominent but controversial township development built with an explicitly international style in mind to see what appeal they hold for their residents. Drawing on interviews with several owners of flats within these buildings, this article highlights the disjuncture between the expectations that owners might have had and the reality of the buildings themselves.

Constructing the Global Indian

I use the term *global Indian* to capture the wide range of identities and appellations used to signify the modern experience of migration from the subcontinent: nonresident Indian (NRI), person of Indian origin (PIO), overseas-born Indian (OBI), desi, East Indian, Indo-Canadian, Indian American, South Asian, and a host of others besides. Yet although the actual experience of emigration has produced multiple variations on the theme of a global Indian—at least in the ways that it is used by the Indian state and many in the culture industries—it refers to a very specific instance: that of primarily elite, transnational subjects living in the countries of the industrialized world. Its power lies in part in being able to subsume the distinct trajectories of migration from the subcontinent under the apparently universal narrative of professionals—doctors, engineers, scientists, and academics among them—who have emigrated to countries such as Britain, the United States, Canada, and Australia. There is also a temporal distinction that is often collapsed in the global Indian formation, between those who are Indian citizens living and working abroad (the NRI) and the Indian diaspora (those with ancestral roots in India).

Thus although there are many other Indians and their descendants who might claim some kind of global identity, the global Indian idea does not necessarily include such diversity. Instead, the narrow view of the elite transnational subject dominates, a perspective made possible only by privileging some migrant experiences at the expense of others. This means, for example, celebrating the lives of the model minority Indians (Prashad 2000) and ignoring or obscuring the histories of indentured migrants to the Caribbean and South Pacific, Sikh labor migrants to North America at the turn of the century, or contemporary Indian workers in the Gulf, to name but three cases. This elite global Indian has become a staple in the popular imagination, the press, and popular discourse, a regular figure in films and television (Desai 2004) and courted by successive Indian governments (Dickinson and Bailey 2007). Indeed, the changing attitude of the Indian state toward its emigrants (both diasporic and short-term) has been remarkable, from one bordering on suspicion and hostility in the postindependence period to one of embracing them more recently (Sahoo, Bass, and Faist 2012), and made possible by the creation of overseas citizenship opportunities, provision of special visas for visits, relaxation of rules on investment, and celebration of the achievements of prominent members of diaspora communities (Bose 2007). Such measures did not create relationships between India and its émigrés; rather, they seek to capitalize on them. When an Indian magazine published the first of what is now an annual global Indian issue, it celebrated the achievements of second-generation "overseas-born Indians (OBIs) . . . [born] in the two great migrations of the 1960s and 1980s to largely educated professionals" ("The OBIs" 2004, 3).

Why the fascination with this particular manifestation of Indian migration by the Indian state and culture industries? What does the global Indian offer the subcontinent? One aspect might be the narrative of success rather than struggle in the lives of software engineers, lawyers, doctors, entrepreneurs, and other professionals living apparently comfortable lifestyles. An Indian government–commissioned study fairly salivated at the prospect of overseas Indians possessing assets roughly equivalent to India's gross domestic product, a Gujarati diaspora in the United States controlling 5 percent of that country's wealth, and Indian information technology (IT) engineers in Silicon Valley ranking second only to Chinese immigrants in terms of numbers of millionaires (Singhvi 2001). But the global Indian must not only be wealthy; he or she must also be a well-behaved and disciplined subject, one whose transnational activities do not extend to, for example, extremist politics or criminal behavior. Instead, he or she must exemplify what one editorial calls a "good NRI," one who "sends their savings back to India, gives money to their poorer relations, sets up schools, colleges and hospitals in towns and villages where they grew up" (Singh 2002). In an interview conducted in 2004, J. C. Sharma, a former senior diplomat and one of the chief authors of the Indian government's influential report on the size and developmental potential of its diaspora (Singhvi 2001), suggested that the success of global Indians rests largely on their ability to behave in accordance with the rules of their new societies while retaining a strong link to their heritage. For Sharma and many others like him, it is this mythologized and idealized global Indian who will help to build the new nation and who therefore holds such appeal for the state and the culture industries alike. In Sharma's view:

> Our destinies are interlinked. The overseas Indians are an excellent resource for us. If today India is shining, then the diaspora is shining too. We say to the diaspora that you have as much to gain from us as we do from you; it is a mutually beneficial relationship (personal communication, interview, 18 March 2004, Ministry of External Affairs, New Delhi).

Global Indians and Kolkata's Rebirth

What kind of future can the global Indian offer Kolkata? The city has been in decline for the better part of a century—due to industrial flight, a large-scale refugee influx, and political turmoil, among other factors. It is no longer a preeminent node in circuits of capital and empire and has been bypassed in stature by Mumbai, Delhi, and Bangalore (Mehta 2004; Dupont 2011; Goldman 2011). The efforts to regain its former prominence have centered on joining the rush to become a world-class city—attempted, for example, by enticing multinationals to relocate, by hosting major spectacles and events, and by engaging in a frenzy of real estate development. Another marker of urban globality has become the physical presence of immigrants (Schiller and Simsek-Caglar 2011). Such sites are becoming locations where global capital and immigrant flows intersect (Sassen 2002). Similarly, a certain kind of "translocality" is remaking urban social space, with migrant agency a key factor in the ongoing redefinition of people, place, and identity (Smith and Eade 2008).

The global Indian can show that Kolkata is a node in ongoing patterns of migration both from and to the subcontinent. In particular, locating the global Indian in Kolkata fits with a narrow view of subcontinental outmigration embodied in the experience of professional classes and transnational elites. The state of West Bengal, of which Kolkata is the capital city, does not have the extensive labor migrant flows or historical diasporas produced by states like Punjab, Gujarat, or Kerala (Raghuram 2007). It is, however, a source of professional emigrants—educated, upper class, often upper caste—the heart of the global Indian image. It is their presence (or memory) in and of the city that can potentially reaffirm its place on the world stage, and indeed it is explicitly to them that local politicians appeal. The state's Chief Minister, Mamata Banerjee, in a recorded message played at the North American Bengali Conference (NABC) in Las Vegas, urged nonresident and diasporic Bengalis to invest in West Bengal:

> I welcome you all and this is your state. Please come here and share your ideas, views and give suggestions to make this state better and I know you all love your state and have sentiments attached to the soil. (Sen Gupta 2012)

Whereas the state government has more recently begun to vigorously recruit its diaspora for developmental purposes, sentimental attachments to the soil have been promoted for much longer in Kolkata. As with other migrant flows and their connections to former homes, much of this has occurred informally, such as the sending of money home, the building of a house, or support for a local charity or initiative. Increasingly, however, real estate developers and returning migrants have begun to influence the reshaping of neighborhoods and regions more purposefully, especially in ways that simulate the experiences—whether long or short term—of being abroad. The story of ethnic enclaves built in destination countries by immigrants is well known, but it has become more evident in recent years that reverse migration can often have the same effect in sending countries. In this sense it is not only a story that Little India is created by expatriates in Dubai but equally that returning migrants might purchase a home in Little Dubai in Delhi, built to remind them of their time in the Middle East (Vora 2008).

In a similar vein, the site of the most significant real estate developments in the city of the past few decades is awash in buildings that conjure the global Indian in lifestyle, vocation, and physical presence. These are the eastern fringes, the only direction in which the city can expand, home to a large expanse of wetlands and a series of satellite townships reclaimed from them. By the late 1990s this area had also begun to attract the IT industry as it reoriented more purposefully to India—not only to primary sites such as Gurgaon, Bangalore, and Hyderabad but also to secondary ones, including Kolkata. Although local environmental protests led to the declaration of protected status for the wetlands, this has done little to stem its attractiveness to developers. Within a few years apartment complexes—many of them public–private partnerships and many others illegal—sprung up throughout the area, especially along a highway that skirts the eastern edges of Kolkata and leads from middle-class residential neighborhoods in the south to the airport in the northeast (Samaddar 2011; Dasgupta 2012).

Boasting names such as Oasis, Bengal Silver Spring, Ideal City, The Empire, Hiland Park, and Technopolis, these projects bring high-rise towers of up to thirty floors onto an urban landscape with few buildings higher than ten stories (Bose 2007). Some—such as Rosedale Gardens, a venture led by nonresident Indians from New Jersey, and Kolkata West International City on the opposite side of town and built with capital and expertise from Indian émigrés in Indonesia—have direct involvement and leadership from global Indians (Bose 2013). Most, however, are domestic enterprises, built in many cases as partnerships between municipal and state agencies and local developers. A rash of multiplexes, shopping malls, golf courses, private schools, medical facilities, amusement parks, and gated green spaces and waterways has burgeoned alongside the new residences. The marketing of such developments is in many ways not unlike what one might find for any condominium tower from Toronto to Berlin to Singapore (King 2004; Lehrer and Laidley 2008; Kern 2010), offering an elite existence replete with all of the amenities deemed necessary for the cosmopolitan life. What sets the Kolkata housing developments apart is the importance of the figure of diaspora in the selling of these buildings, not only to potential inhabitants but also to local politicians, investors, and elites. These housing developments have been a great success, often selling out their units long before ground has been broken, and ensuring yet more to come. During initial fieldwork in 2004 I counted some forty projects marketed toward the global Indian or with their assumed lifestyles in mind, most in the design stage. By 2007 many of these were under construction, with a further forty planned; by 2012 the number had risen to 120, many now no longer on the periphery but rather in the heart of the city itself. Although many have proved to be controversial in their environmental and social impacts, local activists have found them hard to challenge, for to call them into question is to call into question both global Indians and the good they are purported to embody (Vasundhara 2010). I examine how these dynamics play out in the final section of this article by studying a particularly successful township development that straddles the eastern edges of the city.

South City and the Global Life

Flush with the success of selling their condominiums on the edge of town, a consortium of some of the most prominent local and national developers envisioned a new 32-acre minitownship within the boundaries of the city itself. South City was to be built on the grounds of a former industrial site and its promise was grand: the largest development in Kolkata, with the tallest towers, the biggest mall in eastern India, the best school, the most exclusive country club, the most beautifully landscaped pond, and so on—all gated for the enjoyment of its inhabitants (South City Projects 2013). In 2004, the project was little more than a set of plans. By 2007, three of its four towers had begun to rise, and by 2012 it had cemented its place on the cityscape. A promoter (Rakesh[2]) told me that all of his lots had sold out quickly and often at levels far above the original asking price. Indeed, he said that some 1,500 out of 1,740 total units were purchased before a single tower broke ground. In line with the projects in the wetlands, here, too, the lure of a global lifestyle—and perhaps transnational bodies themselves—is key. Gopal, another promoter, boasted that fully 35 percent of apartment owners were NRIs and that some 50 percent of flats were being reserved for overseas Indians—although he also admitted that promoting a building with NRI inhabitants was a great marketing tool. True or not, the perception among some in the area is that projects like South City are for foreigners, not locals. For example, Raju, a construction foreman, said to me in 2007, pointing at the towers inching toward the sky:

> Do you see those flats over there? All of them are owned by NRIs like you. But I will tell you what, come back here at night, come back here at nine or ten o'clock tonight and I'll show you something. A hundred flats and only ten lights will be on. Only ten of them will have people in them. All the rest of you are overseas.

There have been other complaints as well regarding South City. To some it stands as a symbol of the heavy-handed ways in which a desire for global recognition might overwhelm local desires and needs. Its current site was once the home of a prosperous factory now fallen into disrepair and bordered Vikramghar Jheel, one of the largest remaining water bodies within Kolkata. Locals have long sought to clean up the increasingly polluted area so that it might function as a mixed-use landscape for recreation, fishing, and other community needs (Vasundhara 2010). Instead of developing a sustainable space for their own use, however, local residents soon found themselves confronting a scheme that raised global priorities and imagining above their own. Many who live in the neighborhood wondered how the influx of so many newcomers and cars would affect noise and air pollution levels. Others questioned the wisdom of such dense development within an existing residential neighborhood and competition for city services. As the towers took shape, local protests grew on multiple fronts. Trade unions protested the often shoddy and unsafe working conditions that led to several worker deaths during construction (Express News Service 2007). Several residents' groups filed court challenges based on the increase in noise and traffic in the region both during and after construction (Vasundhara 2010). Housing and poverty activists rallied around Sambu Singh, one of the few remaining workers from the previous factory who refused to be displaced for the new developments, although some 7,000 others had been unceremoniously thrown out already. He died under mysterious circumstances but not before drawing attention to the South City Projects (S. Basu 2011). Environmental groups have decried the development's illegal encroachment onto a full acre of the existing water body and filed lawsuits to halt construction. Several state and federal agencies concurred and ordered that the third and fourth towers of South City should be halted (J. Basu 2006). The project, however, with support from civic and state officials who viewed it as a prestigious and landmark development symbolizing Kolkata's renewal, went ahead. And although among some locals South City gained the nickname "Towers of Violation" (Vasundhara 2010), to most it simply became a fact of life.

By 2012, and contrary to prediction, South City was not a ghost town, one of the "migra-villages" common in Central America (Kapur 2005), full of diasporic capital but no diasporas themselves. The township was bustling with people. NRIs, diasporas, and local elites mingle in its glittering mall full of international food chains and franchises, multinational brands and designer labels, supermarkets and luxury items for sale on every floor. A multiplex theater plays Hollywood and Bollywood blockbusters, and South City's International School tells prospective parents that it will "bring out the David Beckham, Bill Gates, or President Obama" in their child (South City International School 2012). As an enclave of consumerist fantasies and middle-class aspirations, South City would not be out of place in any of a thousand locations across the world—tweaked for local idioms and preferences but in essence purveying the same message of elite privilege and the promise of participation in (or perhaps entrance into) the world of cosmopolitan transnationalism.

The developers of projects such as South City target local elites, diasporas, and migrants in their selling of apartments and dreams. Polished model units offer tours for the local market, along with massive billboards that promise the opportunity to "live in the sky," "own a home with India's biggest swimming pool," and to "live the way the world does." Internationally, South City embarked on the roadshows common to many of its fellow developers, often employing celebrities as spokespersons and traveling not only to what salesman Rithwik calls the "London–New York–Toronto triangle" but also to the NABC as well as smaller regional events in places like Liverpool, Sussex, and Birmingham. Rithwik also reported visits to the Persian Gulf; although the Bengali NRI population is much smaller there, South City reasoned that they were even more likely to return to Kolkata than diasporas from the United States and United Kingdom. To diasporas and overseas Indians, the message also concerns "global living," Rather than offering the opportunity to enter an aspirational cosmopolitan lifestyle, however, the focus in such sales pitches has been on providing a dwelling with all the modern conveniences and standards owners might be familiar with and expect from experiences in other parts of the world. Such efforts produced clear results, both at home and away, with the project such a success that the originally planned three towers became five with a substantial number of flats owned by overseas Indians. What motivated these people to buy into South City's vision? What has been their experience with the development since their purchase?

The final part of this article draws on a series of interviews conducted in 2012 with some of the residents to answer such questions. This research is a subsection of a much broader project involving ethnographic fieldwork in Kolkata with residents, planners, politicians, housing advocates, promoters and developers, construction workers, and other key informants regarding changes to the city over the course of the past two decades. In this case I interviewed a total of forty individuals—thirty diasporic and ten resident Indian citizens—recruited via South City's residents' association and snowball sampling methods. All interviewees are between the ages of twenty-eight and fifty-five and included twenty-seven men and twenty-three women. Interviews lasted for approximately one hour and consisted of a series of semistructured questions pertaining to the reasons owners had been attracted to South City and what their experiences with the building had been. Common themes that emerged among respondents in the appeal of the development included the importance of security features, location in the city, and the type and number of amenities—"world class" was an often-repeated phrase. What I also discovered was a clear disjuncture between the image (and appeal) of South City sold to prospective buyers and many of the realities of the flat they actually received.

The Appeal

Of the South City condominium owners interviewed, none were full-time residents in Kolkata but instead treated the flat as a second home to be used for annual visits with friends and family. All were currently based in the United Kingdom, Australia, Canada, or Western Europe; some spent up to six months in India. When asked what drew them back to Kolkata, most gave as their answer some aspect of family ties and connections. Nikhil, a software developer based in Boston, said:

> My parents are getting older and I want to make sure that I can go and see them and make sure that they are well. They used to come and visit me but as they have gotten older the trip is harder on them. So I go out there. It hasn't been easy on me though because the summer is so hot and there's no AC in my parent's house. That's one of the reasons I wanted to buy into South City.

For Mita, an architect in London, there is both obligation and desire to return:

> I go back twice a year—it's expensive but I still do it. It isn't always easy being in Kolkata (or anywhere else in India for that matter) but every time I go back I remember something else I miss.

For most, a flat in South City holds the promise of a return to the city in more comfort than they have been accustomed to in Kolkata. Paroma, an executive with a pharmaceuticals company in Nashville, speaks approvingly of what developments like these portend for the city:

> Yes, I saw the new construction when I was driving in from the airport by taxi. It's about time—finally, the city is beginning to come up again.

Bishnu, a doctor in Toronto, marveled:

> When I first stepped foot into this flat and came up here to the third floor, I opened the windows and felt a breeze I had never felt before in Calcutta. It was cool and refreshing and clean. I couldn't hear the sounds of the city at all. In fact, I felt as though I wasn't in Calcutta at all.

The Reality

Not all had such positive impressions, however. Sonali, a New Jersey–based architect, complains:

> I paid top dollar for this apartment, one of the most expensive in the city. But when I came to inspect the finished product I found construction materials all over the unfinished interior, broken floor tiles, chipped plaster, uneven windows, air conditioners not functioning, and most of the amenities that were supposed to be part of the complex are nonexistent.

Gautam, a London-based marketing executive, was similarly unhappy with the condition of his flat and has received no satisfaction from the property managers—Jones Lang LaSalle Meghraj, the Indian subsidiary of the global real estate services firm in charge of delivering finished flats to the NRIs:

> I have contacted them on multiple occasions and rarely receive a timely response. I'm not in Kolkata enough to stay on top of what's going on with it and I feel that the builders are taking advantage of our absence from the country not to fulfill their promises. Like other Bengali NRIs we bought these flats to maintain our roots in West Bengal. The way we have been treated is shocking and sends the wrong message to us.

Anamitra, a broadcast producer in New York City, was disappointed to find the promised amenities undelivered:

> The lawn and green space that was promised seems now to have been turned into a car park and the clubhouse and pond area remains unfinished in the center of the towers. All in all it just has a feeling of being unfinished.

Perhaps the part of the development that has most fully delivered on its promise of "global living" has been the South City Mall attached to the towers. Unfortunately for some, the poor condition of the apartment towers as opposed to the mall was a sore point. As Toronto-based banker Anjan said:

> Overall I was really surprised at what it looked like inside. While the mall itself is as advertised—gleaming, air conditioned, sparkly, all new and shiny—the apartments are far from it. The hallways are not AC, the windows

seem dirty and the place as a whole is very dusty and really unfinished.

One might read the distance between the promise of South City and the disappointments of owners such as those just cited as not limited to an international-style development in a city of the Global South; we might find such complaints among many a condominium owner in many a place, yet the implications in this context are potentially more significant. South City and similar projects in Kolkata are not simply buildings that might fail to live up to their potential. They are indicative of a particular developmental strategy that relies on successful appeals to transnational subjects and local elites alike. If the space between the promise of global living and that actual life remains so large, what will be the effect on the remade city? Will the projects eventually lose their appeal and become the hollow shells of migra-villages bemoaned elsewhere? For now, residents in South City are not leaving in any significant numbers. Ironically, the disappointment in the lack of promised amenities seems to be one point on which the global Indians in South City and locals might forge commonality—after seeing the shabby state of the pond area, the residents' association within the towers reached out to the same environmental organizations that had fought the construction in the first place to help arrive at a solution.

Conclusion

What can we learn from these complex ways in which migration and development might be manifest in the remaking of cities of the Global South? How do we understand the particular ways in which transnational subjects and locals who wish to live globally might help to transform the cityscape, not only by what they do but also by where they reside? We must look beyond remittances, investments, or even the physical presence of diasporas and overseas Indians. Instead, the invocation of tropes such as the global Indian and "global living" can play an important role in simultaneously catalyzing and affirming connections to international networks and lifestyles. It is also clear that we must look carefully and critically at the types of transformation engendered by such notions of globality. It is necessary, therefore, to interrogate more closely what we mean by the life spaces and migrant experiences we invoke with concepts like the global Indian. I have suggested that this figure as it is popularly imagined is a fetishized, narrow, and exclusionary reading of a much richer and diverse set of migrant flows, returns, and aspirations. Although the appeal of such a celebratory yet limited characterization seems clear, the consequences of a disjuncture between that appeal and the actual reality of the buildings themselves is less apparent.

To some locals who buy in, the buildings might offer an entrance into a transnational life-world as represented by popular culture and promoters alike. For some diasporic and migrant populations they might provide a more comfortable way of visiting and staying rooted to their homeland. In South City's case, its success has led to more projects in other parts of Kolkata and even a $100-million venture to build a township in Sri Lanka (Indian Realty News 2011). But if such housing developments have negative environmental or social effects on their surroundings, or if the construction of the buildings fails to achieve the standards of global living that are such an essential part of their appeal, we could be left with an unsustainable, incomplete, or ultimately abandoned experiment in diaspora-influenced development.

Acknowledgments

The author would like to thank Paroma Roy Chowdhury, Indranil Chakraborty, Mohit Ray, and Arijit Banerjee for their assistance with the fieldwork aspects of the research on which this article is based. The author is indebted to Ananya Ray, Jonah Steinberg, and Tirthankar Bose for numerous discussions, debates, and support for this work. The author would also like to acknowledge Richard Wright and the anonymous reviewers for their constructive feedback on earlier iterations of this article.

Notes

1. A slogan used in the advertising campaign by South City Projects, the housing project examined in the final section of this article.
2. Unless public figures, the names of interview subjects have been altered to preserve confidentiality.

References

Association for India's Development. 2004. AID projects: Rights and social justice, your voice counts. http://www.aidindia.org/hq/misc/vcounts/voice_counts.shtml (last accessed 10 January 2004).

Baecker, D. 2013. Mapping a city onto itself: A note on the reconstruction of Beirut Central District. http://ssrn.com/abstract=2229687 (last accessed 5 September 2013).

Basu, J. 2006. Key curb on mega estate: Law flout alleged, builders deny. *The Telegraph* 1 March. http://www.telegraphindia.com/1060301/asp/calcutta/story_5910083.asp (last accessed 22 November 2013).

Basu, S. 2011. Reeling off the battle against South City Mall. *Statesman (India)* 3 February. http://deshlai.wordpress.com/2011/02/04/reeling-off-the-battle-against-south-city-mall/ (last accessed 22 November 2013).

Bose, P. S. 2007. Dreaming of diasporas: Urban developments and transnational identities in contemporary Kolkata. *Topia: Canadian Journal of Cultural Studies* 17: 111–30.

———. 2013. Bourgeois environmentalism, leftist development and neoliberal urbanism in the city of joy. In *Locating right to the city in the Global South*, ed. T. Samara, S. He, and G. Chen, 127–51. London and New York: Routledge.

Chattopadhyay, S. 2005. *Representing Calcutta: Modernity, nationalism, and the colonial uncanny*. London and New York: Routledge.

Dasgupta, K. 2012. The context of metropolitan planning. In *Transit labour: Circuits, regions, borders*, ed. K. Craig, K. Hepworth, B. Neilson, and N. Rossiter, 30–33. Kolkata, India: Transit Labour.

De Haas, H. 2010. Migration and development: A theoretical perspective. *International Migration Review* 44 (1): 227–64.

Demmers, J. 2002. Diaspora and conflict: Locality, long-distance nationalism, and delocalisation of conflict dynamics. *The Public-Javnost* 9 (1): 85–96.

Desai, J. 2004. *Beyond Bollywood: The cultural politics of South Asian diasporic film*. London and New York: Routledge.

Dey, I. 2011. New Town and labour in transit. In *Transit labour: Circuits, regions, borders*, ed. K. Craig, K. Hepworth, B. Neilson, and N. Rossiter, 10–13. Kolkata, India: Transit Labour.

Dickinson, J., and A. J. Bailey. 2007. (Re)membering diaspora: Uneven geographies of Indian dual citizenship. *Political Geography* 26 (7): 757–74.

Dupont, V. 2011. The dream of Delhi as a global city. *International Journal of Urban and Regional Research* 35 (3): 533–54.

Express News Service. 2007. Another fatal fall at South City. *Express News Service* 17 August 2007.

Fix, M., D. G. Papademetriou, J. Batalova, A. Terrazas, S. Y.-Y. Lin, and M. Mittelstadt. 2009. *Migration and the global recession*. Washington, DC: Migration Policy Institute.

Glasze, G., and A. Alkhayyal. 2002. Gated housing estates in the Arab world: Beirut Lebanon and Riyadh, Saudi Arabia. *Environment and Planning B: Planning and Design* 29:321–36.

Goldman, M. 2011. Speculative urbanism and the making of the next world city. *International Journal of Urban and Regional Research* 35 (3): 555–81.

Hugo, G. 2012. Migration and development in low-income countries: A role for destination country policy? *Migration and Development* 1 (1): 24–49.

Indian Realty News. 2011. Kolkata based South City Projects to foray into Sri Lanka real estate market. http://www.indianrealtynews.com/real-estate-india/kolkata-based-south-city-projects-to-foray-into-sri-lanka-real-estate-market.html (last accessed 24 June 2013).

Kapur, D. 2005. Remittances: The new development mantra? In *Remittances: Development impact and future prospects*, ed. S. M. Maimbo and D. Ratha, 331–60. Washington, DC: World Bank.

Kern, L. 2010. Selling the "scary city": Gendering freedom, fear and condominium development in the neoliberal city. *Social and Cultural Geography* 11 (3): 209–30.

King, A. D. 2004. *Spaces of global cultures: Architecture, urbanism, identity*. London and New York: Routledge.

Lehrer, U., and J. Laidley. 2008. Old mega-projects newly packaged? Waterfront redevelopment in Toronto. *International Journal of Urban and Regional Research* 32 (4): 786–803.

Levitt, P. 2001. *The transnational villagers*. Berkeley: University of California Press.

Mehta, S. 2004. *Maximum city: Bombay lost and found*. New York: Knopf.

Mohan, G. 2006. Embedded cosmopolitanism and the politics of obligation: The Ghanian diaspora and development. *Environment and Planning A* 38 (5): 867–83.

———. 2008. Making neoliberal states of development: The Ghanaian diaspora and the politics of homeland. *Environment and Planning D: Society and Space* 26 (3): 464–79.

The OBIs: A generation of overseas-born Indians that dares to dream. 2004. *India Today* (Special Issue), 19 January.

Perera, S. 2011. Gated communities as packaged fantasies: A meeting of the local and the global and the standardisation of taste in urban Sri Lanka. In *Urban navigations: Politics, space and the city in South Asia*, ed. J. Anjaria and C. McFarlane, 239–64. London and New York: Routledge.

Prashad, V. 2000. *The karma of brown folk*. Minneapolis: University of Minnesota Press.

Raghuram, P. 2007. Conceptualising Indian emigration—The development story. In *Global migration and development*, ed. T. V. Naerssen, E. Spaan, and A. Zoomers, 309–26. London and New York: Routledge.

———. 2009. Which migration, what development? Unsettling the edifice of migration and development. *Population, Place and Space* 15 (2): 103–17.

Roy, A. 2003. *City requiem, Calcutta: Gender and the politics of poverty*. Minneapolis: University of Minnesota Press.

———. 2011. The blockade of the world-class city: Dialectical images of Indian urbanism. In *Worlding cities: Asian experiments and the art of being global*, ed. A. Roy and A. Ong, 259–78. Oxford, UK: Wiley-Blackwell.

Sahoo, A., M. Bass, and T. Faist. 2012. *Indian diaspora and transnationalism*. New Delhi, India: Rawat.

Samaddar, R. 2011. Rajarhat, the urban dystopia. In *Transit labour: Circuits, regions, borders*, ed. K. Craig, K. Hepworth, B. Neilson, and N. Rossiter, 14–16. Kolkata, India: Transit Labour.

———. 2012. The social factory: Production of the new town. In *Transit labour: Circuits, regions, borders*, ed. K. Craig, K. Hepworth, B. Neilson, and N. Rossiter, 6–9. Kolkata, India: Transit Labour.

Sassen, S., ed. 2002. *Global networks, linked cities*. New York and London: Routledge.

Schiller, N. G., and A. Simsek-Caglar. 2011. *Locating migration: Rescaling cities and migrants*. Ithaca, NY: Cornell University Press.

Sen Gupta, T. 2012. Mamata Banerjee to launch NRI cell in West Bengal to woo foreign investment. *Economic Times*. http://articles.economictimes.indiatimes.com/2012-07-10/news/32618403_1_nri-cell-amit-mitra-finance-ministry (last accessed 22 November 2013).

Singh, K. 2002. NRIs, good and bad. *Hindustan Times*. 25 October. http://www.hindustantimes.com/news/181_90809,00300003.htm (last accessed 22 November 2013).

Singhvi, L. 2001. Report of the High Level Committee on the Indian Diaspora. Government of India, Ministry of External Affairs, Non-Resident India and Persons of Indian Origin Division, New Delhi, India.

Smith, M. P., and J. Eade. 2008. *Transnational ties: Cities, migrations, and identities*. New Brunswick, NJ: Transaction.

South City International School. 2012 Home page. http://www.southcityinternationalschool.com (last accessed 24 June 2013).

South City Projects. South City. http://southcityprojects.com/southcity (last accessed 11 September 2013).

Vasundhara. 2010. *South City controversy*. Kolkata, India: Vasundhara.

Vora, N. 2008. Producing diasporas and globalization: Indian middle-class migrants in Dubai. *Anthropological Quarterly* 81 (2): 377–406.

World Bank. 2012. Migration and remittances data. http://go.worldbank.org/092X1CHHD0 (last accessed 24 June 2013).

In the "Service" of Migrants: The Temporary Resident Biometrics Project and the Economization of Migrant Labor in Canada

Rebecca Pero* and Harrison Smith†

*Department of Geography, Queen's University
†Faculty of Information, University of Toronto

Since 1993, the Canadian government has used biometric screening to identify migrants crossing in and out of Canadian territory. Recently, however, the government has sought to enhance the scope of biometric screening through a number of proposed acts and programs. Of particular interest is the 2013 Temporary Resident Biometrics Project. The Project will require that foreign nationals provide enhanced biometric details, including fingerprints and facial capture, which will be shared with other governmental departments such as the Royal Canadian Mounted Police and Canada Border Services Agency. Although the Canadian government has made reference to increasing vulnerabilities in their capacities of identification and verification, it is unclear exactly why the enhanced surveillance and governance of noncitizens is necessary. We argue that the increasing deployment of biometrics is part of a larger global program designed to promote and manage temporary, short-term labor. We explore this use of biometrics by contextualizing the arguments advanced by the Temporary Resident Biometrics Project within the larger neoliberal discourse of market uncertainty and risk management. The working theory draws from Callon's thesis of hybrid forums to investigate the economization of biometric screening and, in turn, how the state's relationship and obligations to noncitizens are increasingly defined through the rhetoric of a market-driven economized service. A brief overview of the Canada–U.S. NEXUS program serves to demonstrate the Project's relevance to understanding how surveillance is being used to manage global flows of labor.

加拿大政府自 1993 年以来，便开始使用生物特徵辨识来识别进出加拿大国境的移民。但近年来，加拿大政府却又提出一系列的行动及计画，寻求提高生物特徵辨识的范围。2013 年提出的 "暂时居民生物特徵识别计画"，便特别值得关注。该计画将要求外籍人士进一步提供详尽的生物特徵，包含指纹及脸部侦测，该资料并将与加拿大皇家骑警以及加拿大边境服务局等其他的政府部门共享。儘管加拿大政府宣称，这项措施与政府辨识并进行确认的能力中，逐渐提高的脆弱性有关，但是加强非公民的监控与管理的必要性为何，却不甚清楚。我们主张，生物特徵识别的加强部署，实际上是设计用来促进并管理暂时、短期劳动力的更为广泛的全球计画的一部分。我们透过脉络化 "暂时居民生物特徵识别计画" 在市场不确定性与风险管理的广泛新自由主义论述中所提出之主张，探讨此般生物特徵识别的使用。我们的理论建构将运用卡龙 (Callon) 的混合论坛理论，以此探讨生物特徵识别的经济化，并回头检视国家之于非公民的关係与义务，如何逐渐透过市场导向的经济化服务之修辞所界定。本文将简要回顾加拿大与美国之间的 NEXUS 计画，以兹证明该计画与理解如何透过监控来管理全球劳动力流动的关联性。

A partir de 1993 el gobierno canadiense utiliza el chequeo biométrico para identificar los migrantes que entran o salen del territorio canadiense. Desde hace poco, sin embargo, el gobierno busca mejorar el alcance de tal chequeo mediante cierto número de propuestas de normas y programas. De interés particular es el Proyecto Biométrico para Residentes Temporales del 2013. Según este Proyecto los extranjeros deberán suministrar detalles biométricos mejorados, inclusive huellas dactilares y capturas faciales, los cuales se compartirían con otros departamentos gubernamentales, como la Real Policía Montada Canadiense y la Agencia de Servicios Fronterizos del Canadá. Aunque el gobierno canadiense se ha referido a las crecientes vulnerabilidades de sus sistemas de identificación y verificación, no es muy claro el porqué se necesita un fortalecimiento adicional en la vigilancia y gobernanza de quienes no son ciudadanos del país. Nosotros arguimos que el creciente despliegue de la biometría hace parte de un programa global de mayores proporciones diseñado para promover y manejar el trabajo temporal, de corto plazo. Exploramos el uso de la biometría contextualizando los argumentos formulados en el Proyecto Biométrico para Residentes Temporales dentro del más amplio discurso neoliberal de incertidumbre del mercado y manejo de riesgo. La hipótesis de trabajo se construye a partir de la tesis de Callon sobre foros híbridos para investigar cómo dar sentido económico al chequeo biométrico ("economización") y, también, cómo cada vez más la relación y las obligaciones del estado para con quienes no son ciudadanos se definen a través de la retórica de un servicio economizante basado en las fuerzas del mercado. Una sucinta mirada al Programa NEXUS, compartido entre Canadá y EE.UU., sirve para demostrar la relevancia del Proyecto para entender cómo está siendo utilizada la vigilancia para manejar los flujos globales del trabajo.

In May 2013, the Canadian government implemented a biometric program for temporary resident visa applicants[1] entering Canada from certain visa-required countries[2] (Citizenship and Immigration Canada [CIC] 2012a). In partnership with the Canada Border Services Agency (CBSA) and the Royal Canadian Mounted Police (RCMP), the CIC will require fingerprints and facial captures prior to an immigrant's arrival to check identities expressed at the time of entry against preexisting data. Immigrants are required to provide this information at their own expense ($85; see Immigration and Refugee Protection Regulations section 315.1). The Temporary Resident Biometric Project (TRBP) is part of the Protecting Canada's Immigration System Act (passed in June 2012). This act is designed to help defend a sovereign Canada from unwanted persons (including criminals, failed refugee claimants, deportees, and would-be terrorists), who are said to exploit weak identity systems, and to strengthen the integrity of Canada's immigration program, increase citizen safety, and assert the nation's attention to migrant bodies and border security to the world (*Canada Gazette* 2012; CIC 2012j).

The focus on temporary residents and the manner in which the nation-state problematizes these populations through the institution of surveillance technologies at border points of entry is of particular interest. The state argues that biometrics support the accurate identification of migrants and organize increased flows of temporary residents (*Canada Gazette* 2012). Biometrics, however, are also tools used to manage global market uncertainties and to justify the (re)construction of spaces of privilege and exclusion for particular migrants (i.e., sought-after temporary laborers). The creation of programs such as the TRBP for regulating the flow of foreign labor, itself not necessarily unique to Canada, occurs within a larger political and economic context wherein the use of advanced information collection and sharing systems are understood as key techniques to manage risk and uncertainty. The result has been a concerted and articulated need by Western powers dependent on foreign labor to develop an infrastructure capable of biometric screening and information sharing to manage foreign labor supplies and respond efficiently to market uncertainty.

The TRBP in Canada demonstrates how geographies are mediated by digital surveillance infrastructures and tied into larger political and economic efforts to contain global flows through a process of identification and singularization. Due the Project's infancy, there is a lack of scholarly literature on the TRBP, and empirical data could not be collected for this research. Instead, this article draws out the larger theoretical understandings to illustrate the Project and employs materials published by the Government of Canada, and other secondary sources, to highlight a larger trend currently unfolding within transnational migration management and the organization of border and migration control mechanisms through digital surveillance infrastructures.

The article is divided into three sections. The first provides an overview of the Canadian government's social construction of the TRBP, paying close attention to the discourses that have sought to problematize existing regimes of migration management, simultaneously characterizing migrants as inherently uncertain in terms of risk management and criminality. This serves to legitimate the TRBP as an effective strategy for containing the uncertainties engendered from migrant flows.

The second section embeds the TRBP within a larger theoretical framework that seeks to address how markets, particularly those deemed uncertain or volatile, are rendered knowable within contemporary global political economies. That is, an understanding of how markets are enacted through practices of calculation and knowledge production. Drawing on Callon's analysis of markets as hybrid forums, we explore how the growth of global biometrics programs can be seen as a politic of calculation wherein migrants are assembled and categorized to allow states to manage the flow of international labor in response to market externalities and overflows.

The final section synthesizes research around geographies of surveillance, particularly digital surveillance infrastructures, to theorize how programs such as the TRBP complement existing regimes for containing global flows and, indeed, that states such as Canada have been significantly invested in developing such infrastructures for managing an assemblage of global flows; in other words, highlighting the complexities of heterogeneous power relations wherein global flows collide with, but sometimes complement, sovereign powers and national political economies. Infrastructures of digital surveillance then are understood as material embodiments of market relations and, moreover, collective desires to manage future market relations. This management of migrants serves to increase the economization of migrant labor in an effort to maximize the liberalization of markets and labor across space while allowing migrant bodies to become calculable within a transnational infrastructure of migration management.

The Temporary Resident Biometric Program

Responding to increased perceptions of identity fraud and vulnerability within the processing of migration applications, the Government of Canada announced the introduction of the use of biometric identifiers through the TRBP to address these deficiencies. This project has also been constructed as a key service that the government can deliver to applicants, offering the promise of increased security and reliability in the application and screening process, presenting biometrics as an effective solution both for securing the Canadian immigration system and simultaneously ensuring a more reliable and efficient processing of migration applications (CIC 2013). The TRBP is set to operate in Visa Application Centres abroad, where third-party service providers under contract or with service agreements will obtain information from applicants, including facial captures and all available fingerprints, that will be sent directly to an RCMP criminal fingerprint repository for comparison with existing records (*Canada Gazette* 2012; CIC 2013). Systems will be put in place and legislative changes made to attempt to ensure applicant privacy in the collection and retention processes, due to the medium-level risk assessment given to biometric technologies in the Interim Privacy Impact Assessment conducted in 2009 (CIC 2012g, 2012h). Technologies will be either enhanced or new technologies established, as necessary, and the Immigration and Refugee Protection Regulations revised to reflect the practices of the Project. A High Value Data Sharing Protocol (HVDS) has also been instituted with the United States, the United Kingdom, Australia, and New Zealand, known as the Five Country Conference (FCC), to share biometric information among themselves (CIC 2012b; New Zealand Visa Bureau 2012).

According to then–Citizenship, Immigration and Multiculturalism Minister Jason Kenney, the move to incorporate biometric technology and digitize migrant bodies will deter abuse of Canada's immigration system, keeping the country's doors open to "legitimate travellers" (CIC 2012f). A glimpse at CIC's Backgrounder page, entitled "Five Reasons Why We Need Biometrics," provides an example of this rationale (CIC 2012c). This page cites five cases of deportation that include an immigrant's full name, the number of deportations that immigrant has faced, and a description of the criminal activity in which the immigrant engaged. Essentially, the information highlights the felonies of five immigrant individuals to justify the need for biometrics. Beyond these five cases, the Government of Canada provides little substantive evidence to support their claims of an increased risk of fraudulent applications and activities. That said, the Canadian government has framed the existing immigration bureaucracy as inefficient, vulnerable, and incompatible with the stated goals and objectives for economic growth. For example, it has sought to problematize and reevaluate the efficiency of the immigration bureaucracy through a series of public hearings and Standing Committee on Citizenship and Immigration meetings to examine immigration processing efficiency. A March 2013 publication, "Standing on Guard for Thee: Ensuring that Canada's Immigration System Is Secure," posits that the system is vulnerable to fraud often due to lack of reliable documentation for identity verification, and that "timely and reliable information is not always available to visa officers when they must make an admissibility determination on an application" (Parliament of Canada 2013). The publication recommends biometric use for the enhancement of identity management capabilities, the acceleration of background checks, and the reduction of visa fraud. Supporters of the program have been quick to articulate the need for authenticating migrant identities through biometrics as a means to deter a rise in global identity fraud, particularly to combat so-called yo-yo bandits, repeat offenders who continue to return to Canada by exploiting bureaucratic loopholes and inefficiencies (Trottier 2012; Koch 2013). Furthermore, the CIC calculates that biometrics could save the government approximately $92 million over a decade by reducing the number of detainees and deportations, as well as $14 million due to a reduction in crime rates (Koch 2013). Last, the TRBP has emerged within a larger economic context that has sought to address domestic labor shortages in high-skilled markets, prompting some to call on the need for a "flood of immigrants" (Friesen 2012).

Such an inflow, however, will be highly managed. The TRBP will be used as a preemptive mechanism to contain populations in the country of origin. It is likewise important to consider how this program might fit into a larger shift in transnational and global regimes of migrant management through surveillance and information sharing agreements. Although the need to contain elements of criminality in migration management arguably provides a key selling point to justify these new programs to the Canadian public, the concern is whether or not the TRBP is exclusively a program to combat fraud while addressing domestic labor shortages. Over the last decade, Western states have begun implementing biometric screening programs to account for and manage the flow of approximately 214 million migrants across international borders (Lyon 2007; Guild 2009; Ackleson 2011; Koslowski 2011). The intensification of global mobility (particularly the shift of migrant bodies from south to north) has prompted investment in techniques to manage migration in a way that is compatible with maximizing the short-term growth of international markets. In the global context, the TRBP is designed with technologies used in other countries in mind and to share information with other privileged partners to further regulate regimes of migration and economic markets (*Canada Gazette* 2012). Tracking, in digital form, the activities of bodies, allows nations to internationally assess movements, (re)frame mobility and allow or deny access for particular purposes.

To further detail the importance of migration management, which encourages the development of global partnerships and strategies, Geiger and Pécoud (2010) noted that (cf. International Organization for Migration [IOM] 2008, 11):

> Migration management is fundamentally about making migration economically beneficial, notably by connecting the supply of labor in the less-developed South with the demand for migrant workers in the North—what the IOM 2008 World Migration report calls "a broad and coherent global strategy to better match demand for migrant workers with supply in a safe, humane and orderly way." (14)

The need for laborers in Western economies is used to justify the investment in biometric technologies and techniques and the use of data globally. The information sharing practices between states contribute to emerging transnational regimes of migration management, a key element in the collection of biometric data (Guild 2009). According to Minister Kenney, the TRBP will "improve our immigration security screening by light years, by orders of magnitude, particularly in the context of the enhanced information sharing agreements that we anticipate with the United States" (Parliament of Canada 2012).

The historical coupling of migration and security further complicates the issue, what Bourbeau (2011, 1) called the "securitization of migration." In Canada, the political conflation of these two themes encourages the use of biometrics (cf. Waever, Buzan, and de Wilde 1993; Bigo 2002). In 1991, Canada first identified migration as a source of concern in the nation's Foreign Policy Themes and Priorities document, where the inclusion of this listing was made "with caution and mostly in terms of the need for the international community to develop cooperative security" (Bourbeau 2011, 20). Notably, this document was prepared before the events of 11 September 2001 (hereinafter 9/11), which set the stage for more regulatory measures in the form of the Immigration and Refugee Protection Act (2002) and the creation of the CBSA (2003). The 2004 Securing an Open Society: Canada's National Security Policy more clearly expresses the need to restrict the movement of people across borders with a focus on the "monitoring [of] illegal drugs and immigration activity" (Bourbeau 2011, 23). The CIC, CBSA, and other federal departments have been integral in these discussions. Ultimately, recent changes to immigration legislation, primarily through the Protecting Canada's Immigration System Act and amendments to the Immigration and Refugee Protection regulations (2012; *Canada Gazette* 2012), further formalize the view that migrant bodies represent a threat to national and individual security, which Bigo (2002) argued reflects a fear of losing sovereign control over territorial boundaries. In this sense, migration is not a self-evident security issue, nor is it exclusive to post-9/11 policies but indeed part of larger neoliberal discourses of global risk and uncertainty (Walters 2010).

Economization as a Calculative Framework for Migration Management

Despite claims that insist on the need for free markets and free trade, it is worthwhile to consider how programs such as the TRBP, when combined with information sharing agreements among an assemblage of Western states, highlight the need to make bodies, identities, populations, markets, and global economic flows calculable and, moreover, to consider how this intersects with larger trends in capitalism. Thrift (2005) identified a shift toward "knowing capitalism" as a means of presenting a new interpretation of capitalism as a field of networks, continuously in a state of mutation based on the contingencies of small events and the necessity of the future as always being uncertain. Drawing from Deleuze's reading of Foucault, Thrift sees capitalism as a diagram, or relations of relation; an "impulse without determinate goals, a functioning which stamps particular forms of conduct on human multiplicity" (2005, 1). From this perspective, uncertainty bears an almost omnipresent role, necessitating a continued investment in instruments and techniques for gathering, processing, and analyzing information. Indeed, concerns for epistemological certainty thus become a key rationale and marketing strategy for biometric investment as a means to accurately target, identify, and render populations knowable.

Markets can be understood as products of collective calculation and investment. Callon (1998, 2007) has devoted significant attention to understanding the theoretical and material practices of calculation for enacting markets and containing uncertainty. Callon's primary interest is in the performative constitution of markets through social and technological investment, particularly as it relates to how actors respond to changes, inefficiencies, and uncertainties in market relations (Callon 1998). Callon has dedicated considerable analysis to understanding what he calls hybrid forums, an analytical concept that denotes situations of market volatility and uncertainty due to a failure in calculation or lack of consensus. What is particularly interesting is that Callon sees the rapid advancement of scientific and technological investment for calculating and enacting markets as paradoxically engendering more uncertainty and volatility (Callon, Lascoumes, and Barthe 2009, 19). In this way, market relations and economic flows are never totalizing or completely stable; markets require strategies of continual calculation, investment, and containment. Hybrid forums in many respects complements Thrift's thesis of knowing capitalism, adding another layer of depth by showing how techniques for calculating or knowing markets likewise engender new conditions of uncertainty.

Callon sees the creation of markets as necessarily produced through performative practices of calculation. In this regard, markets are understood as social frameworks that select, assemble, and in turn reject particular socio-technical practices, behaviors, institutions, and organizations. As such, these practices of framing required to enact markets also in turn generate market externalities and overflows: things that were not otherwise included in the frame or likewise unforeseeably emerged from the enactment of markets. In other words, markets necessarily engender the production of externalities, requiring further practices of calculation and reinvestment. This is precisely why Callon understands markets as socio-technical arrangements for creating powerful mechanisms of social exclusion that create, at least in theory, the possibility of such changes as irreversible courses of action (Callon 2007, 140). Technological investment into markets therefore not only is a process of managing uncertainty but also of generating matters of concern wherein actors are grouped and assembled either inside or outside the frame. This is further reinforced by Callon's interest in what he calls the economization of society, involving the processes by which activities, behaviors, organizations, and institutions are rendered as economic (Çalişkan and Callon 2009).

Externalities and overflows are likewise matters of social concern beyond the exclusive realm of economists (Callon 1998; Callon, Méadel, and Rabeharisoa 2001). One way we can understand this is by examining how the state has increasingly sought to frame migration management as a service that must take into account the calculation of risk and uncertainty as a key priority in policy development, particularly to respond to fluctuations in global markets. Social concerns and actors coalesce into framing migration through its entanglement with institutions of security and economy and necessarily requiring the intervention of a myriad of actors including law enforcement, border security, citizenship, and immigration. This amounts to a singularization of global flows, in this case of migrants, who can be readily identified and scrutinized by these actors as needed. This, in many respects, underscores contemporary social theories of social sorting and the categorization of populations to achieve particular political and economic goals (Bowker and Star 2000; Lyon 2003).

The capacity to individuate applicants, by scrutinizing their identity and assessing relevant skills, education, ethnicity, and other demographic characteristics, while also subjecting them to corporeal identification and verification based on unique physical features, lies at the heart of a surveillance infrastructure that targets, singularizes, and differentiates migrant populations to respond to changes in market demands, such as labor shortages. This suggests that a socio-technical system can potentially provide markets with sources of expendable labor or even a replacement of domestic workers with foreign workers, a possibility that has become a flashpoint in recent political debates in Parliament (Beltrame 2013). In Canada, the TRBP operates by calculating risk, identifying temporary residents from certain visa-required countries only (see Figure 1). According to the Canadian government, a systematic assessment was completed on the basis of "volumes or rates of TRV [Temporary Resident Visas] refusals, removal orders, refugee claims, and nationals arriving without proper documentation or under a false identity" to determine which countries' citizens would undergo biometric data collection (*Canada Gazette* 2012). Moreover, consideration was given to Canada's Global Commerce Strategy and tourism interests in the making of these decisions. There is no need to calculate this same risk for permanent resident applicants applying abroad. In effect, this suggests that far from the diminishing role of the state in an era of neoliberal globalization and so-called free markets, the state is actively managing market uncertainty by reframing their obligations toward externalities through an economization of global labor flows.

Digital Surveillance and the Singularization of Flows

This increasing attention toward market uncertainties has put complex demands on market actors in terms of technological investment (Barry and Slater 2005). Not only do we see a significant increase in technological infrastructure investments to calculate and manage markets but also in increasingly layered, complex, and automated systems of logistics designed to streamline global infrastructures of supply and demand, producing new types of material semiotics and social geographies for managing mobility (Urry 2000; Thrift and French 2002). Such trends have been noted elsewhere; for example, in the emergence of digital surveillance infrastructures capable of automated singularization and delivery of particular services or social resources based on the socio-technological construction of identity. A major concern within infrastructure studies concerns the granularity and extent to which infrastructures can be digitally configured to provide services to those considered economically desirable, effectively requiring complex socio-technical practices of surveillance and digital discrimination of flows (Graham and Marvin 2001; Graham and Wood 2003). With regard to international mobility, a clear example of this has been the rise of trusted traveler programs, where individuals deemed low-risk, frequent flyers can opt to circumvent traditional customs and border clearance queues, passing through automated gates that confirm identity through biometric identifiers. Such automated border clearance mechanisms are also in development for populations at large. Airports, constantly seeking new avenues and greater efficiency in streamlining passenger flows, are increasingly considering using biometrically

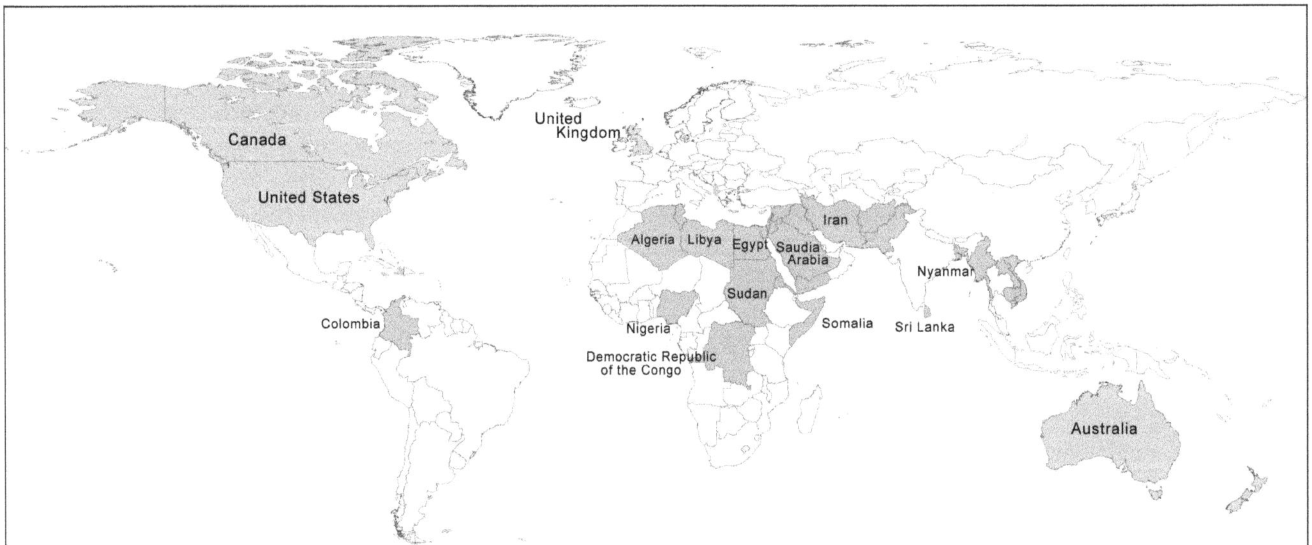

Figure 1. Countries requiring biometric data from citizens for temporary resident visas to Canada (peach) and the Five Country Conference (purple). (Color figure available online.)

enhanced documents and border control devices that could in theory automate passenger flows; populations calculated to be low risk can thus easily cross borders, whereas those calculated to be of higher risk would have to undergo increased scrutiny and human intervention.

By emphasizing the idea that certain unchangeable aspects of the body are in and of themselves stable and unique identifiers of subjectivity, biometric surveillance effectively appeals to a social construction of the body as a socially pure semiotic of identity. As Mordini and Massari (2008) argued, however, every identification system—including biometrics—involves a web of political and economic interests. This presents biometrics as necessarily the embedding of the body within a larger informational database, constructed out of particular social relations. Biometrics renders the body into a readable text (van der Ploeg 2000), suggesting that the context of such legibility will determine the ways in which the body is tied to identity. Biometrics thus becomes valued for its ability to match particular physical features to a larger identity management infrastructure in a database. Trusted traveler programs such as NEXUS, however, emerged as a response to market externalities and overflows and a need to contain (or facilitate) mobility. An increase in border congestion and bottlenecks due to post-9/11 securitization prompted the development of NEXUS. In other words, NEXUS grants increased mobility to a preapproved class of travelers willing to submit to intensified forms of singularization, identification, and surveillance.

Programs such as NEXUS necessitate information collection and sharing agreements, requiring iris and fingerprint biometric identifiers, as well as extensive background checks to access automated self-serve kiosks in designated lanes at both land and air crossings. The primary objective is therefore the construction of an automated identity management system designed exclusively to record low-risk, preapproved travelers (CBSA 2010). More important, the NEXUS program exemplifies the increasing economization of mobility governance, presenting a market-driven service model that automatically discriminates particular socioeconomic groups and classes of flows. Such biometric programs have been instrumental in the creation of legitimate and illegitimate mobilities, effectively implicating everyone in the governance of mobility and ubiquitous profiling:

> It does so via the promise of happiness and security and a world rid of ambiguity and uncertainty. The war on terror not only separates "our war" from "their terror," but also "our globalization" of legitimate and civilized business and leisure travel from "their globalization" of trafficking and illegal migration. (Amoore 2006, 348)

For Sparke (2006), NEXUS demonstrates not only the biopolitical privileging of a global business class but also the macroscale entrenchment of neoliberal policy in transnational regimes of governance introduced through the microscale production of entrepreneurial selfhood. It is therefore of crucial importance to consider how overflows produce matters of political concern, mobilizing market actors into innovative collectivities to participate in undertaking new measures or techniques to reframe and manage market failures.

In terms of singularization and differentiation of flows, a key objective of the TRBP is to use biometrics for reliable identification through unique body identifiers, "thereby fixing the client's identity at the time of application for a visa or for a study or work permit, and allowing verification of that identity when the individual seeks entry at the border" (Treasury Board of Canada Secretariat 2012, 26). Although clearly this technological arrangement seeks in one way to contain externalities created by the perceived threat of visa fraud, at the same time the CIC has emphasized the need to articulate itself as offering an economized service in providing resources, particularly flexible short-term labor, for organizations embedded within a larger global circuit of economic production and distribution facing skills shortages:

> Human Resources and Skills Development Canada (HRSDC) and Citizenship and Immigration Canada (CIC) are continually working on refinements to the Foreign Work Program in order to accommodate the changing needs of workers and employers. This plan includes a commitment to providing service at a speed that will support Canadian employer's efforts to operate effectively in a competitive global environment. (Kustec 2012, 19)

The emphasis on providing flexible sources of labor within a competitive global market effectively creates the need for an infrastructure that can in theory readily mobilize and deploy workers on demand. The Canadian government, however, has been quick to point out that the intensification of global mobility necessarily "brings with it an increase in irregular migration through identity and document fraud, and, therefore, concern—on the part of governments and citizens alike—about the ease with which criminals and would-be terrorists can exploit weak identity systems to travel between countries" (CIC 2008, 1). Nevertheless, the increasing emphasis on governments as providing market services to citizens or foreigners alike suggests that we can understand mobility programs such as NEXUS or now the TRBP as an effort to provide an economized service based on market principles and techniques for evaluating identities based on economic calculations of value, a regime of identity assessment in which the focus is not so much to secure sovereign territory but to manage market externalities and overflows for economic growth. Fundamentally, biometrics and the need for implementing digital surveillance infrastructures to regulate mobility are thus situated within a larger spatial politics of class and access.

Conclusion

Labor migration is a key issue and sometimes a political flashpoint in discussions concerning the future of the Canadian economy. In 2012, Canada settled just over 250,000 permanent residents and permitted the entrance of over 1 million temporary residents,[3] including 491,547 foreign workers (CIC 2012d, 2012e; Statistics Canada 2013a, 2013b). The majority of these workers arrive through the temporary migrant worker programs. These programs allow for a particular, and often exploitative, range of employment relations that domestic workers would resist (Preibisch 2010, 405). Furthermore, these employment practices are not adequately monitored by any government body but "grant employers access to a highly vulnerable group of workers who, once in Canada, are positioned disadvantageously within the labor market through a range of social and political disentitlements" (Preibisch 2010, 405; Preibisch and Encalada Grez 2010, 289). Through technological investment in rapid identification and verification systems, combined with discourses of migrants as always possibly fraudulent criminals, the TRBP will regulate migrant bodies through the stabilization and enhancement of economic interests and, ultimately, through the importation of a "flexible, immobile migrant workforce to respond to market needs" (Gabriel 2011, 138; Sharma 2006; Preibisch 2010; Walia 2010).

To elaborate on Preibisch's point of view, Hennebry (2008, 339) provided a broader perspective in her argument that states:

> "Migrant workers" have become an important resource in the global economy, and not solely for employers and governments. Multilateral agreements, trade liberalization, and advancements in communication and transportation have enabled flows of the world's poor into international labor migration systems, often mediated by a migration industry that profits from providing services to employers and migrants.

The labor migration regime provides a service in the midst of market uncertainty and shapes labor–capital relations and landscapes (Gabriel 2011, 137). In this case, Canada's new legislation works to deter irregular migration, encouraging high levels of short-term economic migration through restrictive policies and biometric technologies. At the same time, it presents Canada's immigration system as fair and balanced (Martin 2011). As Gabriel (2011) argued, the recruitment of immigrants, which was once an important part of the nation-building process, is now a necessary component of the nation's ability to compete in global economic markets. In this global environment, threats to economic stability, through identity theft or fraud as Minister Kenney argues, create anxiety that only technology can manage in any acceptable way (Walters 2010; CIC 2012f).

The TRBP is a particular instance of a larger global trend to manage shifts in migration to address market externalities and overflows. In the last decade, similar programs have materialized across Western states. The qualification of biometric technologies in this way works not simply as a biopolitical technique of securitization or to have migrant bodies to meet labor shortages but as an effort to conceal the control of the market through migration management. Moreover, the categories that biometric technologies justify continue to shift in ways that the nation-state's political and economic system deems necessary and do not necessarily reflect categories of security and risk.

The usefulness of biometric technology, therefore, is thought to speak for itself. As Callon and Muniesa (2005) argued, products are subject to significant metrological work, of qualifications and requalifications of commodities and services to objectify certain properties. This in effect suggests that even economized programs such as the TRBP undergo the same processes as market investments do. The underlying purpose of the TRBP then is not necessarily to manage migrants and repair an inefficient system. Rather, larger socioeconomic conflicts are being manifest through the creation of new spatialities for mobility governance based around the complexities of global economic integration. The rhetoric of delivering an economized service for migration management thus underpins a geopolitics of managing overflows, one that simultaneously invests in the migrant body a series of practices designed to hypostatize and discriminate identity within a larger nexus of risk, uncertainty, and criminality.

Geiger and Pécoud (2010) argued that immigration policies should be questioned by North Americans because they are not necessarily founded on accurate data and expertise. The lack of information available on this Project and the minimal efforts made to consult the general public ultimately make it difficult to discriminate the motivations behind changes to these policies. Although geographic scholarship has developed topics of security and immigration policy in a post-9/11 world (Amoore 2006, 2011; Sparke 2006; Coleman 2009, 2012; Nicol 2011; Salter 2011; Williams 2011) and has worked to delineate the concept of the border, particularly in an age of globalization (Johnson and Jones 2011; Konrad and Nicol 2011; Bauder 2012), the literature has not tended to focus on or reveal motivations for the use of biometrics on immigrant populations other than that of exclusion. Geographers should explore the TRBP and other similar projects to further understand the development of transnational regimes of mobility. This will help unpack the complexities and uncertainties inherent in international migration management and, particularly, how biometrics might serve to enhance the interdependent relationships between markets and global flows of labor.

Acknowledgments

We would like to thank participants of the 2013 Association of American Geographers Annual Meeting in Los Angeles, California, for their feedback and the two anonymous reviewers for the instructive comments, all of which helped to improve the article immensely. We also thank Dr. Richard Wright, 2014 Special Issue Editor, and Robin Maier, Production Editor, for their support and patience.

Notes

1. Including visitors, study, and work permit holders (http://www.cic.gc.ca/english/information/applications/visa.asp; last accessed 31 May 2013).
2. Despite the disclaimer that over 90 percent of immigrants do not require visas to enter Canada and therefore are not affected by this change, it seems that it will only be a matter of time before the state will require biometric data from all immigrants (Parliament of Canada 2012). Countries for biometric screening include the following: Afghanistan, Albania, Algeria, Bangladesh, Burma, Cambodia, Colombia, Congo, Democratic Republic of, Egypt, Eritrea, Haiti, Iran, Iraq, Jamaica, Jordan, Laos, Lebanon, Libya, Nigeria, Pakistan, Saudi Arabia, Somalia, South Sudan, Sri Lanka, Sudan, Syria, Tunisia, Vietnam, Yemen, and the territory of the Palestinian Authority. See Figure 1.
3. This number includes foreign workers, students, refugee and humanitarian cases, and a category defined as other.

References

Ackleson, J. 2011. International cooperation on border security in the developed world: The U.S.–Canada and U.S.–EU cases. In *Global mobility regimes*, ed. R. Koslowski, 95–113. New York: Palgrave Macmillan.

Amoore, L. 2006. Biometric borders: Governing mobilities in the war on terror. *Political Geography* 25:336–51.

———. 2011. Data derivatives: On the emergence of a security risk calculus for our times. *Theory, Culture & Society* 28 (6): 24–43.

Barry, A., and D. Slater, eds. 2005. *The technological economy*. London and New York: Routledge.

Bauder, H., ed. 2012. *Immigration and settlement: Challenges, experiences, and opportunities*. Toronto: Canadian Scholars' Press.

Beltrame, J. 2013. What labor shortage? Number of Canadian job seekers grows, says StatsCan. *The Financial Post* 16 April. http://business.financialpost.com/2013/04/16/canada-economy-job-vacancies/ (last accessed 31 May 2013).

Bigo, D. 2002. Security and immigration: Toward a critique of the governmentality of unease. *Alternatives: Global, Local, Political* 27:63–92.

Bourbeau, P. 2011. *The securitization of migration: A study of movement and order*. London and New York: Routledge.

Bowker, G. C., and S. L. Star. 2000. *Sorting things out: Classifications and its consequences*. Cambridge, MA: MIT Press.

Çalişkan, K., and M. Callon. 2009. Economization, Part 1: Shifting attention from the economy towards processes of economization. *Economy and Society* 38:369–98.

Callon, M., ed. 1998. *The laws of the markets*. Malden, MA: Blackwell.

———. 2007. An essay on the growing contribution of economic markets to the proliferation of the social. *Theory, Culture & Society* 24:139–63.

Callon, M., P. Lascoumes, and Y. Barthe. 2009. *Acting in an uncertain world: An essay on technical democracy*, trans. G. Burchell. Cambridge, MA: MIT Press.

Callon, M., C. Méadel, and V. Rabeharisoa. 2001. The economy of qualities. *Economy and Society* 21:194–217.

Callon, M., and F. Muniesa. 2005. Peripheral vision: Economic markets as calculative collective devices. *Organizational Studies* 26 (8): 1229–50.

Canada Border Services Agency (CBSA). 2010. Programs and services. http://www.cbsa-asfc.gc.ca/agency-agence/program-eng.html (last accessed 30 November 2012).

Canada Gazette. 2012. Regulations amending the Immigration and Refugee Protection Regulations. *Canada Gazette* 146 (49). http://www.gazette.gc.ca/rp-pr/p1/2012/2012-12-08/html/reg2-eng.html#archived (last accessed 31 May 2013).

Citizenship and Immigration Canada (CIC). 2008. Biometrics Planning Project, biometrics field trial:

Evaluation report. http://www.cic.gc.ca/english/pdf/pub/biometrics-trial.pdf (last accessed 30 November 2012).

———. 2012a. Backgrounder—Biometrics in Canada's temporary resident programs. http://www.cic.gc.ca/english/department/media/backgrounders/2012/2012-02-16h.asp (last accessed 28 November 2012).

———. 2012b. Backgrounder—Establishing biometric requirements in Canada's temporary resident immigration program. http://www.cic.gc.ca/english/department/media/backgrounders/2012/2012-12-07.asp (last accessed 20 October 2013).

———. 2012c. Backgrounder—Five reasons why we need biometrics. http://www.cic.gc.ca/english/department/media/backgrounders/2012/2012-02-16g.asp (last accessed 28 November 2012).

———. 2012d. Canada—Permanent residents by gender and category, 1987 to 2011. http://www.cic.gc.ca/english/resources/statistics/facts2011/permanent/01.asp (last accessed 30 November 2012).

———. 2012e. Canada—Temporary residents by yearly status, 1987 to 2011. http://www.cic.gc.ca/english/resources/statistics/facts2011/temporary/01.asp (last accessed 30 November 2012).

———. 2012f. News release—Harper government introduces the Protecting Canada's Immigration System Act. http://www.cic.gc.ca/english/department/media/releases/2012/2012-02-16.asp (last accessed 28 November 2012).

———. 2012g. Temporary Resident Biometrics Project: Stakeholder engagement summary report. http://www.cic.gc.ca/english/resources/publications/biometrics-stakeholders/bio-stake-eng.asp#purpose (last accessed 18 September 2013).

———. 2012h. Temporary Resident Biometrics Project (TRBP) summary. http://www.cic.gc.ca/english/department/atip/pia/trbp.asp (last accessed 18 September 2013).

———. 2012i. Use of biometrics in Canada. http://www.cic.gc.ca/english/department/biometrics-use.asp (last accessed 31 May 2013).

———. 2012j. Why use biometrics? http://www.cic.gc.ca/english/department/biometrics-why.asp (last accessed 28 November 2012).

———. 2013. Canada—Temporary foreign workers present on December 1st by province or territory and urban area, 2008–2012. http://www.cic.gc.ca/english/resources/statistics/facts2012-preliminary/04.asp (last accessed 31 May 2013).

Coleman, M. 2009. What counts as the politics and practice of security, and where? Devolution and immigrant insecurity after 9/11. *Annals of the Association of American Geographers* 99 (5): 904–13.

———. 2012. The "local" migration state: The site-specific devolution of immigration enforcement in the U.S. South. *Law & Policy* 34 (2): 159–90.

Friesen, J. 2012. Why Canada needs a flood of immigrants. *The Globe and Mail.* http://www.theglobeandmail.com/news/national/time-to-lead/why-canada-needs-a-flood-of-immigrants/article4105032/?page=all (last accessed 18 September 2013).

Gabriel, C. 2011. "Advantage Canada" and the contradictions of (im)migration control. In *Constructing and imagining labor migration: Perspectives of control from five continents*, ed. E. Guild and S. Mantu, 137–55. Burlington, VT: Ashgate.

Geiger, M., and A. Pécoud, eds. 2010. *The politics of international migration management.* New York: Palgrave Macmillan.

Graham, S., and S. Marvin. 2001. *Splintering urbanism.* London and New York: Routledge.

Graham, S., and D. Wood. 2003. Digitizing surveillance: Categorization, space, inequality. *Critical Social Policy* 23:227–48.

Guild, E. 2009. *Security and migration in the 21st century.* Malden, MA: Polity Press.

Hennebry, J. L. 2008. Bienvenidos a Canadá? Globalization and the migration industry surrounding temporary agricultural migration in Canada. *Canadian Studies in Population* 35:339–56.

International Organization for Migration. 2008. World migration report 2008: Managing labor mobility in the evolving global economy. http://publications.iom.int/bookstore/free/WMR_1.pdf (last accessed 30 November 2012).

Johnson, C., and R. Jones. 2011. Rethinking "the border" in border studies. *Political Geography* 30 (2): 61–62.

Koch, D. 2013. Critics say biometric borders threaten civil liberties. *Capital News Online.* http://www.capitalnews.ca/index.php/news/biometric-borders-threaten-civil-liberties-activists (last accessed 16 May 2013).

Konrad, V., and H. N. Nicol. 2011. Border culture, the boundary between Canada and the United States of America, and the advancement of borderlands theory. *Geopolitics* 16 (1): 70–90.

Koslowski, R. 2011. The international travel regime. In *Global mobility regimes*, ed. R. Koslowski, 51–72. New York: Palgrave Macmillan.

Kustec, S. 2012. *The role of migrant labor supply in the Canadian labor market.* Ottawa, Canada: Citizenship and Immigration Canada, Research and Evaluation. http://www.cic.gc.ca/english/resources/research/2012-migrant/index.asp (last accessed 30 November 2012).

Lyon, D. 2003. *Surveillance as social sorting: Privacy, risk, and digital discrimination.* London and New York: Routledge.

———. 2007. *Surveillance studies: An overview.* Malden, MA: Polity Press.

Martin, S. F. 2011. International cooperation on migration and the UN system. In *Global mobility regimes*, ed. R. Koslowski, 29–49. New York: Palgrave Macmillan.

Mordini, E., and S. Massari. 2008. Body, biometrics and identity. *Bioethics* 22:488–98.

New Zealand Visa Bureau. 2012. Immigration New Zealand goes biometric. http://www.visabureau.com/newzealand/news/19-10-2011/immigration-new-zealand-goes-biometric.aspx (last accessed 30 November 2012).

Nicol, H. N. 2011. Building borders the hard way: Enforcing North American security post 9/11. In *The Ashgate research companion to border studies*, ed. D. Wastl-Walter, 263–82. Farnham, UK: Ashgate.

Parliament of Canada. 2012. 41st Parliament, 1st Session, Standing Committee on Citizenship and Immigration. http://www.parl.gc.ca/HousePublications/Publication.aspx?DocId=5525470&Language=E&Mode=1#Int-7538283 (last accessed 16 November 2012).

———. 2013. Standing on guard for thee: Ensuring that Canada's immigration system is secure. http://parl.gc.ca/HousePublications/Publication.aspx?DocId=6038280&Language=E&Mode=1&Parl=41&Ses=1&File=57 (last accessed 31 May 2013).

Preibisch, K. 2010. Pick-your-own labor: Migrant workers and flexibility in Canadian agriculture. *International Migration Review* 44:404–41.

Preibisch, K., and E. Encalada Grez. 2010. The other side of El Otro Lado: Mexican migrant women and labor flexibility in Canadian agriculture. *Signs* 35:289–316.

Salter, M. 2011. Places everyone! Studying to performativity of the border. *Political Geography* 30 (2): 66–67.

Sharma, N. R. 2006. *Home economics: Nationalism and the making of "migrant workers" in Canada*. Toronto: University of Toronto Press.

Sparke, M. B. 2006. A neoliberal nexus: Economy, security and the biopolitics of citizenship on the border. *Political Geography* 25:151–80.

Statistics Canada. 2013a. Canada—Permanent residents by gender and category, 1988 to 2012. http://www.cic.gc.ca/english/resources/statistics/facts2012/permanent/01.asp (last accessed 25 October 2013).

Statistics Canada. 2013b. Canada—Temporary residents by yearly status, 1988 to 2012. http://www.cic.gc.ca/english/resources/statistics/facts2012/temporary/01.asp (last accessed 25 October 2013).

Thrift, N. 2005. *Knowing Capitalism*. London: Sage.

Thrift, N., and S. French. 2002. The automatic production of space. *Transactions of the Institute of British Geographers* 27 (3): 309–35.

Treasury Board of Canada Secretariat. 2012. Citizenship and Immigration Canada—Supplementary tables: 2012–13 report on plans and priorities. http://www.tbs-sct.gc.ca/rpp/2012-2013/inst/imc/imc00-eng.asp (last accessed 30 November 2012).

Trottier, B. 2012. Bernard Trottier on protecting Canada's Immigration System Act. http://openparliament.ca/debates/2012/5/17/bernard-trottier-1/only/ (last accessed 16 May 2013).

Urry, J. 2000. Mobile sociology. *British Journal of Sociology* 51 (1): 185–203.

van der Ploeg, I. 2000. The illegal body: "Eurodac" and the politics of biometric identification. *Ethics and Information Technology* 1:295–302.

Waever, O., B. Buzan, and J. de Wilde, eds. 1993. *Identity, migration, and the new security agenda in Europe*. New York: St. Martin's.

Walia, H. 2010. Transient servitude: Migrant labor in Canada and the apartheid of citizenship. *Race & Class* 52:71–84.

Walters, W. 2010. Migration and security. In *The Routledge handbook of new security studies*, ed. J. P. Burgess, 217–28. London and New York: Routledge.

Williams, A. M. 2011. Migration, risk, and uncertainty: Theoretical perspectives. *Population, Space and Place* 18 (2): 167–80.

Index

Note: Page numbers in *italics* represent figures
Page numbers in **bold** represent tables

287(g) program, the 129, 130, 132n1

accessibility and net migration 92
adaptive migration 71, *72* *see also* comparative analysis of political economy and adaptive systems framework to environmental change in Honduras and Peru
advocating for undocumented immigrants 101–3, 157 *see also* case study on the demands of immigrant advocacy on the boundaries of liberal democracy
AFL-CIO (American Federation of Labor and Congress of Industrial Organizations) 110
Agamben's mobilization of zones of exception 161
airports' use of biometrics 184–5
amenity index for local areas of nonmetropolitan southeastern Australia 85–9, **86,** *87, 88,* **89**
amenity migration 59–60, 83–4 *see also* case study on temporal and spatial dimensions of urban migration to rural Australia; case study on the impact of environmental hazards on migration
amenity premium and farmland value 93–5, *94*
amenity premiums and farmland values in Australia 93–5, *94*
anti-immigration policy in the U.S. 97–8, 100
Anti-Slavery International 55
appeal for NRIs of owning exclusive property in Kolkata, India 175
arrests of undocumented immigrants 109
Association or India's Development 170
asylum controls in the US and the UK 151, 152, 155–7, 167
Australian agriculture and migration 86, **89**
Australian treatment of asylum seekers 165

Banerjee, Mamata 172
Berlusconi, Silvio 165
biometric screening of migrants in Canada 1, 180–2, 183–7, *184,* 187n2
blending among different cultures 15
boats as symbols of chaotic migration 164
borders and national identity 163, 164, 167
border walls 164; and digital surveillance technology 180–1, 184
branding of communities in Africa as 'migrant' or 'foreign' 31
Bush, George W. 25, 109

Canadian legislation on migration 165, 182, 186
capital accumulation through construction 119
capitalism and migration management 182–3, 186
CASA de Maryland 130
case study of migration patterns of gay people in Ottawa and Washington D.C. 4–10, *7, 8*
case study of the integration of children of Hong Kong immigrants in Canada 12–19, **13**
case study on how labor unions shape state immigration policy 108, 109–14
case study on Palestinian migrants on family life in Britain 22–3, 24–8
case study on Sudanese and Nigerian migration to the EU 42–6
case study on temporal and spatial dimensions of urban migration to rural Australia 84–5, 92–3, 95; amenity index for local areas of nonmetropolitan southeastern Australia 85–9, **86,** *87, 88,* **89,** 91; amenity premiums and farmland values 93–5, *94*; relationship between rural amenity and migration **89,** 89–92, **91,** *93, 94*
case study on the construction labor market and urbanization in Dubai 119–20, 121–3; feminist and intersectionalist approach to immigrants in construction 119–21
case study on the demands of immigrant advocacy on the boundaries of liberal democracy 126–32
case study on the impact of environmental hazards on migration 60–5, **62,** *62,* **63, 64, 65,** 67–8; job growth and environmental hazards 65–7, *66, 67*
case study on the spatial and temporal dimensions in defining internal migration in Sweden 135–6, 138–40, **139,** *139, 140,* 148–9; by administrative region 141–5, **142,** *143,* **144,** *145*; by education **146,** 146–8, **147**; by gender and age 145–7, **146, 147**; individual characteristics of migrants 140–1, **141,** 148–9
case study on trafficking of women in North Korea 51–2
CBC (Canadian-born Chinese) 14 *see also* case study of the children of Hong Kong immigrants in Canada
CBSA (Canada Border Services Agency), the 182
Centreville Immigration Forum 130
Chinese involvement in the trafficking of North Korean women 53–5, 56n2
Christianity and the welcoming of strangers 98–9, 103–4 *see also* churches in the U.S. South and treatment of immigrants
churches in the U.S. South and treatment of immigrants 98, 99–104
CIC (Citizenship and Immigration Canada) 180, 181, 185
citizenship, conceptions of 128; insurgent citizenship 128–9, 130–2
climate change scenarios 71
Clinton, President Bill 164
collective bargaining and protections for immigrant workers 111–14
coming out process by gay individuals and relocating 5, 6
commodity relations of urbanization 121–2
comparative analysis of political economy and adaptive systems framework to environmental change in Honduras and Peru 72–5, *73, 75,* 78–80; water provision and glacier recession in Peru *76,* 76–9, **78**

connections between émigré communities and their former homes 170
construction labor markets and urbanization 116–23
construction sector in Canada, the 121
contiguous and noncontiguous migration 138
court-watching as a tactic of support for asylum seekers 154
crime and human trafficking 51–2, 121
criminalization of migrants and asylum seekers 167
crisis narratives and the criminalization of migrants 162–3
cultures of relatedness in migrant families 23

day laborer centers 130
decentralized administrative units in Africa 34
Democracy and the Foreigner (book) 127
deportations 166
deregulation in the construction sector 118
detention facilities 163, 164, 165–6
DHS (Department of Homeland Security), the 109–10, 111, 133n4
diaspora-influenced housing development in Kolkata, India for cosmopolitan émigrés 170–1, 172–6
digital surveillance technologies at border points 180–1, 184
"disaster capitalism" 163
disclosure of gay identities 5, 9
dissatisfaction of NRIs with exclusive property bought in Kolkata, India 175–6
distance in the definition of migration *see* case study on the spatial and temporal dimensions in defining internal migration in Sweden
DOL (US Department of Labor), the 109–10, 111
DREAM Act, the 127, 129

ecological fallacy 137
economic disparity and migration 162
economic growth in the American South 98
economic spatial equilibrium 59, 68
education and distance relationships in the definition of migration 137–8, **146,** 146–8, **147**
Engle's Law 60
environmental hazards and migration 58–9, 60 *see also* case study on the impact of environmental hazards on migration
Equal Employment Opportunity Commission, the 109
ethnicity and political support in Africa 33, 34–5, 36
Euclidean migration distances **139,** *139*, 139–40
E-Verify 112, 114
exceptionalism and border control 164
exteriorization of migrant controls, the 151–2

faith-based organizations and legal advocacy for immigrants 130–1
families and migration 22–4
farm family incomes in Australia 85
farmland values and amenity premiums in Australia 93–5, *94*
fear and insecurity and immigrant securitization 163, 164
federal immigration enforcement activity in the U.S. 108, 126–7, 129, 131, 132n1, 166
feminist and intersectional approach to immigrants in construction work 119–20
feminist geopolitics 51–2
flooding and population change at Santa Rosa de Aguán, Honduras 75, *75*
FOB (fresh off the boat) immigrants 14, 17
from-ness of place 14, 18

Garifuna, the 72, 73, 74, 79
gerrymandering through demographics 35
glacier recession and water provision in Peru *76*, 76–9, **78**
global Indian, the 171–2, 176
global networks and transnational migration 163–4
GLOFs (glacier lake outburst floods) 78, 79
governance on African migration effects 33–4

health of the urban poor in Africa 36
HII (environmental hazards impact index) 61, 62
HIV/AIDS activism 8
Hoffman Plastic Compounds, Inc. v. National Labor Relations Board 109
"homosexual" as deviant identity 4
horizontal interdependence between MASGs 156–7
Howard, John 165
HRC (Human Rights Campaign) 9
HRSDC (Human Resources and Skills Development Canada) 185
Human Rights Watch 121
human trafficking 51–2, 53–5, 121
Hungarian crime syndicate in human trafficking 121
Hurricane Mitch (1998) 75, *75*

I-9 form, the 112
ICE (U.S. Immigration and Customs Enforcement) 109, 113, 132, 133n4, 153, 156
Illegal Immigration Enforcement Ordinance of northern Virginia 131
Immigrant Freedom Ride, the 111
immigrants as threats to the national fabric 162
Immigration and Refugee Protection Act (Canada, 2002) 182
Immigration Reform and Control Act, the (US, 1986) 108, 112
Indian diaspora, the 171
individual characteristics of migrants to Sweden 140–1, **141**
institutional access and the life events of gay men 8–9
institutional change and impact on migrants and the urban poor 31, 36–7
insurgent citizenship 128–9, 130–2
Internet, the 8, 9
intervening opportunities 137
irrigation and amenity pursuits 90, **90, 91,** 92
islands as detention centers 165
Israeli detention policy on migrants 165
Italian legislation on migration 165

Kenney, Jason 181, 182
Kent, Karen 113
Kenyatta, Uhuru 35
kinship studies 23 *see also* case study on Palestinian migrants on family life in Britain

labor actions by construction workers in Dubai 121–2
labor costs in the construction industry 119
labor unions and the workplace 110
land tenure and use in Honduras 73–4
land tenure and water governance in Peru 76
language as a barrier 15, 16
Latina immigrants and churches in the U.S. South 100–1
Law for the Modernization and Development of the Agricultural Sector (Honduras) 74
life course events in gay/queer lives 4, 10 *see also* study of migration patterns of gay people in Ottawa and Washington D.C.
life course theory 3–4
livelihood resilience and environmental change 72
Lynch, Austin 111, 112, 113–14

makeshift accommodation for migrants in Dubai 120
manipulation of the voting process in Africa 34–5
marketing of exclusive homes to global communities 170 *see also* diaspora-influenced housing development in Kolkata, India for cosmopolitan émigrés

markets as products of collective calculation 183
Marxian theorizations of urbanization 120–1, 122
MASG (irregular migrant support groups) spatial tactics against migrant control 151, 152, 153–8
MAUP (modifiable area 1 unit problem) 137, 148
media portrayal of migrant journeys 164
Mexicans without Borders 131
migrant incorporation into the construction sector 117–18 *see also* case study on the construction labor market and urbanization in Dubai
migration management as policy to compete on global economic markets 183, 186
migration selectivity 60
Minato, Susan 112, 113
mobility and African migration 32
mobility and migration studies 40–2, 45–6, 163 *see also* case study on Sudanese and Nigerian migration to the EU
multiculturalism of CBCs (Canadian-born Chinese), the 18

NABC (North American Bengali Conference) 172
natural amenities as a superior good 60
natural disasters and population migration in Peru 78
neoliberalization of immigration enforcement, the 156
Netanyahu, Benjamin 165
NEXUS traveler program 185, 186
NGOs 50, 51, 52
Nigeria and migration *see* case study on Sudanese and Nigerian migration to the EU
noncontiguous migration 148
North Korea and human trafficking 51–2, 53–5 *see* case study on trafficking of women in North Korea
North Korean Human Rights Act (2004, US), the 50, 52
NRIs (non-resident Indians) 171, 172, 173, 174

OBIs (overseas-born Indians) 171
"outsourcing" of detention 165–6

Pacific Solution, the 165
pluriactivity 85
political economy and the adaptive systems framework to environmental change 71–2 *see also* comparative analysis of political economy and adaptive systems framework to environmental change in Honduras and Peru
political geography and migration 161
principle of least effort, the 137
privatization of immigrant detention 163
Protecting Canada's Immigration System Act 180
protests in Kolkata against real estate developments for émigré Indians 174

queer, as term 10n1

racism 16–18, 162
remittances from émigrés 17, 170
Rudd, Kevin 165
rural amenity migration in Australia 2, 84 *see also* case study on temporal and spatial dimensions of urban migration to rural Australia
rural and urban bias in Africa 33–4, 35

SAPs (structural adjustment policies) 33
seasonal climatic variables and measurement of rural amenity 84–5
Secure Communities program, the 126, 129, 130, 131, 132, 132n1
Senate Bill 1070 (Arizona) 97–8, 131, 133n3
Sharma, J.C. 172
SHELDUS (Spatial Hazard Events and Losses Database for the United States) 60–1
Snowden, Edward 1
social support in the gay community 5–8, 10
socio-ecological impact of Topical Depression Gamma at Batalla, Honduras *73*
Southern Baptist Convention, the 99
sovereignty and disciplinary power 153, 166
spatial clustering in rural Africa 34
spatiotemporal logics of chaos and crisis in global migration 161–6
spiritual assistance for undocumented immigrants 101–3
state attitude towards emigrants in India 171–3
strikes by migrants in Dubai 120
Sudan and migration *see* case study on Sudanese and Nigerian migration to the EU
Sweeney, John 110
systems approach to human adaptation to climate change 71 *see also* comparative analysis of political economy and adaptive systems framework to environmental change in Honduras and Peru

Tanzania Association of Informal Construction Workers 121–2
technological infrastrauctural investments to manage markets 184, 185–6
technological tactics in support of asylum seekers 154–5
Temporary Foreign Workers Program (Canada) 121, 186
Trafficking in Persons Report 52, 53
trafficking of women in North Korea 50
transnational kinship among families 24, 25, 26–7
transnational migration research 41–2, 45–6 *see also* case study on Sudanese and Nigerian migration to the EU
transnational turn, the 40–1
TRBP (Temporary Resident Biometric Project), the 180, 181–3, 184, 185, 186, 187
Tropical Depression Gamma 73
TRVs (temporary resident visas) 184, *184*

UN, the 51, 80
UN-HABITAT projects 36
Unite Here 110 *see also* case study on how labor unions shape state immigration policy
urbanisation and the political process in Africa 32–3, 34–6
urban spatial fix, the 119
US State Department, the 50, 52–3

Vikramghar Jheel water body, Kolkata 173–4

water provision and glacier recession in Peru *76*, 76–8, **78**
wealth and natural amenities 60; income changes and migration due to environmental hazard 61–5, **62**, **63**, **64**, **65**
white supremacy as a liberal Western ideal, belief in 162
Wilhelm, John 110
Workers' Defense Project, the 122
workplace as site for immigration enforcement, the *see* case study on how labor unions shape state immigration policy

ZANU-PF 35